Agrarian Crisis in India

Published by
ACADEMIC FOUNDATION
in association with

NATIONAL INSTITUTE OF RURAL DEVELOPMENT (NIRD)
(Ministry of Rural Development, Govt. of India)
Hyderabad

Agrarian Crisis in India

The Way Out

Editors

K. Suman Chandra

V. Suresh Babu

Pradip Kumar Nath

ACADEMIC FOUNDATION

NEW DELHI

www.academicfoundation.com

First published in 2013
by

ACADEMIC FOUNDATION
4772-73 / 23 Bharat Ram Road, (23 Ansari Road),
Darya Ganj, New Delhi - 110 002 (India).
Phones : 23245001 / 02 / 03 / 04.
Fax : +91-11-23245005.
E-mail : books@academicfoundation.com
www.academicfoundation.com

in association with the
National Institute of Rural Development (NIRD), Hyderabad.

Disclaimer:
The findings/views/opinions expressed in this book are solely those of the author and do not necessarily reflect the views of the publisher.

Cataloging in Publication Data--DK
Courtesy: D.K. Agencies (P) Ltd. <docinfo@dkagencies.com>

Agrarian crisis in India : the way out / editors, K. Suman Chandra, V. Suresh Babu, Pradip Kumar Nath.
p. cm.
Contributed articles.
ISBN 9789332700321

1. Agriculture--Economic aspects--India. 2. Agriculture and state--India. 3. Farmers--Suicidal behavior--India. I. Suresh Babu, V. II. Chandra, K. Suman. III. Nath, Pradip Kumar.

DDC 362.280888630954 23

Typeset by Italics India, New Delhi.

Printed and bound by The Book Mint, New Delhi.
www.thebookmint.in

Contents

Technological Dimensions in Agriculture

Emerging Trends in Agriculture and Food Security

Trends in Costs and Returns in Agriculture

Economic Reforms, Agricultural Credit and Indebtedness

List of Tables and Figures

Tables

Figures

About the Editors/Contributors

Editors:

K. Suman Chandra is Professor & Head, Centre for Agrarian Studies and Disaster Mitigation, National Institute of Rural Development (NIRD), Hyderabad. His areas of specialisation include agrarian reforms in general and tenancy reforms in particular. A sociologist by training, he has extensive field work experience which served as the basis for training rural development functionaries over the last two decades. To his credit there are four books and several articles published in reputed journals.

V. Suresh Babu is Assistant Professor, Centre for Agrarian Studies and Disaster Mitigation, NIRD, Hyderabad. His areas of research include: agriculture, agro-forestry, bio-inputs, watersheds, wage and self-employment. Dr Babu has published several papers in reputed journals.

Pradip Kumar Nath is Adjunct Faculty, Centre for Agrarian Studies and Disaster Mitigation, NIRD, Hyderabad. Presently, he is actively engaged in research and training in the areas of disaster mitigation and agrarian issues. Trained in economics, he has to his credit few publications on regional disparity, folklore, and issues related to marginalised communities (SC & ST).

Contributors:

Sudipta Bhattacharyya is currently Associate Professor, Department of Economics and Politics, Visva-Bharati, Shantiniketan, West Bengal. He served as Visiting Professor at the Asia Research Centre, Copenhagen Business School, Denmark (2009-10), Commonwealth Fellow at the Department of Economics and International Development, University of Bath, UK (2003-04) and Visiting Scholar at the Queen Elizabeth House, Oxford (1997-98). He has to his credit two books and many papers. His recently published edited book is entitled: *Two Decades of Market Reform in India: Some Dissenting View* (Anthem Press London).

Binod K. Das, Visiting Faculty, Nirma University, Ahmedabad, Gujarat, teaches social entrepreneurship. His present area of research is social exclusion and inclusion in India and Central Asia. Construction of deprivation index of excluded groups is the focus of his research.

K.K. Datta is Professor, Agricultural Economics, Division of Dairy Economics, Statistics and Management, National Dairy Research Institute, Karnal, Haryana. He post graduated from Calcutta University and did his PhD in Economics from Kurukshetra University. In 2010, he was awarded ICAR Award for Outstanding Interdisciplinary Team Research 2007-08 in Social Science. He has more than 100 research publications in various peer reviewed and refereed research journals of national and international repute. He has received several awards for best paper as well as best presentation in well-reputed leading journals like *Indian Journal of Agricultural Economics* and *Agricultural Economics Research Review Association.*

Rajiv Dey, Research Scholar, Visva-Bharati, Shantiniketan, West Bengal, is presently pursuing PhD from Visva-Bharati, Shantiniketan. The topic of his doctoral thesis is "Crop Diversification, De-peasantisation and Problems of Farm Viability in the Post-Liberalisation Era". He has published articles in reputed journals and attended national seminars on the topics of de-peasentisation and problems of farm viability.

Shrikant Shaligram Kalamkar is currently Director and Professor, Agro-Economic Research Centre (sponsored by Ministry of Agriculture, Government of India), Sardar Patel University, Anand, Gujarat. He has made important and useful contributions to the subject of agricultural economics. He has published more than 55 research papers/articles in reputed journals and 5 books.

Elumalai Kannan, Associate Professor, Institute for Social and Economic Change (ISEC), Bangalore, holds a Master's degree in Agricultural Economics from Haryana Agricultural University, Hisar, Haryana and PhD in Economics from Jawaharlal Nehru University (JNU), New Delhi. He has worked at various reputed national and international research institutes like International Maize and Wheat Improvement Centre, International Livestock Research Institute, National Council of Applied Economic Research and Madras School of Economics. He has published on issues related to India's agricultural growth, agricultural trade and markets, credit and agrarian issues.

E. Naveen Kumar is presently pursuing PhD in Economics from University of Hyderabad.

N. Kishore Kumar is PhD from College of Agriculture, Vellayani, Kerala. His specialisation is in the discipline of agricultural extension and his research work is in the fields of print and electronic media.

R.V. Ramana Murthy is currently Professor, School of Economics, University of Hyderabad. He works in the area of development economics, political economy and agrarian studies.

Ghanshyam Kumar Pandey, Research Scholar, Department of Economics, Punjabi University, Punjab, is presently pursuing PhD at Department of Economics, Punjabi Univestity, Patiala.

P.A. Lakshmi Prasanna graduated from Andhra Pradesh Agricultural University in 1990 and completed MSc (Ag) from the same university in 1993. She completed her PhD in Agricultural Economics in 2008 from Tamil Nadu Agricultural University. She joined National Centre for Agricultural Economics and Policy Research, New Delhi, as

Scientist in July 1996. Currently, she is working as Senior Scientist and focusing on the research theme 'institutional change'.

P. Prashanth is presently pursuing PhD from College of Agriculture, Acharya N.G. Ranga Agricultural University, Hyderabad.

G.K. Rajesh, Sericulture Officer, Commissionerate of Rural Development, Department of Rural Development, Government of Kerala, was educated at Centre for Development Studies, JNU; University of Agricultural Sciences, Bangalore and Mysore University. He has two decades of professional experience in silk industry.

Chirala Sankar Rao, Research Scholar, Centre for the Study of Regional Development, School of Social Sciences, JNU, New Delhi, is presently pursuing PhD in economics at JNU, under the supervision of Prof Ravi S. Srivastava. His research interests are: agriculture, labour and rural economics both at theoretical and empirical analysis.

Shiv Raj Singh Rathore, after graduating from S.K.N College of Agriculture, Jobner, Rajasthan, completed his Masters in Dairy Economics from National Dairy Research Institute, Karnal, Haryana. His research interests are in the areas of food security, livelihood security, agro-industry and dairy sector policy. Since 2012, he is working as Assistant Professor (Dairy & Food Business Management) at Shri G.N. Patel Dairy Science & Food Technology College, Gujarat.

M. Jagan Mohan Reddy, Professor, Department of Agricultural Extension has been working in ANGRAU for past 15 years in the fields of research, extension and teaching, and specilising in information communication technology, management and behavioural science research. He has authored 2 books and 70 research articles in various national and international journals. Presently, he is working in College of Agriculture, Rajendranagar, and has guided more than 10 MSc and PhD students in the field of agricultural extension.

Anil Kumar Roy is presently Associate Professor, Faculty of Planning and Public Policy, CEPT University, Ahmedabad. He holds a PhD degree from JNU, New Delhi. He was awarded Canadian Studies Doctoral Fellowship for his PhD work and was affiliated with Dalhousie University, Halifax, Canada. Dr Roy's research papers have been published in various journals. His recent book: *Port City Planning and Management: Social and Environmental Concerns in India and Canada*, has been published by Lambert Academic Publishing, Germany.

Partha Saha is currently Assistant Professor, Jindal School of Government and Public Policy, O.P. Jindal Global University, Haryana. His primary areas of interest lies in understanding agrarian relations, socioeconomic development in backward regions, skill development and enhancing employability of workers, enhancing rural non-farm sectors, and determinants of participation of females in the labour force. He has published edited books and contributed to journals on issues related to agrarian relations, socioeconomic development and employment creation. He has also contributed to the second *Human Development Report* for the country.

Swatahsiddha Sarkar is currently Assistant Professor, Department of Sociology, University of North Bengal, West Bengal. He was recently awarded PhD for his doctoral thesis on "Gorkha Ethnicity and Movement in the Darjeeling Hills". Besides having published research articles in academic journals and edited volumes, he keeps keen interest in the field of identity politics and ethnicity. He has authored a book on *Gorkhaland Movement* (2013) and co-edited a volume on *Ethnicity in India* (2013).

T. Satyanarayana is presently pursuing PhD in Economics from University of Hyderabad.

Sangeeta Shroff is Associate Professor, Agro-Economic Research Centre, Gokhale Institute of Politics and Economics, Pune, and also Officer-in-charge of the Centre. She has published several research papers in reputed national and international journals. She has worked in large number of projects sponsored by Ministry of Agriculture, Government of India, Planning Commission, Government of

Maharashtra and World Bank. She teaches at the post-graduate level and also guides PhD students.

Mamata Swain is currently Professor of Economics, Ravenshaw University, Cuttack, Odisha. She has published 5 books and more than 70 research papers in national and international journals, and edited books. She has completed 15 research projects sponsored by the World Bank, USAID, DFID, SANDEE, ICAR, ICSSR, Planning Commission etc. She has done Post Doctoral Research Work in the University of British Columbia, Vancouver and University of Tsukuba, Japan.

Mrutyunjay Swain is currently working as Research Officer (Economics) at Agro-Economic Research Centre, Sardar Patel University, Vallabh Vidyanagar, Gujarat. His areas of research interests are agricultural and resource economics, ecological economics, rural finance, participatory irrigation management, climate change, disaster management, food security and sustainable development. He has published 2 books and more than 30 research papers in referred journals and edited books.

Amarnath Tripathi, PhD in Economics from Banaras Hindu University, Varanasi, is presently Assistant Professor, Institute of Economic Growth, Delhi. His research interest includes climate change vulnerability and adaptation, issues related to Indian agriculture and the Uttar Pradesh economy. He has published many research articles in the area of climate change, Indian agriculture and economy of Uttar Pradesh.

Foreword

India is a vast country of sub-continental proportions with varied agronomic, economic and climatic regions. Excessive generalisations or conclusions based on aggregate data for the country as a whole would not only be inappropriate but also misleading. Diversity of conditions require special attention. In this context, it is heartening to see that the contributions to this volume have focus upon empirical, region-specific and state-specific studies and hence the conclusions drawn from these studies are of immense policy relevance. It gives us confidence that the subject is still of continuing interest to many of our young scholars who have done very useful studies upon the multifaceted nature of the farming systems favouring communities and agrarian relations. It would be a good idea for the NIRD authorities to network with these scholars and their institutions and bring out periodically state-specific, region-specific and issue-specific analysis and trends, policy prescriptions, and the implementation aspects.

When I was a Member of the Planning Commission in-charge of rural development, it was a matter of great concern for me to notice the extent to which the cosmos of agriculture at the Centre was segmented, fragmented and divided. Although agriculture and allied activities include a wide range of subjects like crop production, fisheries, animal husbandry, forestry, etc., it is unfortunate to witness the functional fragmentation of the universe of agriculture in the form of departmental silos. And this practice has spread to every level from the Centre to the state, district and sub-district levels. The absence of horizontal coordination amongst all the sub-systems relevant for agricultural investment and development is a serious lacuna that has adversely affected farming and the farmers' interest, resulting in sub-critical and sub-optimal results.

Agriculture moves forward only when all its constituent sub-systems are in right place and are functioning successfully. The main constituent sub-systems of agriculture are R&D, marketing, credit, input supply, extension, soil health, land and water conservation, and protection. The erstwhile Green Revolution of the 1960s and 1970s succeeded because these relevant sub-systems were decentralised, dynamically coordinated and integrated in implementation. Even now, certain regions and states, where agricultural growth has significantly increased, careful analysis shows their efficiency in understanding the role of the sub-systems and harmoniously ensuring their coordination and synchronised implementation especially at the district and at the farm levels. Some of the major concerns in agriculture and allied sectors are:

- Degradation of natural resource base.
- Uneven and slow development of technology.
- Inefficient use of available technology and inputs.
- Lack of adequate incentives and appropriate institutions.
- Widening economic disparities between irrigated and rainfed areas.
- Adverse impact on small and marginal farmers due to lack of timely access to credit and other inputs like seeds and fertilisers; and
- Declining groundwater table.

Amongst the immediate causes for the stagnating agricultural productivity, and the deep distress experienced by the farming community, I wish to stress three causes. First, the Indian agriculture is increasingly becoming small and marginal farmer dominant. The average size of land holding has declined to 1.30 hectare. Small and marginal farmers now comprise 83 per cent of the total land holdings and account for about 41 per cent of the cultivated area. These farmers are mostly outside the coverage and reach of the institutions meant to extend support to their livelihood activities. All agricultural sub-systems in most of the country bypass them. Raising the productivity of these small farms and the incomes of the small farmers is essential for the sustainability and security of agriculture in

India. Unfortunately, in the transformation that is taking place through migration, small and marginal farmers are also quitting agriculture instead of being a modernising force as in the countries in the East. In the absence of input supply, extension, marketing support and the credit support they are unable to take any risk to enhance the productivity and have become extremely vulnerable without succour or support from any source. They are caught in a vicious circle of low income, low investment and low productivity. Most of the farmers' suicides are on account of these failures. The great transformation in Europe lucidly explained by Karl Polanyi is not happening in India in the absence of adequate opportunities of employment in non-farm and industrial activities.

The second important issue contributing to agrarian crisis in India is the preponderance of non-production interests dominating agricultural assets. Agrarian structure and agrarian relationships have undergone serious changes which require deeper study. People are quitting agriculture in hordes without suitable or visible alternatives. They are forced to become circulating labour. For the first time in the Indian history, the growth of urban population is more than the rural population between 2001 and 2011. The urban population increased by 91 million and rural population by 90 million. Actually these are all rural people circulating in urban areas, and may or may not include small and marginal farmers. A new form of 'neo *zamindari*' is emerging in rural areas. Land under tenancy varies between 30 to 60 per cent. Neither the NSSO nor the agricultural holding census is able to capture this phenomenon correctly since the bulk of tenancy is unrecorded, unregistered and underground. A new class of landholders have emerged who have no interest in agriculture but would make investments in land for speculative and rentier purposes. Many of these could be the erstwhile farming communities who have reaped the fruits of Green Revolution, but now have migrated into trade, real estate, fertiliser and pesticide dealerships, marketing, service provisions, but yet do not want to leave the land. This is the group which has benefited directly both by the failure of the government to operate the important agricultural sub-systems and by implementing some intervention programmes. They appropriate all the interventions like institutional credit and insurance, and they gain from the State failure by becoming dealers or brokers, supplying agricultural inputs. In several countries and in a few regions in

India, there are legal provisions prohibiting ownership of agricultural land by non-agricultural interests. However, this law and other land reform measures such as tenancy reform, effective implementation of land ceilings and redistribution of ceiling surplus land, have become more or less dead letters, and the capture of the commanding heights of agriculture by corporate giants, traders and speculators is more or less complete especially in areas where the land is fertile and valuable.

The third important crisis and the challenge is rainfed agriculture and watershed management. Rainfed agriculture which consists of 60 per cent of our cultivated land suffered neglect in the past in not having received differentiated technological, institutional, infrastructural and investment support. Under the Department of Land Resources, new guidelines of integrated watershed development are reputed to have become popular and useful to the farmers in the erstwhile drought prone areas. However, it would not be an exaggeration to suggest that hardly 20 per cent or less of the area requiring watershed treatment has so far been covered under any watershed development programme activity. On account of the heavy incentive structure in favour of cereal crops such as rice and wheat, and the absence of agricultural support systems to the dryland farmers, mono-cropping systems which require least investment and effort are becoming extensive especially amongst those who have withdrawn from intensive cultivation. The importance of comprehensive bottom-up district agricultural plans cannot be underestimated and the launch of schemes like RKVY, BRGF etc., which are predicated upon such plans require to be internalised. There are several other issues and concerns which have led to the current crisis in different parts of the country.

Various studies in this important volume show the variegated nature of the challenges and problems encountered by farmers and their impact on livelihoods and agricultural production. Taken together, these papers presented at a National Seminar, make a valuable contribution by demonstrating the need for rigorous analysis relevant to different regions of the country, and towards the development of disaggregated approach for dealing with the principal issues to enable agriculture to move on to the high road of growth. The matrix of such development can and has to be different. It is worthwhile to work on this matrix comprising of not one uniform approach of mere distribution of subsidies but reconstructing and reinvigorating the totality of sub-systems which are vital for the forward movement of agriculture to a higher plateau.

— B.N. Yugandhar

Former Member

Planning Commission of India

Acknowledgements

There is a growing sense of despair that agriculture is becoming unremunerative as well as unattractive as a livelihood option in rural India. Given an option, more than one-third of the farmers in India would like to quit agriculture and move to other sectors. Lack of opportunities and skills required for new livelihoods keep majority in the primary sector only. Notwithstanding the importance and critical areas of food security and employment generation, the sector is being neglected by both policymakers and programme implementers. In this juncture, it is apt and need of the hour to understand different areas of concern in agriculture to bring out feasible recommendations. This in view, a seminar was organised and opportunity was given to the young and energetic scholars to shed light on areas which require deep understanding. Several persons and institutions contributed and helped for successful conduct of the seminar.

First and foremost, we are greatly indebted to Dr M.V. Rao, Director-General, National Institute of Rural Development (NIRD) for the encouragement in organising the seminar and enabling the participation of young scholars from across the country. We are also grateful for the financial support for organising the seminar and the publication of this volume. We are highly indebted to Shri B.N. Yugandhar, Former Member, Planning Commission and Prof D. Narasimha Reddy, Chair Professor, S.R. Sankaran Chair for their contributions and constant support. We would like to express special gratitude to the paper writers for enriching understanding of the issues set out for the seminar.

We are also grateful to several resource persons, especially Prof K.P. Kannan, Prof Praveen Jha, Dr Jayaranjan, Dr Marcella D'Souza, Prof S.N. Tripathy, Prof Sushanta Kumar Nayak, who chaired the technical sessions and steered the proceedings towards stimulating and fruitful discussion on the present conditions of agriculture.

We are also profoundly grateful to our NIRD colleagues (seminar team members and others) for their dedication and commitment in bringing out this volume.

— K. Suman Chandra

Prof and Head (CAS&DM)
NIRD,
Hyderabad

Introduction

Agrarian Crisis in India: The Way Out

K. Suman Chandra, V. Suresh Babu and P.K. Nath

Rates of Growth of agriculture in the last decade have been poor and are a major cause of rural distress. Farming is increasingly becoming an unviable activity

– Manmohan Singh, Prime Minister of India, 2007

Introduction

Agriculture, the most important sector in Indian economy, is presently at the crossroad. Although, the share of agriculture in the national GDP had decreased to 18 per cent in 2007, and subsequently 14 per cent in 2011, yet the sector continues to provide livelihood support to more than two-third of the total population in India. The slowdown in performance of the sector during the past couple of years is a cause of concern in the overall agenda for food and income security. The aggregate productivity has been low, growth of production decelerating and more importantly, the sector is exposed to risk and uncertainty. More than 80 per cent of the farmers, mostly with marginal and small holdings, depend on agriculture as the primary activity and are highly vulnerable to multiple risks. The downtrend in growth of total factor productivity (a measure of contribution of technology) has increased the vulnerability and induced unsustainability in farm income. At the same time, the escalating cost of production and unremunerative output prices have made farming increasingly unviable.

Several factors have contributed to the imbalance, of which the low agricultural productivity and its stagnation are notable. Among the supply side factors, declining factor productivity, degradation of soil and water resources due to intensive agriculture, inequitable access to input and imperfection in output market are prominent. The combined effect of these factors has resulted in escalation of the cost of production, low and uncertain production and falling farm income. The failure of the farming sector to absorb the rapidly growing pool of manpower in rural areas has also caused the decline in household income to add fuel to fire in the distress scenario.

Agrarian Distress in India: Evidences

The agrarian crisis and distress of the rural population which is the outcome of accumulated eventualities and adversity, has emerged as a nationwide phenomenon and a leading policy issue. Decelerating agricultural growth, stagnating yields, erosion of farm income, shrinking access and entitlements to food are the manifestations of agrarian crisis. The distress among the rural peasant community, allegedly manifested in farmers' suicide, is commonly attributed to debt trap, crop failure and/or yield loss. The increased incidences of farmers' suicides that have hogged the media limelight, are undoubtedly the most unfortunate reflection of agrarian distress which cannot be brushed aside as an event merely associated with drought or other natural disasters.

Declining Agricultural Output and Decelerating Growth

The macroeconomic scenario and growth trends show loss of dynamism in agriculture and allied sectors during the past 10-15 years. Growth of agricultural GDP decelerated from over 3.5 per cent per year during 1981-82 to 1996-97 to only around 2 per cent during 1997-98 to 2004-05. The deceleration is observed in almost all major sub-sectors, including horticulture, livestock and fisheries where, growth was expected to be high. Under the circumstances, achieving targeted growth rate of 4 per cent fixed by the Government of India (GoI) will be a herculean task.

Another crucial indicator of looming agrarian crisis is the steep decline in the growth of total factor productivity (TFP) during 1990s as compared to the 1980s, indicating the decreasing importance of technological contribution to food production. The TFP of paddy, for instance, has decreased in Punjab and Haryana during the period 1985-2000, remained unchanged in Assam, West Bengal, Karnataka and Uttar Pradesh and increased marginally (1-2% per annum) in Andhra Pradesh, Bihar, Madhya Pradesh and Tamil Nadu. The decadal analysis of the productivity growth of this most important cereal crop has shown that compound growth rate (CGR) of paddy yield after peaking during the 1980s (2.97%), decelerated rapidly since the early 1990s, depicting an inverted bowl shaped growth pattern. The downside growth pattern is a worrisome issue and further continuation of the trend is likely to have deteriorating impact on rural livelihoods.

Eroding Farm Profitability

In addition to decelerating or stagnating growth trend in production and productivity, the available figures on returns to farming are the striking evidence of simmering distress in the countryside. Looking at the NSSO and cost of cultivation statistics, it is apparent that the returns from agriculture for small and marginal farmers are not adequate to take them above the poverty line, in case agriculture is the only source of income.

Although there are inter-regional variations in farm profitability due to inter-crop variations in net returns per hectare, in general, scenario of income generation from crop production is not very encouraging. In five out of six major rice-growing states, in the triennium ending 2003-04, the share of returns to entrepreneurship in value added by the output was negative, which indicates economic unavailability of rice cultivation.

High Incidence of Indebtedness

As millions of small farmers have very little disposable production and income, borrowing is unavoidable to carry out the farm activity and to meet the basic household needs. This has become a gradual drag factor pushing the farmers into the deep debt trap. The impact of the phenomenon on the subsistence-oriented small and marginal farmers is disastrous. The available literature shows that growing indebtedness has increased the probability of suicides. Radhakrishna Committee on Agricultural Indebtedness has stated that the credit-related problems of farmers' indebtedness have assumed serious magnitude across the country in the recent years.

Farmers Suicides

Let us focus on the most affected regions of agrarian crisis. The most affected region has been found in the Vidharbha in Maharashtra; Deccan and Hyderabad Karnataka in Karnataka, i.e., Northern Karnataka;

Telangana and Rayalaseema in Andhra Pradesh; Wayanad, Idduki and Kannur in Kerala. It has been found by most of the studies that these regions (except Kerala region) are dry regions and had poor irrigation facilities, hence faced adverse climatic conditions. The regions of Kerala are mainly hilly regions; however during the period of crisis even these regions reported to have deficit rainfall. The crops cultivated in these regions were mainly commercial crops or cash crops. In Northern Karnataka regions the main crops were *jowar*, *bajra*, cotton, *ragi*, wheat, paddy, and *tur*. Commercial crops also have a significant presence in the cropping pattern. The presence of at least one/two such cash crops can be seen across the size classes. In Kerala the major crops cultivated were coconut, rubber, tea, coffee, areca nut, spices, which are mainly commercial crops or export-oriented crops. It has also been noticed that, taking a cue from the trends in commodity prices over time farmers have been switching from cultivation of multiple crops to mono crops. The major crops of Vidharbha were: cotton, oilseeds, sugarcane, soyabean. However, extensive cultivation of cotton has been reported. Shift towards cash crop production has also been seen in case of Andhra Pradesh, where major crops cultivated were tobacco, cotton, chillies, and groundnut. All these studies reveal that the victims have been mostly marginal and small farmers in most of these regions, who were highly indebted. The victims were mainly male and very young. In Karnataka, it is observed that, victims belong to ages between 25 to 35 and are mainly OBCs, though there are also cases of farmers committing suicide, hailing from dominant castes such as Lingayats and Vokkaligas. In Maharashtra (Yavatmal), it is found that the age group of victims are 30 to 50 and are mainly Hindus. In case of Kerala, majority were male (90%) and belonging to the age group of 41-60 across all caste and religious groups. A survey in Andhra Pradesh on socioeconomic characteristics of farmers who committed suicide in 2003 and 2004 in four districts shows that the majority of the victims were male in the age group of 30-50 and majority of them belonged to backward communities.

Analysing Farmer Suicides: Major Causes

Indebtedness

The primary cause that evolves from most studies is `indebtedness' which again can be linked to mismatching of 'expectation'. A farmer gets indebted when his/her return from the field is not as much as he expected and he becomes unable to pay the loan, which he might have taken for productive purpose, social purpose or to meet his immediate consumption needs. Indebtedness is quite evident in most of the affected districts. In Kerala, indebtedness is higher compared to national average due to availability of good informal and formal credit facilities, cash crop production and higher value of assets per households.

The failure of cooperative institutions has further made the large number of farmers to fall back on the moneylenders who charges exorbitant interest. This charge varies from 36 per cent to 60 per cent. Still most of the victims were found to prefer taking loan from moneylenders, although they charge exorbitant interest rate due to cumbersome process followed by institutional sectors. In case of Maharashtra and Andhra Pradesh also indebtedness were reported to be the major factor.

Agriculturist moneylender unlike the professional moneylender is a native of the village and powerful in the structure of village economy also. In many cases, he is the landowner who lease out land to the peasant. It is this combination of functions of landowner, moneylender and trader that makes him a powerful agent in the village, thus making him able to exploit. Besides this, commission agents, dealers and sub-dealers of pesticides and private source also provides credit and generally at very high interest rate and exploit farmers.

Indebtedness only shows us the tip of iceberg! The secondary factors which led to indebtedness and eventually agrarian crisis can be cited as: climatic factors and misuse of resources, low yield, neo liberal policies, seed sector liberalisation and GM crops, lack of alternative livelihood as well as social factors. These factors get interrelated together to form a complex process and causes agrarian distress.

Neo-liberal Policies and their Impact

Neo-liberal policies were also cited to have aggravated the problem. Impact of liberalisation has been felt in terms of declining output prices and increasing input prices. Due to withdrawal of supports by Government from agriculture in terms of declining subsidy, investment in irrigation and encouragement of private seed companies, has led to increase in the costs of inputs for production. The overall public capital formation in agriculture is declining over time. Dryland agriculture has been long neglected. The public investments in these regions are relatively low when compared to endowed regions. It is true that the magnitude of investment both in public and private investment in money terms has slowed down. But the bulk of public investment is on surface irrigation and concentrated on projects, which have taken an unconscionably long time to complete. Costs have risen and outlays have not resulted in any significant additions to irrigated area. The other, much smaller component of public investment has been in land and water conservation in rainfed areas through watershed development.

Non-Availability of Seeds

In the present context, the multinational seed industries are seeking total control over seeds. This has made farmers feel threatened as the barriers are raised in terms of growing, distributing and exchange of seeds. Again the shift towards cash crop production, which requires monetised inputs, including fertilisers and subsidies has further increased the cost of cultivation. Coupled with these the output prices falling since mid-1990s caused huge loss and indebtedness among farmers. Even when the output prices started increasing, there was no gain because of increasing cost of cultivation and falling yield.

Sociocultural Factors

Sociocultural factors also play catalytic role in the process of agrarian crisis. Low quality of public education and health facilities drives rural poor to private provisions rendering high cost. In some cases, the productive loan has been used in meeting social obligations such as dowry, community functions etc. Some studies have also pointed out alcoholism as another factor for incurring debt. In case of Punjab, it has been noticed that maintaining status (e.g., having tractor) and keeping up with neighbour has led to unproductive spending and hence indebtedness. Hence, many of these debts have been incurred due to non-productive purposes. In Andhra Pradesh, leasing-in land for cotton cultivation by small farmers also aggravates the condition of cultivators. It has also been found in some studies that, support from family, friends, relatives, *panchayat* members, etc., helps one to relieve the stress they face due to distress. Not discussing one's problem with others leads to closing an avenue for letting out ones pent up feelings and frustration. Societal support, public participation and close knit intra-family relationship helps in averting such incidents by providing with confidence. However, in recent times it is found that, the support system that was being provided by the family and the village system in the earlier days has been fast disappearing.

Conclusion

It is quite true that the problem cannot be solved through economic packages alone. What is needed is social interventions so that the farmers realise that suicide is not the way out. They should understand that they need to develop self confidence. The future generation should have the mental strength to face life's challenges. What can be done is: social support should be provided as a buffering effect for stress. The greater the support from friends, family, local communities, national policies and society as a whole, the less impact stress will have on individuals and families. It is crucial for governments and NGOs to work together, on a local, national and global level, to address and solve this critical issue.

Structural, Institutional and Resource Dimensions of Indian Agriculture

1 Changes in Landholding Structure in Indian Agriculture

Implications for Input Use, Cropping Pattern and Productivity

P.A. Lakshmi Prasanna

Introduction

Indian agriculture is characterised by predominance of smallholders. This structural feature has implications for input use and cropping pattern and in turn agricultural output growth, productivity and efficiency at macro-level. At micro-level, smallholders livelihood status depends on agricultural growth and its incidence (i.e., distribution of benefits of growth) which in turn is dependent on farm holding structure. The nature and strength of this two way relation is dynamic and changes with, changes in agricultural technology, liberalisation etc.

In this backdrop in the present study an attempt has been made to examine farm size—input use—cropping pattern trends over the years in Indian context and probe into farm size productivity relationship in recent period. The study is organised into five sections. In section one, a brief account of changes in landholding structure is presented. In section two, empirical evidences on changing input use, adoption of improved technology etc., by farm size wise are presented. In section three, a brief discussion about farm size wise (aggregate crop) productivity and efficiency are presented. In section four, ways to strengthen smallholder's productivity and improve their livelihood are explored. In the last section of the study conclusions are presented.

Structural Changes in Landholdings in India

Between 1970-71 and 2010-11, the total number of operational holdings in India increased from 71.01 million to 137.75 million and operational holdings area declined from 162.13 million ha to 159.18 million ha. As a result of this structural change, the average size of landholdings in India decreased to 1.16 ha. During this period the share of smallholders (farmers with operational area less than 2 ha) in number of operational holdings increased from 70 per cent to 85 per cent and their share in operational area increased from 21 per cent to 44 per cent (Table 1.1).

Table 1.1

Structural Change in Landholdings in India

	Marginal	*Small*	*Semi-Medium*	*Medium*	*Large*	*All Size*	*Smallholders*
			Percentage share in number of holdings				
1970-71	51	19	15	11	4	100	70
2005-06	65	19	11	5	1	100	83
2010-11	67	18	10	4	1	100	85
			Percentage share in operational area				
1970-71	9	12	18	30	31	100	21
2005-06	21	21	24	23	11	100	42
2010-11	22	22	24	21	11	100	44
			Average farm size (Ha)				
1970-71	0.40	1.44	2.81	6.08	18.10	2.28	0.68
2005-06	0.38	1.38	2.68	5.74	16.22	1.21	0.61
2010-11	0.38	1.42	2.71	5.76	17.38	1.16	0.60

Source: *All India Report on Agricultural Census*, various years.

Input Use

Crop Intensity

Crop intensity has been a major source of productivity increase and growth in agriculture in the country. At all-India level, crop intensity increased from 124 per cent in 1981-82 to 130 per cent in 2006-07 (Table 1.2). The results clearly reveal that crop intensity declined as farm size increased in all the years. On marginal farms it increased from 134 to 142 per cent, on small farms it increased from 128 to 130 per cent. This indicates that marginal and small farmers in India have been making much efficient use of land compared to upper size categories.

Table 1.2

Cropping Intensity

Year	Cropping intensity of farms of different size categories (%)					
	Marginal	Small	Semi-medium	Medium	Large	All Size Category
1981-82	134	128	125	120	116	124
1991-92	137	130	124	121	118	126
2001-02	139	128	126	125	121	128
2006-07	142	130	126	126	125	130

Source: *All India Report on Input Survey*, various years.

Use of Irrigation and Access to Irrigation

During 1980-81, 40 per cent of the net sown area (NSA) under the marginal category and 33 per cent of NSA on small farms was under irrigation (Table 1.3). On the other hand medium and large farms had irrigated only 24 per cent and 16 per cent of their NSA, respectively. Over the years share of irrigated area increased on different size category farms. In the year 2005-06, around 58 per cent of NSA of marginal farms was irrigated and on small farms it was 47 per cent. On large farms 37 per cent of NSA was irrigated. As landholding size increased, the percentage of area under irrigation declined in all the selected years.

Across different size category farms, in 1980-81 around 40 per cent of the irrigated area was under canal irrigation, but in 2005-06, tube wells constituted major source of irrigation. On marginal farms, area irrigated under tube wells constituted 46 per cent of net irrigated area in 2005-06, on other size category farms, its share ranged between 41 to 43 per cent. Over the years across different farm size groups, share of area irrigated by canals and tanks declined. This trend indicates increased tendency of individualised access to water through private investment in irrigation across different farm size groups. In 2005-06, share of area irrigated under wells was directly related to farm size (with the exception of large size farm), share of area irrigated under tanks and tube wells was inversely related to farm size.

Table 1.3

Proportion of Irrigated Area and Share Area Irrigated by Different Sources on Different Size Farms

	Year	Marginal	Small	Semi-medium	Medium	Large	All Categories
		Share of irrigated area in NSA (%)					
	1980-81	40	33	29	24	16	27
	1990-91	44	36	33	30	22	33
	2000-01	51	39	37	36	31	39
	2005-06	58	47	45	45	37	47
		Share of different irrigation sources in total irrigated area (%)					
	Year	*Marginal*	*Small*	*Semi-medium*	*Medium*	*Large*	*All Categories*
Canals							
	1980-81	39	40	39	38	40	39
	1990-91	35	34	33	34	36	34
	2000-01	28	27	27	30	32	28
	2005-06	28	26	26	28	31	27
Tanks							
	1980-81	14	11	9	6	6	9
	1990-91	10	8	6	4	4	6
	2000-01	7	5	4	2	2	4
	2005-06	7	5	3	2	2	4
Wells							
	1980-81	12	16	19	23	26	19
	1990-91	11	18	21	25	27	20
	2000-01	10	18	21	22	20	18
	2005-06	9	17	20	21	19	17
Tubewells	1980-81	26	26	27	27	24	26
	1990-91	35	33	32	31	28	32
	2000-01	44	40	40	39	35	40
	2005-06	46	43	43	42	41	44
Other sources							
	1980-81	9	7	7	6	5	7
	1990-91	9	8	7	6	6	7
	2000-01	11	10	8	7	11	9
	2005-06	10	9	8	7	7	8

Source: *All India Report on Agricultural Census*, various years.

High Yielding Varieties (HYV)

In 1996-97, the share of area under HYV on marginal farms constituted 59 per cent, it increased to 75 per cent in 2006-07 (Table 1.4). In 1996-97 share of area under HYV on small farms constituted 55 per cent, it increased to 70 per cent in 2006-07. On the other hand in 1996-97 share of area under HYV on large farms constituted only 42 per cent, it increased to 52 per cent in 2006-07. The results clearly reveal the share of area under HYVs decreased as farm size increased in all the years studied. Further, the difference in adoption of HYVs between large and other categories was much larger on unirrigated land than on irrigated land.

Table 1.4

Share of Area under HYV in Different Categories of Farms (%)

	Year	*Marginal*	*Small*	*Semi-medium*	*Medium*	*Large*	*All Size Category*
Irrigated							
	1996-97	80	76	76	76	75	77
	2001-02	89	86	85	82	78	85
	2006-07	91	89	86	83	81	87
Unirrigated							
	1996-97	37	37	38	36	25	35
	2001-02	52	54	52	46	30	48
	2006-07	58	52	50	46	31	49
Total							
	1996-97	59	55	54	53	42	54
	2001-02	72	68	65	61	47	64
	2006-07	75	70	67	64	52	67

Source: *All India Report on Input Survey*, various years

Fertiliser Use

Analysis of fertiliser use per hectare of NSA on different size categories of farms revealed the following trends. During 1981-82, marginal farms applied 55 kg fertiliser per hectare, whereas large farms applied only 27 kg per hectare which is only half of rate applied on marginal farms (Table 1.5). Thus, fertiliser use was inversely related to farm size in 1981-82, and the same trend continued in 1991-92, 2001-02 and 2006-07 also. In 2006-07, marginal farms used 2.3 times the fertiliser used by large farmers on per hectare basis. Similarly, small farms used 1.96 times the per hectare fertiliser used by large farms. Further disaggregated analysis of fertiliser use in irrigated *versus* unirrigated area revealed that: (i) the difference in fertiliser use between irrigated and unirrigated land was lowest on marginal farms and (ii) marginal farms were using higher fertiliser per hectare compared to large farms even under unfavourable conditions. The share of fertiliser treated area in gross cropped area (GCA) increased from 50 per cent in 1981-82 to 77 per cent in 2001-02, but declined to 74 per cent in 2006-07 on marginal farms. In all size category it increased from 40 per cent to 73 per cent during this period indicating that all-India level, in 2006-07, 27 per cent of GCA was not treated with any chemical fertiliser (Table 1.6). Subsidised fertiliser policy could be one underlying factor of this increase in share of fertiliser treated area.

Table 1.5

Fertiliser Use on Farms of Different Sizes

		Fertiliser consumption per ha of NSA according to farm size class (Kg)					
	Year	*Marginal*	*Small*	*Semi-medium*	*Medium*	*Large*	*All Size Category*
Total	1981-82	55	48	42	36	27	40
	1991-92	99	85	77	68	54	76
	2001-02	175	129	112	95	68	119
	2006-07	198	167	137	120	85	147
Irrigated	1981-82	100	98	99	97	101	99
	1991-92	145	140	147	144	157	146
	2001-02	252	206	190	174	171	203
	2006-07	276	256	209	195	173	229
Unirrigated	1981-82	24	22	18	14	9	16
	1991-92	58	48	39	32	19	38
	2001-02	96	76	63	46	22	61
	2006-07	114	87	75	56	29	75

Source: *All India Report on Input Survey*, various years.

Farmyard Manure (FYM) and Other Organic Manure Use

It is observed that share of area treated with FYM increased with farm size upto small category, later declined with farm size in all the studied years (Table 1.6). However, no definite trend was observed over years with regard to share of area treated with FYM in each size category farm. In

1981-82 on marginal farms, 34 per cent of GCA was treated with FYM as against only 19 per cent on large farms. In 2006-07, 28 per cent of cropped area was treated with FYM on marginal farms, as against 19 per cent on large farms. Variation in share of area treated with FYM over years was lower on large category farms compared to all other size category farms. With respect to green manure, oil cake and other organic manure also, the analysis results revealed that share of area treated in total cropped area decreased with farm size. In case of green manure, on marginal farms, small farms and semi-medium farms, share of area treated increased over years. But with respect to proportion of area treated with oil cake and other organic manure, no consistent trend over years was observed on each size farm category.

Table 1.6

Share of Area Treated with Fertilisers and Other Different Manures in GCA on Different Farms (%)

		Marginal	*Small*	*Semi-medium*	*Medium*	*Large*	*All Size*
Fertilisers							
	1981-82	50	45	42	37	29	40
	1996-97	64	63	61	57	45	59
	2001-02	77	75	72	67	52	70
	2006-07	74	77	76	71	58	73
FYM							
	1981-82	34	36	31	27	19	29
	1996-97	25	26	24	20	16	23
	2001-02	30	34	32	26	18	29
	2006-07	28	30	29	23	19	27
Green Manure							
	1996-97	1.26	0.84	0.68	0.50	0.43	0.74
	2001-02	1.44	0.97	0.78	0.61	0.42	0.88
	2006-07	1.53	1.28	0.89	0.60	0.60	1.03
Oil Cake							
	1996-97	1.64	1.38	0.90	0.46	0.19	0.93
	2001-02	1.26	1.02	0.80	0.62	0.28	0.84
	2006-07	1.88	1.15	0.83	0.43	0.25	0.99
Other Organic Manure							
	1996-97	3.32	2.30	1.53	0.86	0.48	1.71
	2001-02	2.15	1.07	0.66	0.36	0.17	0.93
	2006-07	2.30	1.48	1.22	0.72	0.31	1.32

Source: *All India Report on Input Survey*, various years.

Pesticide Use

Pesticide treated area share on different size category farms increased over the years (Table 1.7). In 1981-82 it ranged between 8 to 10 per cent across different farm size categories, in 2006-07 it ranged between 38 to 43 per cent. In 2006-07, the share of area treated with pesticides increased with farm size (with the exception of small farms and large farms), similar trend was observed in other years also (with the exception of large farms). But further disaggregated analysis, indicated that positive relation between proportion of area treated with pesticides and farm size was there in 1981-82, 1991-92, and 2001-02 in irrigated area but not in unirrigated area. Cropping pattern together with irrigation pattern could be the underlying factors of the observed trend in share of area treated with pesticides. In 1981-82, on marginal farms ratio of pesticide treated area under irrigation to un-irrigation worked out to be 6, it declined to 1.7 in 2006-07. On small farms the ratio declined from 4.2 to 1.4, and on large farms the ratio declined from 5 to 3. Thus, over the years disparity in proportion of area treated with pesticide under irrigated condition *versus* unirrigated condition on different farms declined.

Table 1.7

Share of Area Treated with Pesticides on Different Size Farms (%)

		Marginal	*Small*	*Semi-Medium*	*Medium*	*Large*	*All Size*
Irrigated							
	1981-82	18	21	21	21	21	20
	1991-92	25	27	33	37	41	32
	2001-02	40	40	43	48	57	44
	2006-07	49	45	52	57	65	52
Unirrigated							
	1981-82	3	5	4	5	4	4
	1991-92	8	12	12	12	7	11
	2001-02	23	27	26	21	12	23
	2006-07	28	32	32	30	22	30
Total							
	1981-82	9	10	10	10	8	9
	1991-92	16	18	20	21	18	19
	2001-02	32	33	33	33	29	32
	2006-07	39	38	41	43	41	40

Source: All India Report on Input Survey, various years.

Institutional Credit use

Analysis of farm size wise institutional credit availment yielded the following insights. In all the selected years, share of holdings availing institutional credit increased with farm size (Table 1.8). In 2006-07, only 20 per cent of marginal farms availed institutional credit as against 40 per cent on large farms. In 2001-02 only 20 per cent of all holdings category availed institutional credit, it increased to 25 per cent in 2006-07. On the other hand according to NSSO Situation Assessment Survey, 48.6 per cent of farmer households in India were indebted (in 2002-03). This indicates the persistent importance of non-institutional credit sources in credit to farmer households. Prevalence of indebtedness increased with farm size. Over the years institutional credit amount per holding increased with farm size. According to NSSO survey, amount of outstanding loan also increased with farm size.

In 1986-87 on marginal farms share of short term credit in institutional credit worked out to be 57 per cent and it increased to 71 per cent in 2006-07. This is in accordance with the observed increased input use on these farms. In other size category farms no consistent trend over years was observed in the share of short-term credit. However in all the years, short-term credit constituted the major credit in credit availed on different size category farms. In 1986-87 and 1991-92, long-term credit was the second major credit item across different size category farms, but later the trend changed. In 2001-02 and 2006-07, only on medium size and large size category farms, long-term credit was the second major credit item, in all other cases medium-term credit was the second major item of credit. It is worth to note that in 2006-07 on different size category farms, medium-term credit share worked out to be around 20 per cent. In 2006-07, share of long term credit on different farm size category ranged between 10 to 33 per cent, it increased with farm size.

In the years 1986-87 and 1991-92, share of fertiliser value in short-term institutional credit did not follow any definitive trend with farm size but in later years i.e., 2001-02 and 2006-07, it increased with farm size. However, over the years share of fertiliser value in short-term credit declined following an inconsistent trend across different size category farms. In 1986-87, across different farm size category farms, share of fertiliser constituted 30 to 45 per cent of short-term credit. In contrast it ranged between 18-29 per cent only, in 2006-07. In 1986-87 and 1991-92, share of fertiliser value in short-term credit was maximum in the case of medium farms, but in 2001-02 and 2006-07, the share was maximum in the case of large farms.

Table 1.8

Institutional Credit Use and Extent of Indebtedness among Farmers in India

	Share of holdings availing institutional credit (%)					
Year	*Marginal*	*Small*	*Semi-Medium*	*Medium*	*Large*	*All*
1981-82	18	27	31	33	33	23
1991-92	16	18	22	22	22	18
2001-02	14	28	32	33	29	20
2006-07	20	33	35	39	40	25
	Prevalence of indebtedness among farmer households in 2002-03(%)					
	45	51	58.2	65.1	66.4	48.6
	Institutional Credit per holding (₹)					
Year	*Marginal*	*Small*	*Semi-Medium*	*Medium*	*Large*	*All*
1981-82	592	1111	1384	2248	3332	1162
1986-87	1020	1446	1965	3088	4929	1682
1991-92	1240	2571	3694	5921	11554	2572
2001-02	10817	16781	24972	42637	72519	19264
2006-07	21847	31086	46015	74163	114043	33673
	Amount of outstanding loan per farmer household in 2002-03 (₹)					
	7549	13762	23456	42532	76232	12585
	Share of different types of institutional credit on different size category farms					
Year	*Marginal*	*Small*	*Semi-Medium*	*Medium*	*Large*	*All*
			Short-term credit			
1986-87	57	54	49	45	47	51
1991-92	63	68	64	61	58	63
2001-02	69	71	67	60	53	66
2006-07	71	67	63	53	47	64
			Medium-term credit			
1986-87	17	21	20	16	11	18
1991-92	16	11	11	10	8	12
2001-02	16	15	17	17	20	17
2006-07	19	20	19	20	20	20

contd...

...contd...

Year	*Share of different types of institutional credit on different size category farms*					
	Marginal	*Small*	*Semi-Medium*	*Medium*	*Large*	*All*
			Long-term credit			
1986-87	26	25	31	39	41	31
1991-92	21	20	25	30	34	25
2001-02	14	14	16	23	27	17
2006-07	10	13	18	28	33	17

Year	*Share of fertiliser value in short term institutional credit on different size category farms*					
	Marginal	*Small*	*Semi-Medium*	*Medium*	*Large*	*All*
1986-87	31	30	40	45	37	36
1991-92	16	26	33	33	28	27
2001-02	21	26	30	36	45	28
2006-07	18	20	22	25	29	21

Source: (i) *All India Reports on Input Surveys* and (ii) NSSO 59th Round Survey, Farmers Situation Assessment Survey Reports.

One argument that is being often put forth for lower institutional credit availment by marginal and small farmers, is their limited credit absorption capacity. But the present analysis results of input use pattern and increasing share of short-term credit in total credit, clearly shows that credit absorption capacity of smallholders is not limited. Hence, the other reasons for lower availment of institutional credit by smallholders can be risk-averse behaviour of smallholders and inaccessibility of institutional credit due to higher transaction cost. Higher prevalence of indebtedness among smallholders coupled with their higher dependence on informal credit sources support the reason of inaccessibility of institutional credit. Innovation in institutional credit delivery *viz.*, Kisan Credit Card (KCC) is reported to increase the availability of institutional credit to smallholders. Samantara (2010) based on a study on this aspect, reported that KCC scheme has made dent in the horizontal growth of credit in terms of coverage of farmers, but there is lack of depth in credit flow. Kumar *et al.* (2011) reported that in 2003, share of households having KCC increased with farm size.

Smallholders Share in Different Inputs in Indian Agriculture

A look into the smallholders share in different inputs indicates that: (i) Smallholders share in number of holdings availing institutional credit increased over the years, but still their share was 75 per cent only against their share of 83 per cent in number of holdings in 2006-07. (ii) Smallholders share in total fertilisers and FYM used, and total institutional agricultural credit increased over the years and is matching with their share in land (Table 1.9). However as observed in previous tables, there is still scope to increase share of smallholders (area) using different inputs in total smallholders (area).

Table 1.9

Smallholders Share in Different Input Use (%)

	1981-82	*1986-87*	*1991-92*	*1996-97*	*2001-02*	*2006-07*
Total number of holdings	75	76	77	80	81	83
Number of holdings who availed institutional credit	65	64	72	68	69	75
Total holdings area	28	30	33	36	39	41
Total NSA	29	31	35	37	40	42
Total GCA	31	33	37	38	42	44
Total gross irrigated area	36	37	42	42	46	46
Total net irrigated area	37	38	42	42	45	45
Area treated with fertilisers	37	36	39	41	45	46
Area treated with FYM	37	38	44	43	46	49
Area treated with pesticides	32	29	33	45	42	42
Total nitrogen used	39	38	42	45	50	52
Total phosphorous used	33	39	39	44	50	52
Total potassium used	39	47	50	59	62	63
Total fertiliser used	37	39	42	46	51	53
Quantity of FYM used	44	39	54	63	53	58
Total Institutional credit	43	45	45	48	47	56

Source: *All India Reports on Input Surveys*, various years.

Cropping Pattern

Cropping pattern is also one of the determinants of productivity in agriculture. But cropping pattern of different size category farmers may be influenced by different motivations like ensuring family food security, risk spreading, reaping benefits of scale economy, responding to changes in market demand etc. In the Indian context based on micro-level studies, availability of irrigation (Saleth, 1997), access to infrastructural facilities (Chand, 1996; De and Chattopadhyay, 2010) were identified as major determinants of cropping pattern rather than farm size. Birthal *et al.* (using NSSO 54th round data pertaining to year 1998) showed that provided enabling factors are there, smallholder's participation in high value crop cultivation was more, particularly in vegetables due to higher availability of family labour with them.

Table 1.10

Cropping Patten on Different Size Farm Categories

		Share of different crop groups in total cropped area (%)					
		Marginal	*Small*	*Semi-medium*	*Medium*	*Large*	*All Size Category*
Cereals							
	1996-97	71	62	57	51	45	57
	2001-02	70	61	56	53	50	58
	2006-07	72	63	60	55	52	61
Pulses							
	1996-97	4	5	5	6	6	5
	2001-02	7	10	11	12	13	10
	2006-07	8	9	9	10	13	9
Sugar crops							
	1996-97	2	2	2	2	1	2
	2001-02	3	3	3	2	1	3
	2006-07	3	3	3	3	1	3
Spices and condiments							
	1996-97	0	0	0	0	0	0
	2001-02	1	1	1	1	1	1
	2006-07	1	1	1	1	1	1
Fruits							
	1996-97	1	1	1	0	0	0
	2001-02	1	1	1	1	1	1
	2006-07	1	1	1	1	1	1
Vegetables							
	1996-97	1	1	1	0	0	0
	2001-02	3	2	2	1	1	2
	2006-07	2	4	1	1	1	2
Oilseeds							
	1996-97	6	10	12	15	14	12
	2001-02	9	12	14	16	14	13
	2006-07	8	13	15	17	15	13
Fibres							
	1996-97	2	3	4	4	4	3
	2001-02	4	7	7	7	6	6
	2006-07	3	4	5	6	5	4

Source: *All India Report on Input Survey*, various years.

In the present study, cropping pattern across different farm sizes at all-India level was analysed using data from input surveys in three years i.e., 1996-97, 2001-02 and 2006-07 and the results were presented (Table 1.10). The analysis results revealed that the area allocated to cereals decreased with farm size in all the three years. But in the case of marginal, small and semi-medium categories, share of area allocated to cereals declined between 1996-97 and 2001-02, increased between 2001-02 and 2006-07. On the other hand, on medium and large size farms, proportion of area allocated to cereals increased continuously. In 2006-07, share of area allocated to cereals stood at 72 per cent in the case of marginal farms, 63 per cent in the case of small farms and 52 per cent in the case of large farms. The share of area under pulses increased with farm size in all the years. The share of area allocated to vegetables was maximum in the case of marginal farms in 2001-02, in the case of small farms in 2006-07. In all the three years, share of area allocated for oilseeds was maximum in the case of medium farms. Smallholders allocated lower share of their area for oilseeds, fibres and pulses compared to large farmers. The results indicate that the production of staple food crops is dominant consideration in all size categories.

Farm Size-Productivity-Efficiency-Livelihood

Analysing NSSO 59th round farmer's situation assessment survey data (pertaining to the year 2002-03) inverse relation between farm size and land productivity was reported with respect to gross receipts by Chand *et al.* (2011) and net receipts by Gaurav and Mishra (2011). The results from further probe into the data are presented in the Table 1.11.

From Table 1.11 it is evident that net receipts per rupee of total paid out cultivation expenditure increased with farm size upto semi-medium class but then declined with farm size. Similar trend was observed with respect to net receipts per rupee labour expenditure. However, per person net receipts from cultivation (when accounted for family size) increased with farm size. Per capita net receipts on marginal holdings was one-tenth of that on large farms and per capita net receipts on small farms was one-fourth of that of large farms. This low per capita receipts on small farms is making them unviable.

Table 1.11

Productivity and Efficiency on Different Size Category Farms in 2002-03

	Unit	*Marginal*	*Small*	*Semi-Medium*	*Medium*	*Large*	*All Size*
Total paid out cultivation expenses	₹/ha	9608	7727	6787	6452	3845	7194
Gross receipts from cultivation (including value of by products)	₹/ha	22232	18128	16219	14451	8205	16618
Net receipts over paid out costs	₹/ha	12624	10401	9432	8000	4361	9424
Net receipts per rupee of total paid out cultivation expenditure	₹/₹	1.31	1.35	1.39	1.24	1.13	1.31
Net receipts per rupees of labour expenditure	₹/₹	5.97	6.10	6.20	5.40	4.63	5.83
Per person net receipts from cultivation	₹	1001	2498	4031	6586	10141	2093
Per capita availability of land	ha	0.08	0.24	0.43	0.82	2.2	0.22

Source: Author's computation using NSSO 59th round survey farmers situation assessment survey data.

Strengthening Smallholder's Productivity and Improving their Livelihood

Literature is more inconclusive about relationship between farm size and labour productivity (Fan and Chan-Kang, 2005; Brigham, 2003). But given the established relation between labour productivity, land productivity and land to labour ratio, and clear inverse land productivity relation at all-India level (observed in the present study also), the way to improve efficiency of agriculture together with improved livelihood of smallholders is to improve land to labour ratio on these farms. This can be brought about by either increasing availability of land through tenancy or surplus land redistribution or by reducing the number of people dependent on agriculture. Vijay (2012) reporting increasing share of non-cultivating peasant households i.e., households retaining interest in land but not cultivating, declining share of cultivating households and stable proportion of agricultural labour households in India, opined that going back to land reforms is one option for dealing with agrarian crisis. However in the context of widely documented decreasing tenancy transactions and limitations on surplus land distribution, labour productivity can be improved only by reducing the number of people dependent on agriculture on small farms. Initially this can be on part time basis, but on permanent basis in the long run.

Positive effect of shifting the surplus labour from cropping to non-cropping sector was reported in the context of China (Lin, 1994). But in the Indian context this shifting/mobility is lower. There appear two underlying factors for this. As per 2006-07 input survey, average age of Indian farmer worked out to be 48 years. Share of farmers of above 50 years age was 38 per cent on marginal farms, and 40 per cent on small farms. On the other hand share of illiterate farmers worked out to be 33 per cent on marginal farms, and 32 per cent on small farms. This age factor coupled with illiteracy, reflecting as limitation on skill of farmers may limit their ability to shift completely to a non-farm job of high remuneration. Hence, creating non-farm job opportunities for smallholder cultivators outside agriculture but within countryside can bc one policy option. Creation of part time non-farm employment opportunities will aid in improving livelihood of smallholders.

The second underlying factor pertains to intergenerational occupational mobility. In a recent study Motiram and Singh (2012) analysing the data of India Human Development Survey (IHDS), 2005, using transitional matrix approach, reported that socioeconomic limitations and inequity in opportunities is leading to lower intergenerational occupational mobility in India. Hence, these constraints need to be addressed for moving surplus labour from agriculture to non-agriculture and improve their livelihood in the long run. Improving opportunities for education and development of skills of children of farming population can aid in this intergenerational mobility for improved livelihood.

Observed high cost of cultivation on marginal and small farms (Table 1.11), indicates the possibility of improving net returns by cost management. Addressing resource constraints faced by smallholder farmers and improving their access to modern inputs also will aid in strengthening their productivity. Increasing smallholder's access to institutional credit, increasing share of fertiliser value component of short-term credit on marginal and small farms which stood at 18 and 20 per cent respectively in 2006-07, can be one option in fertiliser cost management. Providing enabling conditions for acceleration in diversification in farming by smallholders will also help in improving their livelihoods.

Conclusions

Smallholders are ahead of other farm size categories in adoption of improved technologies and use of fertiliser, organic manure and irrigation. But they are lagging behind in availing institutional credit and crop-diversification more specifically in diversification towards oilseeds and fibre crops. Despite this limitation, present study results revealed that smallholders in India are superior in terms of land productivity performance, but they are weak in terms of generating adequate income and sustaining livelihood. This is mainly due to lower per capita availability of land on these farms besides limited diversification. Given the limitations recorded regarding earlier land reforms, the option of increasing per capita availability of land may not be that much easier. The other option is to shift some members in smallholder families to non-farm sector. There are some constraints in doing this (like illiteracy and inequality in opportunity) and these needs to be addressed. Besides this, improved cost management on small farms can also aid in improving their income. Hence, multipronged strategy is needed for improving livelihood of smallholders and harnessing their strengths in improving Indian agricultural performance.

References

Birthal, P.S, P.K. Joshi, Devesh Roy and Amit Thorat (2007). "Diversification in Indian Agriculture Towards High-value Crops: The Role of Smallholders", *International Food Policy Research Institute Discussion paper* 00727. IFPRI.

Brigham, Anne Margrethe (2003). "Just Hungry? Analyzing the Relationship between Land Concentration and Food Insecurity". Dr. Polit dissertation in Political Science. Department of Sociology and Political Science, Norwegian University of Science and Technology.

Chand (1996). "Diversification through High Value Crops in Western Himalayan Region: Evidence from Himachal Pradesh", *Indian Journal of Agricultural Economics* 51(4): 652-63.

Chand, Ramesh, P.A. Lakshmi Prasanna and Aruna Singh (2011). "Farm Size and Productivity: Understanding the Strengths of Smallholders and Improving their Livelihoods", *Economic and Political Weekly* XLVI(26 and 27): 5-11.

De, Utpal Kumar and Manabendu Chattopadhyay (2010). "Crop Diversification by Poor Peasants and Role of Infrastructure: Evidence from West Bengal", *Journal of Development and Agricultural Economics* 2(10): 340-50.

Fan, Shenggen and Connie Chan-Kang (2005). "Is Small Beautiful? Farm Size, Productivity and Poverty in Asian Agriculture", *Agricultural Economics 2005* 32(S1): 135-46.

Gaurav, Sarthak and Srijit Mishra (2011). "Size-Class and Retuns to Cultivation in India: A Cold Case Reopened", *Working Paper* WP-2011-027. Mumbai: Indira Gandhi Institute of Development Research.

Kumar, Anjani, Chitra Yadav, Shiv Jee, Sant Kumar and Sonia Chauhan (2011). "Financial Innovation in Indian Agricultural Credit Market: Progress and Performance of Kisan Credit Card", *Indian Journal of Agricultural Economics and Policy Research* 66(3): 418-28.

Lin, Bo.Q (1994). "Rural Reforms, Structural Change and Agricultural Growth in the People's Republic of China", *Asian Development Bank Report* No.62.

Motiram, Sripad and Singh Ashish (2012). "How Close Does the Apple Falls to the Tree? Some Evidence from India on Intergenerational Occupational Mobility", *Economic and Political Weekly* XLVII(40): 56-65.

Saleth, Maria (1997). "Diversification Strategy for Small Farmers and Landless: Some Evidences from Tamil Nadu", *Indian Journal of Agricultural Economics* 52(1): 73-86.

Samantara, Samir (2010). "Kisan Credit Card: A Study", *Occasional Paper* No.52, Mumbai NABARD.

Vijay, R. (2012). "Structural Retrogression and Rise of 'New Landlords' in Indian Agriculture: An Empirical Exercise", *Economic and Political Weekly* XLVII(5): 37-45.

2 Agrarian Reforms and the Precarious State of Affairs in the Darjeeling Hills

A Review

Swatahsiddha Sarkar[1]

It is by now a well established fact that the ecological and geomorphological factors largely determine the rather agriculture unfriendly character of the hill region. Such issues did not however, diminish the significance of land and for that matter of land reform for the agrarian regions located in the hills. In fact, one can hardly deny the very fact that land and agriculture still contributes, in no less significant manner, towards the maintenance of livelihood of a sizeable section of population in the hill region of the district of Darjeeling. Conspicuously, studies available on the social implication of agrarian reform measures in West Bengal, by and large, focused their otherwise scholarly attention to the ground realities of the plain regions. Studies on hill agrarian situation, in general and on Darjeeling hills, in particular, are not merely few but rare indeed. Only sociological works available on the issue having reference to Darjeeling hills were produced almost single handedly by Subba (1983, 1985, 1986, 1989), and to some extent by Chaudhury (1985). The present study is based on the presumption that issues like agrarian reform measures are capable of ushering in equally significant social consequences for the hill areas too, notwithstanding the fact that such regions are ecologically not so suitable for tilling purposes. However, as the present review will show, the actual reality in the Darjeeling hills with regard to the question of land reform is far from being impressive.

---

1. The author is indebted to Dr Anirban Dasgupta, Associate Professor, South Asian University, New Delhi, for his generous permission to use the data set of his research for the present paper.

The Area

Darjeeling, the northern most district of West Bengal, is constituted of both hill and plain areas. While the hill region of the district comprises the three hill subdivions namely Darjeeling Sadar, Kurseong and Kalimpong, the Siliguri subdivision is made up of the plain areas of the district, usually referred to as Terai. The district of Darjeeling, besides having this rather rigid geomorphological distinctiveness, is also earmarked by its own historical idiosyncrasies. Available official historical sources on Darjeeling's past pinpoint that the area was made a part of Bengal only in 1835 and finally in 1865. With the inclusion of Kalimpong (in 1865), the entire region took the shape of a district as of now. The territories which now constitute the whole district earlier belonged to the three Himalayan kingdoms of Sikkim, Nepal and Bhutan. Darjeeling, during the time of its first colonial encounter, was an uninhabitable bushy tract thinly populated by about hundred souls 'probably' of Lepcha origin. It deserves mention that tea plantation has been the staple industry of Darjeeling hills since the middle of 19th century. The rapid development of tea plantation in the region was the result of direct penetration of colonial capital and colonial patronage. Contrary to this the issue of agriculture in Darjeeling hills did flourish, particularly in the Kalimpong subdivision, through indigenous efforts alone. The tea plantation has had its huge concentration in the two hill subdivisions of Darjeeling Sadar and Kurseong located in the western part of the river Tista. Consequently, the agrarian issues of the hills receieved its due consideration only after the inclusion of Kalimpong region, which took place in 1865. Although the Lepchas, the so-called autochthones of the region, had been ignorant in the practice of settled cultivation, agriculture floursihed in the Darjeeling hills with the incoming of the Nepalis in the region. The Nepalis taught the Lepchas and to some extent the Bhutias too (who were familiar only with subsistence based shifting cultivation) the art of settled cultivation following the method of terrace cultivation. All these perhaps explain why the Nepalis enjoy, still today, the pivotal position in the overall hill agrarian social system.

Evolution of Agrarian Social Structure in Colonial Darjeeling Hills

The historical evolution of agrarian relations in Darjeeling hills and its distinctiveness was upheld by the provision that the government remained to be the sole proprietor of all estates within the tract and there existed practically none in between the government and the *raiyat* (tenant), the actual tiller of the soil. Neither the Bengal Tenancy Act (1885) was promulgated in Darjeeling hills nor was even the entire hilly tract of the district brought under the provisions of permanent settlement. In place of permanent settlement or the Bengal Tenancy Act, several regulations were promulgated for the administration of 'Waste Land' regions like: Waste Land Rule of 1825, Regulation III of 1828, Grant Rules of 1830, Waste Lands Rules of 1853, Fee Sample Grant and Redemption Rules of 1863. However, the rights and obligations of all classes of tenants and land revenue were settled according to the provisions of Bengal Rent Act, 1859 and Bengal Rent Settlement Act, 1879, respectively. This arrangement remained intact till the implementation of West Bengal Estate Acquisition Act of 1953 (Todarmal, 2002: 77–78).

The uniqueness of Darjeeling's agrarian structure was largely circumscribed by such differential administrative arrangements. As an immediate consequence of such provisions *zamindars* or of *zamindari* system had not been encouraged in Darjeeling hills, the *raiyats* since the first settlement were not permitted to sublet their lands (which they received for 10 years lease on the basis of payment of rent) and only local hillmen *viz.*, Lepchas, Bhutias and Nepalis were entitled to become *raiyats*. Above the *raiyats* there existed another category of people who were termed as *mandals* (village headman). *Mandals* were entrusted with the responsibility to collect the rent from the *raiyats* according to the rent rolls and remit them to the Manager, who remained responsible to the Deputy Commissioner. Apart from this the *mandals* had also the responsibility to make arrangements for the free labour needed for the purpose of construction of roads and they also played the instrumental role of being the intermediary between the government and the *raiyats*. Being in possession of larger income, resources, power and privileges, the *mandals* were able to place themselves in the highest echelon of hill agrarian hierarchy. During 1880–1900 (i.e., the period ranging between the enactment of the first settlement to the third) the *mandals* had exerted tremendous amount of power and control upon the *raiyats* in general and the new coming settlers in particular through their capacity to settle the new *raiyats* and to realise the amount of rent whatsoever being charged against such new settlements. The fallout of such provisions, particularly for a larger body of a mass that used to arrive there in search of livelihood, could be well surmised. The possession of large amount of land and the power of supervision of the entire agrarian belt capacitated the *mandals* to maintain a process of subinfeudation—disregarding the provisions of law—in which the illiterate and ignorant cultivators remained always under their absolute control who offered at times new leases from their storage of land defying the provisions of the *patta* (legal sanction of holding a plot) which they received from the government.

As pointed out earlier, the provisions of revenue administration during colonial period has represented the entire agrarian set up as the abode of *raiyats*, the real proprietors of all the estates of the tract, who held land on lease for tilling purposes on payment of regular rents charged by the government. However, among the *raiyats* some were entrusted with special responsibilities of collection of rents (along with certain other duties for which they enjoyed special privileges too), who were termed as the *mandal*. Thus, in Government records the entire agrarian population of Darjeeling hills was categorised generally as the *raiyats* or specifically as two categories of the *raiyats*: one group was represented simply as the assemblage of those *raiyats* who actually tilled the land, paid rents on a regular basis, and renewed the lease on usual intervals; and the other group was represented by a smaller section of privileged *raiyats*—the *mandals*, who had to perform other responsibilities laid on them by the government, apart from paying rents, or renewing the lease of the tract they possessed on a regular basis.

More often than not, social life in agrarian societies as opposed to their industrial counterpart is governed more by personal ties and loyalties than by impersonal rules. Srinivas and Béteille (1964) had well reminded us that social networks in agrarian societies are usually close-knit and social relations have a multiplex character so that one's partner in work is also related to oneself in a number of other ways. The concern for agrarian social structure as a system incorporating the relations of mutual cooperation and loyalties can help us understand other crucial processes of social significance which often fuse the realm of culture with that of political economy. The social history of agrarian social structure of the hills sheds enough light on all these issues and processes. The provisions of cooperation and mutuality has enabled the hill agrarian social system acquire a resilient structure that maintained almost the same categories bearing almost the similar connotations even in the post independent phase too. While taking into consideration the hill agrarian situation it was observed that amongst the various structural units, which provided the basis for the kind of mutual role-relationship and cooperation that existed historically, the structural bases of kinship, locality[2] (in the sense of *char chhimeki* or neighbourhood), and forms of exchanging labour (*viz.*, *parma* and *hoori*)[3] had reinforced each other.

---

2. The locality or neighbourhood happens to be a vital component of hill agrarian social structure that provides the basis for the networks of cooperation and mutual role responsibility spread amongst the incumbents having varied class backgrounds. In a region like Darjeeling hills, where the frequent movements of people—due to migration and immigration—have been historically acknowledged, the significance of *char chhimeki* (a local term whose connotation involves at least three interrelated meanings *viz.*, neighbourhood, belongingingness to the community and a commonly shared past of co-residentiality, in developing the relationship of cooperation in non-economic terms) can hardly be disregarded. The concept of *char chhimeki* has reinforced not only the forms of labour exchange but also brought the subordinated rural folk closer to their *maliks* or *pattadars* (landlords) at times of distress, economic or otherwise. In such situations, the relations between landowners and the landless were not always restricted to land or cultivation, but covered other spheres of life like providing aid—in cash or kind—in matters of familial occasions or festivities. Moreover, the linguistic cleavage and the frequency of intermarriage among different communities have helped the consolidation of *char chhimeki* in the rural society of Darjeeling hills.
3. The nature of labour exchange has been the crucial domain that explains why cooperation and mutuality was a likely phenomenon in the hierarchically arranged rural social structure of the hills. *Parma* and *hoori* were two such forms of labour exchange commonly practised throughout the hill agrarian belt until recently. While the *parma* system guarantees the rotational exchange of individual labour to each other's household (generally two households were involved) the *hoori* scheme assures the rotational exchange of group labour against those households which are involved in such a scheme. Although both the systems are out of the purview of cash payment but in case of *hoori* arrangements the services of the surplus labour remained marketable. This happens in such situations when the recipient household of *hoori* services cannot make use of the labour provided by the entire group (composed of as many as 20 persons) and have the authority both to sell the services of surplus labour in either cash or kind and to enjoy the returns of the sold out labour. Both these forms of labour exchanges underlie the strong cooperative spirit of the local communities who generally belong to the subordinate categories of rural hierarchy like the marginal farmers and/or the sharecroppers. These forms of labour exchange had been the ingenious creation of the earlier generation in the region when cultivation remained heavily labour intensive due to unfavourable soil type which contained rugged terrain, undulating surface and often required clear felling so as to prepare the grounds for cultivation.

It is true that during colonial period the intimately knit personalised networks of hill agrarian social structure had helped maintain the organic unity of the primary producers with the means of production. However, this hardly prevented it from experiencing the challenges of subinfeudation, land alienation, declining size of average per capita holding, and economic disparity. It had been possible to resolve such crises amicably during colonial times as long as tea plantation provided the scope for alternative livelihood arrangement, on the one hand and Kalimpong contained reasonable space to absorb the growing size of population economically, on the other. Consequently, the grievances of the rural mass during colonial times had never reached the level of unaccommodating conflicts of interests. But with the passage of time, particularly since the post-reform period, the rural social formation of the hills generated in itself the forces of its own adversity.

The Beginning of Land Reform in Darjeeling Hills

With India's independence, Darjeeling hills has been brought under the uniform legislative measures promulgated by the Government of West Bengal throughout the state on the issue of land. Two such legislations were of enormous significance which brought fundamental changes in the overall rural social structure of the state of West Bengal in general. The first such an Act was the West Bengal Estates Acquisition Act (WBEAA) of 1953, passed on February 12, 1954 and extended by stages to cover the entire state, which attempted to eliminate—among other things—the interests of the *zamindars* and other intermediaries by acquisition on payment of compensation. Although the WBEAA of 1953 had successfully met the need of arresting the growth of a feudal seeming social order throughout the countryside of the state, it was largely ineffective in Darjeeling hills. Darjeeling hills was described in colonial revenue administration as 'waste land' and stayed outside the purview of both Permanent Settlement of 1793 and Bengal Tenancy Act, 1885. Under Regulation III of 1828 all 'waste lands' were declared as Crown's property, hence the question of *zaminadari* did not arise in case of Darjeeling hills. Consequently, the provisions of Estate Acquisition Act, which might have brought significant changes in rural West Bengal, failed to bring forth similar consequences for the hill region of the district of Darjeeling. Perhaps this was due to the underlying factor that both the tea gardens and agricultural lands in hill areas of Darjeeling district were exempted from the stipulation for a ceiling on existing holdings at 25 acres in respect of agricultural lands as per the provisions of the WBEAA of 1953 which was applied uniformly throughout the state (GoI, 1966: 151). The exemption of Darjeeling hills from the ceiling rider maintained up to the late 1970s until the provisions of West Bengal Land Reforms Act (WBLRA), 1955 were promulgated for the first time on January 1, 1978.

The measures promulgated under the Estates Acquisition Act of 1953 to do away with the *zamindari* system necessitated corresponding changes in the tenural and land ceiling arrangements. This has been done through the promulgation of WBLRA of 1955 which contained provisions for the fixation of the upper limit of the ceiling. As per the stipulations of the said Act (Section 14 M of the WBLRA, 1955) the ceiling area ranges from 2.5 standard hectares for an unmarried adult to a maximum of 7.0 standard hectares for an owner with a nine-member family. It needs mention here that the ceiling limit had been unanimously applied in all districts but the state had the authority to allow the landowners to hold more lands than the ceiling limit if they intend to establish tea gardens (Clause 14Y of WBLRA, 1955, inserted by West Bengal Land Reforms Amendment Act 1996, dated 27.8.1996, w.e.f. 27.8.1996). Besides all these, the Government of West Bengal since 1978 had given special emphasis (on Clause 50 of WBLRA, 1955) to record the name of the *bargadars* (sharecroppers) in the record-of-rights so as to prevent their frequent evictions and thereby to guarantee tenural security to the *bargadars* under the programme called operation *barga*. The land reform measures adopted by the Left Front Government (LFG) since 1978 also consisted of the programme of redistributing ceiling surplus land among rural landless/land poor households. Although the successful implementation of both the reform measures (i.e., tenural security and land redistribution) throughout the state had enabled West Bengal acquire global recognition for setting off a ruling machinery that carried out the principle of redistribution of resources amidst a liberal bourgeois democratic national set up, the situation in Darjeeling hills is not that rewarding. Table 2.1 gives us the impression that could hardly be considered as very encouraging. So far as the reform programmes involving the issues like the area of land redistributed and the quantity of land recorded under the Operation Barga are concerned Darjeeling's performance is worst among the six districts of North Bengal. Moreover, the percentage share of Darjeeling district in both the spheres fall far short of the state average too.

Table 2.1

District-wise Progress of the Agrarian Reform Programmes in Terms of Area (up to September 2003) in North Bengal

Districts	*Area of Land Redistributed (in ha.)*	*Area Recorded under OB (in ha.)*	*Net Sown Area (in ha.)*	*% of Area under OB*	*% of Area under Re-distribution*	*% of Area under Reform*
Darjeeling	12575	7012	143108	4.90	8.79	13.69
Jalpaiguri	40189	39089	329846	11.85	12.18	24.03
Coochbehar	25675	33785	253539	13.33	10.13	23.45
Dinajpur (N)	28973	9037	275213	3.28	10.53	13.81
Dinajpur (S)	19258	21634	194609	11.12	9.90	21.01
Malda	31122	31999	228485	14.00	13.62	27.63
West Bengal	439585	452370	5354196	8.45	8.21	16.66

Source: Adapted from Dasgupta, 2006: 27.

The difficulty to get hold of authentic quantitative data on land is well known. Moreover, another problem was that the available data—no matter how scanty it may be—on Darjeeling maintained as a whole the district picture. It is proposed here that in the absence of hills-plains distinction of data sources on the issue of land any attempt to assess the implication of land reform in Darjeeling hills would be both an unrealistic and unfair exercise. While maintaining a closer look at actual facts of agrarian reforms in West Bengal since 1977 scholars have provided a gloomy picture of land reform particularly in the district of Darjeeling. Data regarding the district-wise progress of agrarian reforms programme in Dasgupta's work reveal that up to 2003, the percentage of area under reform in Darjeeling district was 13.69. While the area recorded under operation *barga* was as low as 7012 hectares (4.90%), the area of land redistributed was moderately high at 12575 hectares (8.79%). The number of *bargadars* recorded under Operation Barga in Darjeeling district—the same source maintained—is 12879 (9.17% of the total 140506 number of agricultural workers) and the figure for land reform beneficiaries is 51991 (37% of the total number of agricultural workers). Dasgupta also noted that in the figures on perceived security, Darjeeling (and also Jalpaiguri) district occupied the alarmingly high proportion (47%) of *Bargadars*, who feel insecure about their rights as tenants. Table 2.2 is helpful in comprehending the situation of Darjeeling hills in a comparative manner with the performances of other five districts of North Bengal.

Table 2.2

District-wise Progress of the Agrarian Reform Programmes in terms of Beneficiaries (up to September 2003) in North Bengal

Districts	*Land Reform Beneficiaries*	*Bargadars Recorded under OB*	*Agricultural Workers= Cultivators+ Agricultural Workers*	*Land Reform Beneficiaries as % of Agricultural Workers*	*OB Beneficiaries as % of Agricultural Workers*	*Agrarian Reform Beneficiaries as % of Agricultural Workers*
Darjeeling	51991	12879	140506	37.00	9.17	46.17
Jalpaiguri	129637	61333	495462	22.16	12.38	38.54
Coochbihar	128023	84799	647285	19.78	13.10	32.88
Dinajpur (N)	147235	31942	647706	22.73	4.93	27.66
Dinajpur (S)	75812	71836	411516	18.42	17.46	35.88
Malda	153178	81693	689178	22.23	11.85	34.08
West Bengal	2,744,415	1,514,457	12964104	21.16	11.68	32.84

Source: Adapted from Dasgupta, 2006: 28.

It appears to be a surprising contrast that while West Bengal has shown an impressive performance in registering the tenants as *bargadars*, the performance of Darjeeling district in this regard is worst amongst all the districts of North Bengal, if not among all the districts of the state as a whole. Regarding the poorest performance of the district of Darjeeling in matters of registration of *bargadars*, the West Bengal Human Development Report (WBHDR), 2004 maintains that this might be because of the prevalence of tea estates on the one hand and the less practice of tenant-based cultivation in the region on the other. WBHDR also maintains that the role of local self government institutions like the panchayati raj institutions (PRI) has been very significant in determining the differntial performances of the districts in this regard.

The analysis maintained so far showed in more than one ways the fact that Darjeeling district had practically made modest progress in implementing agrarian reform measures. Although the overall district data in many cases represent a reasonably gleaming picture regarding the impact of reform measures, the state of affairs in the hill region of the district is not that wholesome as is reflected in the overall district profile. In fact, any attempt to make an appraisal of the reform measures in the context of Darjeeling hills, that too in the absence of hills-plains division of data sources, appeared to be challenging but not an impossible task. Even the findings of the WBHDR is not helpful in explaining the actual state of affairs in the Darjeeling hills. Some qualitative information, gleaned out from some secondary sources, may help us deduce the prospect of agrarian reforms in Darjeeling hills, if any.

Post–Reform Hill Agrarian Society: Consequences and Constraints

In fact, the scope for agrarian reforms remained largely limited in Darjeeling hills due to several innate constrains to which the hill people do not have any control. For example, the large forest coverage[4] of the district happens to be a natural barrier to usher in faster agricultural growth in the hills. Moreover, a sizeable amount of agricultural land had been brought under non-agricultural uses (i.e., plantation). The remaining portion of agrarian belt of the hill region is not also free from rugged terrains, undulating soil types etc. To bring more lands under cultivation is also not ecologically feasible. Besides all these, the uninterrupted migration had resulted into significant decrease in the average size of land holding particularly in the hills. Moreover, migration happened to be one of those crucial factors which restricted the proper implementation of operation *barga* in a significant way. Apart from these historical factors and ecological limits which hampered the smooth progress of reform measures promulgated by the former Left Front Government in West Bengal, some

4. As per 2001 Census data more than 38 per cent of the total geographical area of the district is under forest, in earlier Census decades the percentage share of forest coverage remained higher and if we consider only the case of the hill subdivisions forest coverage amounts to be around 49 per cent of the total geographical space.

structural changes definitely have taken place in the region under study. For example, the significance of *mandal* (village headman) system of earlier days has faded away to a great extent. Although *mandals* can still be found in some pockets of hill rural belt, but at the moment they lost their erstwhile power, privileges and positions in the rural hierarchy. However, this needs to be stressed out that the insignificance of the *mandals* in the hill agrarian situation is not due to the effective implementation of WBEAA. The institution called *mandal* system was not the result of permanent settlement and that is why it failed to gain the reformatory attention under the provisions of the said Act. Yet the spread of *panchayati raj* did affect the significance of the *mandals* in the rural social structure of the hills.

The Role of PRI

It deserves mentioning that while PRIs have effectively curtailed down the significance of the *mandals*, the PRIs in the hills unlike their counterpart in the plains did fail to prepare the ground for smooth functioning of land reform measures in the region. In fact, the evolution of PRIs in Darjeeling hills did reveal a chequred history of its own. West Bengal Panchayat Act, 1956, came into force in this region in 1961. Initial grooming of PRIs in the hills took place at three levels having *Gram Sabhas*, *Gram Panchayat* and *Anchalik Parishad/Anchal Panchayat* at the village level, *Panchayat Samiti* at the sub-divisional level and the District Board at the district level. In the 1970s, the structure of *panchayat* in the hills became somewhat coherent. According to the West Bengal Panchayat Act, 1973, the scheme of *panchayati raj* became three-tier system—*Gram Panchayat* at the village level, *Panchayat Samiti* at block level and *Zilla Parishad* at district level. The West Bengal Panchayat Act, 1973, came into operation in Darjeeling hills only in 1978. Within the passage of just one decade there has occurred again a remarkable shift that in fact transformed the the entire idea of three-tier arrangement of decentralised power that PRIs upheld throughout the country.

As is well-known, the decade of 1980s was perhaps the most disturbed decade in the history of Darjeeling hills. The deacde of 1980s earmarked the movement of the Gorkhas to have a state of their own, which soon attained extremist overtones. The movement of Gorkhaland after enough bloodshed, loss of property and life was ultimately resolved through the signing of a tripartite agreement in the fashion of Darjeeling Gorkha Hill Council (DGHC) in 1988. The formation of DGHC was unique in the sense that it attempted to adjudicate the aspiration of self-rule keeping in tact national integrity and was designed to provide all the benefits that a politico-administrative arrangement at the sub-state level could bestow. It is remarkable to note that the DGHC Act of 1988 has initiated a new mode of local governance that actually invalidated the already existing three-tier *panchayati raj* structure in the hills. The said Act mentioned in clear terms that there will be no *Zila Parishad* in the district of Darjeeling. The hill Council will be a substitute for the *Zila Parishad* in the hill areas, the existing HADC and the District Planning and Coordinating Council, and for the plains areas of the District there shall be the Siliguri Mahakuma Parishad as a substitute for *Zila Parishad*. According to the provisions of DGHC Act, Darjeeling hills has enjoyed the atypical two-tier system of local governance that too was opposed by Ghising, the Chairman of the DGHC, and which ultimately ceased to exist from 2006. The 73rd Amendment Act promulgated three-tier system of *panchayat* throughout the country with apex body (*Zilla Parishad*) at the district level followed by *Panchayat Samiti* at the block level and *Gram Panchayat* at the village level. However, the said 73rd Amendment Act did contain statutory provisions for maintaining two-tier structure (without detailing out what would be the composition of this two-tier structure) for the Darjeeling hills. This is why three-tier *panchayati raj* election in the hills did not take place particularly after the formation of the DGHC. In the recently held tripartite agreement of Gorkhaland Territorial Administration (GTA) designed to pacify the second round of Gorkha agitation, signed on July 18, 2011, there is a refernce to three-tier *panchayati raj* structure to be introduced once again in the hills. However, the GTA Act even after its due ratification by the government, fails to indicate in clear cut terms how this three-tier structure would be operationalised in the hills.

The brief outline of the evolution of *panchayati raj* is indicative of the very fact that the efficacy of the PRIs in Darjeeling hills was limited particularly with reference to the question of being a supportive agency to cater the needs of agraian reforms. The point is that the effective dissemination of Left political activism through PRIs contributed a lot towards the smooth functioning of the agraian reform measures throughout the state of West Bengal during the three decades of Left Front rule. The same however, was not true in case of Darjeeling hills. In its intial phase the *panchayati raj* bodies, in whatever form they existed in the hills, acted more as a development agency of some sort. Even after 1978, the Left was not able to establish its strong hold at the grassroot level power structures of the hills. As a result the rural power structure in the hills (or the hill PRIs), unlike the plains, were hardly integrated with the political activism and mobilisation associated with the agrarian reforms measures (involving such crucial issues, which are germane to the cause of successful implementation of reform measures, like identification of the beneficiaries, location and distribution of ceiling surplus land, raising consciousness among the tenants regarding the benefit of being a government notified *bargadar* etc.).

Agrarian Relations in Darjeeling Hills in Post Reform Period

After independence, in general and in post-reform era in particular, hill agrarian hierarchy contained mainly two categories of population landowners (locally known as *pattadars* or as *maliks*) at one extreme and landless agricultural workers (*pakhurey*) at another. Although the attempt to classify the hill agrarian situation in two-fold categories is based on the criterion of land ownership, it defies the three-tier model expounded by such scholars like Mukherjee (1957), Béteille (1969) or Thorner (1976). The *pattadars* were basically the landowning groups who often acted as non-cultivating owners, as cultivators, as cultivator-cum-*adhiwals* (sharecroppers), or even partly as a *khetala* (agricultural labourer). The *pakhurey* category was represented by the landless agricultural labourers or the *bargadars*. The *pakhureys* either have cultivated the plot of the *pattadars* or were not cultivating at all. The non-cultivating *pakhureys* used to serve as daily labourers at the *malik's* (landowner's) house located within or outside the village. Given this two-fold classification of agrarian social structure, it was not feasible that a *pattadar* could become a pakhurey at the same time in the same village. Howeever, a *pattadar* of a village could be a *pakhurey* of a nearby village and vice-versa. In the absence of any fixed local terminology for sharecroppers, a *pattadar* can engage in sharecropping and when he does so he is termed as *adhiwal* and when he leases out his lands on sharecropping he is referred to as *malik* by his lessee(s). But when a landless engages himself in sharecropping, he is called as *pakhurey*. The category of *pakhurey* also includes those non-cultivating landless labourers who used to serve (virtually as servants) in their *malik's* household. Subba's (1958: 53) observation is very significant in this regard who pinpoints that the degree of involvement of the *pakhureys* in the houses of their *maliks* depends partly on the amount of land held by the latter and partly on his personal nature. The non-cultivating *pakhureys* have to attend their *maliks* throughout the year and fulfill whatever is required out of them including the fulfilment of such menial works like repairing the cowshed, goatpen or a pigsty. This is particulalry the case when the *maliks* are big cultivators and in possession of sizeable amount of land. It deserves mention that the agrarian categories mentioned so far did not exist as discrete bodies of people. They are in fact, interlinked with each other in the overall operation of the agrarian production process itself. Some of them are linked with each other in terms of land (like the *pattadars* and *pakhureys*), some others through labour exchange (like the *parma* and *hoori* arrangements as described in footnote 3). Still others are linked by agricultural employment, oligation or bondage (like big cultivators and non-cultivating *pakhureys*). The point is that any attempt to analyse the evolution of agrairan social structure in the Darjeeling hills and the corresponding changes that has occurred has to be contextualised within this network of relationships.

While there has been a sharp increase in the volume of the agricultural labourers in post-independence period, the category of of agricultural labour was non-existent in the Census records till 1951. This is perhaps because of the close knit network of agraian social structure that the Darjeeling hills has maintained historically, which in fact provided the space needed for the absorption of the various agrarian categories within the available tenancy arrangements. The hill agrarian economy, argues Subba (1985: 34), is characterised by large tracts of unirrigated land, very small size of landholdings, single cropping pattern, large uncutlivable lands and forest areas make it altogether difficult for agricultural labourers to exist as a separate category. Not to mention that these observations go against the prognosis of *West Bengal Human Development Report*, which mentioned that less practice of tenant-based cultivation in Darjeeling might have accounted for the lower proportion of *bargadars* in the district (WBHDR 2004: 33).

While examining the causes of sharp increase in agricultural labourers in the northern districts (including Darjeeling) of West Bengal during 1961-1971, Bandyopadhyaya (1977: A-125) maintained that "(. . .) *pakhureys* of Darjeeling hills are generally poor migrants from adjoining hills and mostly belonging to the Nepali community and in the overwhelming number of cases the *pakhureys* had enumerated themselves as agricultural labourers and not as sharecroppers." Chaudhuri (1985: 101-02) likewise, has pointed out that governmental measures and comprehensive area development programme (CADP) initiatives have grossly failed to encourage the *pakhureys* to record themselves as sharecroppers. Subba (1985: 82) has also noted the abrupt emergence of sharecroppers' association in the style of Pakhurey Sangh during the late 1970s in some villages of Kalimpong like Echhay and Pudung where a sizeable amount of evictions took place and also the sudden death of such associations too.

It is proposed here that apart from the close knit network of the rural society, the migratory background of the *pakhureys* perhaps discouraged them from becoming government notified *bargadars*. Since being a migrant, the *pakhureys* more often than not were landless and resource less poorer lots and they had to depend—in most cases—thoroughly upon their *pattadars* even for the basic minimum inputs needed for the purpose of tilling. Amidst such situations the surfacing of a patrimonial bond amongst the *pattadar* and *pakhurey*—particularly when both of them belonged to the same community and sharing the same historical past—was not unusual, which in fact led them to consider the position of *bargadar* as an incompatible status to the spirit of their everyday life. Another noticeable change that occurred in the post-reform era is the process of labour commodification. That is to say the issue of labour, which remained entwined with the community earlier, has been brought to market. Highland tilling in Darjeeling now-a-days solely depended upon the agricultural labourers. This also explains why there has been a sharp increase in the category of agricultural labourers in the last few decades. Besides the unprecedented growth of agricultural labourers, the bleak agrarian scenario in the hills is also characterised by a sharp decline in the volume of cultivators during the past few decades. In fact, the percentage of cultivators is declining faster than the percentage growth of agricultural labourers in Darjeeling hills (see Figure 2.1). Agriculture became less rewarding and the significance of primary sector in the hill economy has been taken over by secondary and tertiary sectors. On the whole, the earlier organic unity of the primary producers with the means of production has been thoroughly altered in contemporary rural Darjeeling hills.

Figure 2.1

Agricultural Work Situation in Darjeeling Hills (1961-2001)

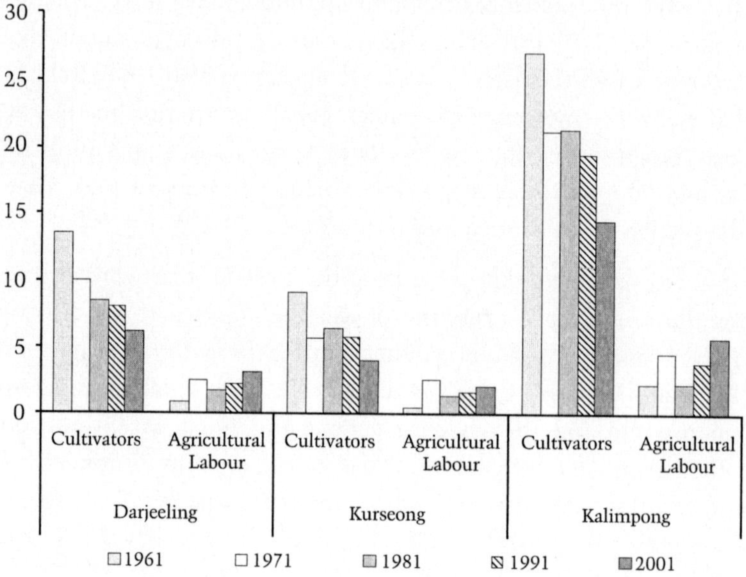

Source: *District Census Handbook: Darjeeling, Census 1961*, Calcutta: Government Printing; *District Census Handbook: Darjeeling, Census 1971*, Calcutta: Government Printing; *District Statistical Handbooks: Darjeeling (1981–2001)*, Government of West Bengal, Bureau of Applied Economics & Statistics.

Impact of Declining Land-Man Ratio

Although the declining trend of per capita landholding in the hill agrarian belt has had a colonial legacy of its own, but the vulnerability of the process had never been as acute as it is now. The ceaseless migration in post-colonial Darjeeling hills had deteriorated the agrarian situation to the extent of no return. In the absence of labour absorbing capacity of the plantations coupled with the land-locked nature of the agrarian sector of the hills, the rural mass have been experiencing the heavy yoke of decreasing land-man ratio. This becomes crystal clear when we see that rural areas though contained a sizeable section of hill population (72.81% of the total hill population as per 2001 Census), had actually showed a declining trend of growth.[5] Such developments do suggest that people are moving out from the villages in search of livelihood and one possible reason behind such a process of rural out-migration in the hills is the scarcity of arable plots for the growing number of population.

The downsizing rate of average landholding in the district during the last 30 years is another disturbing trend that affected the smooth functioning of reform measures. The average size of landholding has been slimming down from 3.06 hectares during 1970-71 to 1.66 hectares during 2000-01 (see Figure 2.2). The mounting pressure of migration had in fact furthered the decline of land-man ratio in the hills and this process has resulted into the acutely felt need of homestead land and consequently the large scale encroachment of government *Khasmahals* (land in which the control of the Government is direct); illegal clear felling etc., became common even in the urban areas too. In the rural areas, apart from the dire need of homestead lands, declining land-man ratio has given birth to an unprecedented growth of landless agricultural labourers. In fact, the declining trend of average landholding size accompanied the gravity of contradiction in agrarian situation of the hills. The smaller size of the holdings has narrowed down the scope of land distribution in the hills and the extent of benefit to be accrued therefrom. Consequently, the likelihood of surplus land extraction from those who hold more than Government stipulated ceiling of 7 standard hectares of land and their redistribution is extremely limited.

---

5. Although the size of rural population in the hills as per the 2001 Census data contains more than two-third of the total hill population of the district, in earlier Census decades the percentage share of hill rural population happened to be much higher than its urban counterpart. During the last three Census decades (1971-1991), rural population stood out to be 76.29 per cent (1991 Census), 77.56 per cent (1981 Census) and 82.76 per cent (1971 Census) of the total hill population.

Figure 2.2

Average Size of Landholding in Darjeeling District (1970-2001)

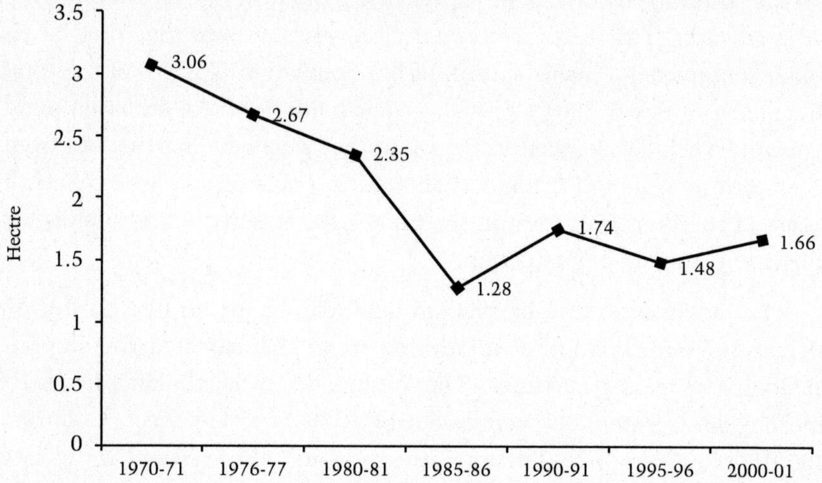

Source: *District Statistical Handbooks: Darjeeling (1970–2001)*, Government of West Bengal, Bureau of Applied Economics & Statistics.

Towards Conclusion

The agrarian relations in Darjeeling hills has been changing in response to the forces like urbanisation, politicisation, spread of education and migration. The hill agrarian economy has been gradually shifting from being a self-consumption based peasant economy towards a market driven cash crop oriented capitalist economy, therby putting the men of capital and businessmen instead of the peasants at the centre stage of contemporary rural economy of the hills. Even in the remotest corner of the hills the steady increase of cash crops like ginger in place of seasonal crops is indicative of such a process of change. This is however, true in the whole of India where the growing concern for cash and the penetration of capital has destabilised the existing rural arrangements. Extensive studies of Subba (1985) also endorse such a change. Subba (1985: 21) even went to the extreme in claiming that in the wake of such a qualitative change the most benefited category of population in the hills were neither the *pattadars* nor even the *pakhureys* but the men of capital—the businessmen (like the *Marwaris*, *Beharis* and the *Nepali* business caste *Newars*) in particular.

As noted earlier the rapid increase in the numbers of the agricultural labourers has been another significant change that has taken place in the hill rural social system over the years. The incessant flow of the migrants in the hills have been largely responsible for the continuous swelling of the category of agriculatural labourer. The new migrants were usually thrown to the bottom of agraian hierachy where they have nothing to offer except their labour power. The presence of this section of population who were always ready to sell out their labour power, under whatever terms and conditions being set by their employer, has created a new body of 'rural reserve army'. This has resulted in two direct consequences. On the one hand it has increased ceaseless competition and hankering to grab the scarce resources like land among the landless agricultural labourers and on the other hand their significant presence has paved the way for an enviable situation for the landowners who can recruit, transfer and even terminate any of his *pakhureys* and replace them by new ones drwan from the 'rural reserve army' category. Subba (1985) has noted that such a change has severly curtailed down the bragaining power of the rural labour. This also explains why rural wage structure in Darjeeling hills has remained relatively stable till the late 1980s (Ibid: 119).

In the wake of Operation Barga, the landowners of the hills, much like their plains counterparts, have become overcautious while leasing out their lands to the *pakhureys*. A trend of 'reverse tenancy' has also taken place in the region in which much of the non-cultivating owners, who used to lease out their lands earlier, have taken back their lands from their sharecroppers for self-cultivation. Some others have either sold out their lands or started changing their tenants every now and then. The availability of the new tenants were not at all a problem. As noted earlier the 'rural reserve army' has provided opportune moments to the landlords which ultimately impaired the steady growth of the rural economy. Consequently, an unstable agrarian social structure has ensued in the hills that implicated the breakdown of traditional harmony between the landlords and the tenants partially as a result of the land reform meaures. The most sordid feature of such a change, argues Subba (1985: 117), is the fact that a large number of *pakhureys* are turning into landless labourers.

While the challenges of subinfeudation, land alienation and economic disparity crept within the hill agrarian situation in colonial times were harmoniously resolved within the organic unity of the primary producers with the means of production, the post-colonial agrarian situation in the hills showed a declining trend not only in the number of cultivators but also led towards the growth of landless labourers, besides diminishing the size of per capita landholding in every subsequent Census decades. All these processes have had their cumulative effects in breaking down the earlier organic unity of the primary producers with the means of production, which characterised the colonial agrarian social formation of Darjeeling hills to a large extent and are indicative of the fact that the rural areas in the hills have remained crisis ridden over the years. The analysis maintained so far has attempted to delineate the deteriorating status of the hill agrarian situation—one of the major foundations of hill social formation—which caused marginality as an unavoidable reality for a sizeable section of the hill population. The point is that, the roots of marginality or deprivation in the hill agrarian sector, if any, is embedded in the political economy of the hills itself. And the land reform measures instead of ameliorating the agrarian impasse have remained largely irresponsive to the innate peculiarities of the hill rural social situation. This is why the consequences of land reform measures for Darjeeling hills can hardly be compared with the overall impressive picture of the state of West Bengal.

However, it needs to be stressed out that the successful operation of land reform measures and their significance can be brought into practice in the hills not on the premise as they were operationalised till date. Basically the 'land to the tiller' principle or for that matter the *Bargadari* system, the land ceiling and land redistribution measure or any such programme of action needs to be assimilated with the overall socio-political realities of the region. Special consideration shoud be made on such grave issues which indicate the difference of hill agrarian social system from its plains counterpart. For example, the issues like land fragmentation, small holding size, soil type, ecological limits to increase the amount of arable lands, little prospect for multiple cropping or crop diversification, huge institutional land ownership and the demand of legal entitlement to the piece of land possessed by the plantation workers for generations, are some such crucial parameters which may help us work out a more realistic pattern of land reform measures for the hill region of the district of Darjeeling. In fact, the problem lied not with the conception of land reform but perhaps in the mode of their articulation and operation. The hill regions of the district of Darjeeling had differential exposure to the forces of colonial governmentality particularly in the matters related to land revenue administration. Besides the historical factor, the ecological and topographical factors largely determined the bases of difference still maintained by the agrarian social system of Darjeeling hills. Hence, it is desirable that the issues of local significance should have their proper reflection in the policy perspectives too, which however, had hardly been the case. Neither the uniform approach of land reforms for both the hills and the plains has been able to generate similar social consequences nor even the present system of land reform measure could address justifiably the pangs of rural poor or accommodate the aspirations of the rural hill folk either.

References

Bandyopadhyay, N. (1977). "Causes of Sharp Increase in Agricultural Labourers, 1961–71: A Case Study of Social–Existence Forms of Labour in North Bengal", *Economic and Political Weekly* 12(53): A-111–A-126.

Béteille, André (1969). *Caste, Class and Power: Changing Patterns of Stratification in a Tanjore Village*. Calcutta: Oxford University Press.

Chaudhuri, B. (1985). "Agrarian Relations in A Hill Region", in S.K. Chaube (ed.), *The Himalayas: Profiles of Modernisation and Adaptation*. New Delhi: Sterling Publishers.

Dasgupta, A. (2006). "Agrarian Reforms in West Bengal since 1977", Phd Dissertation, University of California, Riverside. (Abridged on-line version). Available at: *http://economics.ucr.edu/seminars/winter05/ped/ped03092005.pdf* [Retrieved on December 12, 2009].

District Census Handbook: Darjeeling, Census 1961. Calcutta: Government Printing.

District Census Handbook: Darjeeling, Census 1971. Calcutta: Government Printing.

District Statistical Handbooks: Darjeeling (1970–2001). Government of West Bengal, Bureau of Applied Economics & Statistics.

Government of India (1966). *Implementation of Land Reforms: A Review by the Land Reforms Implementation Committee of the National Development Council*. New Delhi: Planning Commission.

Government of West Bengal (2004). *West Bengal Human Development Report (WBHDR)*, Development and Planning Department.

Mukherjee, Ramkrishna (1957). *The Dynamics of Rural Society: A Study of the Economic Structure in Bengal Villages*. Berlin: Akademie-Verlag.

Srinivas, M.N. and André Béteille (1964). "Networks in Indian Social Structure", *Man N.S.* 64: 165–8.

Subba, T.B. (1983). "Caste, Class and Agrarian Relations in the Nepalese Society". Unpublished Ph. D. Thesis. Department of Sociology and Social Anthropology, University of North Bengal.

———. (1985). *The Quiet Hills: A Study of Agrarian Relations in Hill Darjeeling*. Delhi: ISPCK.

———. (1986). "Agrarian Relations and Development in Sikkim", Paper presented in National Seminar on *Rural Development in Eastern and North Eastern India*. May 8-10, Kolkata: IIM Joka Campus.

———. (1989). *Dynamics of a Hill Society: The Nepalis in Darjeeling and Sikkim Himalayas*. Delhi: Mittal.

The West Bengal Land Reforms Act, 1955; 2005. Kolkata: Kamal Law House.

Thorner, Daniel (1976). *The Agrarian Prospect in India*. New Delhi: Allied Publishers.

Todarmal (2002). *Evolution of the Indian Land Revenue System: Analysis of Land in West Bengal*. Kolkata: Anima Prakashani.

3

Participation in Tenancy Market

An Analysis of Asset Ownership across Caste and Class in Two Villages in Uttar Pradesh

Partha Saha[1]

Introduction

The process of agricultural production in a peasant economy involves use of different productive resources. However, pattern of ownership of means of production varies across households, which necessarily implies that agricultural production is organised in different ways by different households. Further, households with different resource endowments will enter into different types of relations with each other in the process of agricultural production. The pattern of asset ownership in general, and ownership of means of production in particular, is an important determinant of the location of a household in the system of agrarian relations. Whether workers in a household sell their labour power or deploy it in their own household enterprise depends crucially on the extent to which they own various means of production.

Ownership of means of production influences the process through which tenancy operates. This process varies from village to village depending on various socioeconomic factors. This paper explores the differences in tenancy relations due to differences in asset ownership in the two villages. In Harevli, the ownership of means of production played

1. This work is a part of Project on Agrarian Relations in India (PARI) initiated by Foundation for Agrarian Studies in Kolkata. The author would also like to thank Dr Vikas Rawal, Prof V.K. Ramachandran and Prof Madhura Swaminathan for their guidance all through out the work. The field work for this study was conducted in May 2006.

a major role in determining the specific terms and conditions of tenancy contracts. In Mahatwar, even though the terms and conditions of tenancy contracts were more or less generalised, ownership of certain means of production was an important determinant of incidence of tenancy.[2]

Some Features of the Study Villages

In Harevli, Dalits formed the largest social group (with Chamar being the largest community). However, Other Caste Hindu households (to be more specific Tyagi households, which constituted 30% of village households) had absolute dominance over all socioeconomic aspects of the village society. Agriculture was the mainstay of economy of Harevli. Income from crop production, rental income from agricultural land and orchards and income from wage labour in agriculture accounted for about 64 per cent of average household income in the village.[3] Sugarcane was by far the most important crop grown in the village and it accounted for about 53 per cent of gross cropped area (GCA). Ownership of agricultural land in Harevli was highly unequal with 73 per cent land being owned by the top 16 per cent per cent of households while 33 per cent of households did not own any agricultural land. All households within this top 16 per cent were Tyagi households, and in all, Tyagi households owned 85 per cent of land in the village. Dalits, on the other hand, owned only 6 per cent of land in the village.

In Mahatwar, Other Caste Hindus constituted only 7 per cent of the total population, while Dalits constituted the largest caste group with a population share of 59 per cent. Other backward classes (OBCs) had a

2. Harevli is a village in Bijnor District of Western Uttar Pradesh, and Mahatwar is a village in Ballia District of Eastern Uttar Pradesh.
3. Household income refers to the total income accruing to the household and aggregate earnings of the members of the household. The measurement of income in the study villages was associated with several difficulties primarily because of the fact that most rural households were simultaneously involved in production, consumption and investment, and none of the households maintained any accounts of the transactions undertaken during the reference period. As a result, income had to be derived through a complex process of calculation based on detailed information (collected through household survey) on values of output produced and costs incurred in the economic activities during the reference period. For details of the methodology involved in calculation of household income please see Rawal (2008b), *http://www.agrarianstudies.org*.

significant presence in the village and accounted for 34 per cent of the total population. Eastern Uttar Pradesh is noted for small size of landholdings.[4] This was also the case in Mahatwar. Mahatwar did not have very large land owners. The largest ownership holding in the village was 10.33 acres while the average was only 0.9 acres. Even though Other Caste Hindu households accounted for only 7 per cent of households, they owned 45 per cent of land in the village. The major crops grown in the village were paddy (*kharif* season) and wheat (*rabi* season). These two crops together accounted for about 88 per cent of the GCA. For the village as a whole, income from agriculture accounted for only 10 per cent of total income.

In both the study villages, there was a very strong congruence between caste and class. The economic classification of households in the study village was done on the basis of ownership of assets, principal source of income, and whether the members of the households participated in any manual operation on their operational holding. The households were categorised into 11 economic classes (landlord, rich peasant, upper middle peasant, lower middle peasant, small peasant, manual labour, artisans, households whose main sources of incomes were from business activities other than direct crop production, households whose major earnings were from salaried employment (private or government), households whose main incomes were from remittances, pensions and scholarships, households whose main incomes were from rent or money lending).[5] Table 3.1 and Table 3.2 clearly brings out the very strong caste-class correlation in the two study villages. In both villages, Dalit households were concentrated in small peasant and manual labour economic classes. Other Caste Hindu households, on the other hand, were concentrated within the top three economic classes in Harevli, while they were spread across all peasant households in Mahatwar.

4. According to AIDIS, 2002-03, among all the four regions of Uttar Pradesh, average size of operational holding was the smallest in Eastern Uttar Pradesh (1.3 acres) while Southern region had the largest average size of operational holding (3.5 acres). Smallness of landholdings in eastern UP has been noted in other studies as well. See, for example, Singh (1987).
5. In Harevli, an important additional feature of landlord, rich peasant, and upper middle peasant households was that in addition to substantial rent extraction, they maintained unfree labour through the process of inter-relationship established between tenancy market and labour market.

Table 3.1

Distribution of Household Belonging to Different Caste and Class, Harevli, 2006

	Dalit	*OBC*	*Muslim*	*Other Caste Hindu*	*All Social Groups*
Landlord	0	0	0	2	2
Rich Peasant	0	1	0	9	10
Upper Middle Peasant	0	0	2	13	15
Lower Middle Peasant	5	3	3	6	17
Small Peasant	15	11	2	0	28
Manual Labour	18	4	4	0	26
Artisans	0	0	0	2	2
Business/Self-employment	3	0	1	0	4
Households Primarily Dependent on Salary	3	1	0	0	4
Households Primarily Dependent on Remittances/Pensions	1	0	1	0	2
Households Primarily Dependent on Rent/Money Lending	1	0	0	1	2
All Economic Classes	46	20	13	33	112

Source: Survey Data.

Table 3.2

Distribution of Household Belonging to Different Caste and Class, Mahatwar, 2006

	Dalit	*OBC*	*Other Caste Hindu*	*All Social Groups*
Landlord	0	0	1	1
Rich Peasant	0	0	1	1
Upper Middle Peasant	0	7	1	8
Lower Middle Peasant	9	23	1	33
Small Peasant	30	6	1	37
Manual Labour	26	2	0	28
Artisans	0	1	1	2
Business/Self-employment	15	9	1	25
Households Primarily Dependent on Salary	5	0	1	6
Households Primarily Dependent on Remittances/Pensions	8	1	3	12
All Economic Classes	93	49	11	153

Source: Survey Data.

Asset Ownership in Study Villages

Distribution of assets in Harevli was extremely skewed. The Gini coefficient for total value of assets was 0.75. The bottom 70 per cent of households owned only 10.5 per cent of total value of assets, while the top 5 per cent of households owned 43 per cent of total value of assets.[6] Table 3.3 shows that Dalit, Muslim, OBC households constituting 70.5 per cent of households in this village owned only 17.6 per cent of total assets. On the other hand, Other Caste Hindu households constituting 29.5 per cent of households owned 82.4 per cent of total assets in the village.

Table 3.3

Proportion of Households and Assets Owned by Social Group, Harevli, 2006

Social Group	*% of Households*	*% of Assets Owned*
Dalit	41.1	7.2
Muslim	11.6	3.8
Other Backward Classes	17.8	6.6
Other Caste Hindu	29.5	82.4
All Social Groups	100	100

Source: Survey Data.

Analysis of variation in asset holdings across socioeconomic classes shows that landlord and rich peasants owned disproportionately high share of assets in the village. These two classes together constituted 11 per cent of households but owned 61 per cent of total assets (Table 3.4). On the other extreme, small peasant and manual labour classes constituting 48 per cent of households in this village owned only 3.4 per cent of total assets.

Likewise, in Mahatwar also, the extreme form of wealth inequality was reflected in the high value of Gini coefficient (0.69) of ownership of total assets in the village. Dalit households in this village were located mostly in the bottom four wealth deciles, while OBC households were

6. More than 90 per cent of households in the two top-most wealth deciles belonged to Tyagis. The majority of households in the bottom seven deciles in the village belonged to Dalit, Muslim and OBC social groups. Half of the Dalit households owned assets worth less than ₹ 0.6 lakh, while almost three-fourth of Other Caste Hindu households owned assets worth more than ₹ 10 lakh.

mostly distributed across the middle and upper wealth deciles (Table 3.5). Other caste Hindu (Tyagi) households were located in the two top-most wealth deciles.

Table 3.4

Proprtion of Households and Assets Owned by Economic Class, Harevli, 2006

Economic Class	*% of Households*	*% of Assets Owned*
Landlord	1.8	22.3
Rich Peasant	8.9	38.3
Upper Middle Peasant	13.4	22.1
Lower Middle Peasant	15.2	9.8
Small Peasant	25.0	2.4
Manual Labour	23.2	1.0
Artisan	1.8	0.04
Business/Self Employed	3.6	0.4
Households Primarily Dependent on Salary	3.6	2.9
Households Primarily Dependent on Remittances/Pensions	1.8	0.1
Households Primarily Dependent on Rent/Money Lending	1.8	0.7
All Economic Classes	100.0	100.0

Source: Survey Data.

Table 3.5

Percentage Distribution of Households by Wealth Decile by Social Group, Mahatwar, 2006

Wealth Decile	*Dalit*	*OBC*	*Other Caste Hindu*
0 - 10	13	6	0
10 - 20	15	2	0
20 - 30	15	2	0
30 - 40	15	4	0
40 - 50	11	8	9
50 - 60	11	10	0
60 - 70	12	10	0
70 - 80	4	20	9
80 - 90	4	18	18
90 - 100	0	18	64
All Deciles	100	100	100

Source: Survey Data.

Land was by far the most important asset owned by the households in both the study villages. In Harevli, it constituted 78 per cent of the total value of assets, and in Mahatwar, it constituted 73 per cent of total value of assets.[7] In both the villages, land and other forms of assets were concentrated among other caste Hindu households and the value of 'access index' for all categories of assets was higher as compared to households belonging to Dalit, Muslim, OBC households (Tables 3.6 and 3.7). [8]

Table 3.6

Access Index of Different Social Groups by Items of Assets, Harevli, 2006

Item of Asset	*Dalit*	*Muslim*	*OBC*	*Other Caste Hindu*
Land	0.15	0.34	0.29	2.88
Building	0.29	0.15	0.49	2.64
Animal Resources	0.40	0.48	0.92	2.09
Other Means of Agricultural Production	0.24	0.48	0.92	2.32
Means of Transport	0.14	0.21	0.42	2.86
Furniture	0.18	0.37	0.49	2.68
Electrical Equipment	0.12	0.33	0.45	2.82
Others	0.16	0.29	0.57	2.70
All Assets	0.17	0.33	0.37	2.79

Source: Survey Data.

In a nutshell, given the strong caste-class congruence in both the study villages, distribution of assets was characterised by asset concentration among upper caste households belonging to landlord, rich peasant and upper middle peasant economic classes, while Dalit households mostly belonging to small peasant and manual labour economic classes were characterised by lack of ownership of assets. This skewed distribution of asset ownership across households belonging to different social groups and economic classes had its impact in the way households participated in tenancy contracts. In Harevli, with very high participation of households

7. Concentration of land in the hands of Tyagi households was the fundamental feature of landownership in Harevli, and Tyagi households owned 85 per cent of total value of land in the village. In Mahatwar, other caste Hindu households constituting 7 per cent of all households accounted for 45 per cent of land ownership.
8. 'Access index' is due to K. Nagaraj, as cited in Ramachandran (1990). In the present context access index is defined as the ratio of proportion of assets owned by a social group to the proportion of households belonging to that particular social group.

in tenancy market, ownership of assets had an impact on the specific terms of the contract. In Mahatwar, where non-agricultural labouring out was wide-spread among asset poor Dalit households, ownership of means of production determined their participation or non-participation in the tenancy market.

Table 3.7

Access Index of Different Social Groups by Items of Assets, Mahatwar, 2006

Item of Asset	*Dalit*	*OBC*	*Other Caste Hindu*
Land	0.28	1.18	6.28
Building	0.50	1.58	2.58
Animal Resources	0.69	1.55	1.14
Other Means of Agricultural Production	0.12	2.38	2.28
Means of Transport	0.20	0.82	8.53
Furniture	0.75	1.10	2.60
Electrical Equipment	0.73	1.26	2.10
Others	0.61	1.30	2.92
All Assets	0.34	1.28	5.33

Source: Survey Data.

Tenancy in Harevli

Tenancy was an important institution in Harevli and it played a major role in the village economy. A large number of households (69%) in Harevli accessed land through different forms of tenancy (Table 3.8). An interesting feature of land tenancy in Harevli was the presence of some households simultaneously as tenants and lessors in the lease market. Simultaneous leasing-in and leasing-out was a characteristic present among other Caste Hindu Households, 27 per cent of which were involved in such tenancy contracts.

Tenancy in Harevli took various forms and the only common feature in all lease contracts was the fact that labour was provided entirely by the tenants (whether it was the labour of the family members of the tenants' household or the casual labourers hired in by the tenants). In Harevli, both share tenancy contracts and fixed tenancy contracts existed. Between the two broad classifications of tenancy contracts, share tenancy was clearly the predominant form and the incidence of share tenancy was twice that of fixed rent tenancy. Within the category of share contracts, several patterns of crop sharing and sharing of input were observed, and to a very large extent it depended on the socioeconomic status of landowners and tenants. Among the various forms of tenancy contracts, the two most prominent forms of tenancy contracts will be discussed in the following two sub-sections.

Table 3.8

Incidence of Tenancy, Harevli, 2006

Land leased in	
Proportion of households	46.4
Area leased in as a proportion of total operated area	29.3
Land leased out	
Proportion of households	35.7
Area leased out as a proportion of total owned area	20.7

Source: Survey Data.

Fixed Rent Tenancy between Non-resident Tyagi Lessors and Resident Tyagi Tenants (Type 1 Contract)

An important feature of Harevli was ownership of agricultural land by non-resident Tyagi households which had migrated to Delhi and other urban areas for non-agricultural businesses and salaried jobs. Such Tyagi non-resident landowners leased out their land to resident Tyagi households and typically had kinship ties with the tenants. In such cases, the tenants belonged to landlord, rich peasant and upper middle peasant classes and owned substantial means of production. These tenancy contracts were fixed rent contracts payable in cash.[9] The landowners had no real interest in agriculture in the village. Their sole motivation was to retain control over land, and so they were not really bothered with the fact that they were getting only one-fourth share of the produce.[10] Such long-term

9. The average rental income of the landowners from such tenancy contracts was roughly one-fourth of gross value of output from cultivation of sugarcane, which was the only crop cultivated under such type of tenancy contracts.

10. It may be noted here that the resident Tyagi tenants paid a fixed amount of money as rent to the non-resident Tyagi landowners. After calculating the gross value of output of the resident Tyagi tenants from sugarcane cultivation from such leased-in land it was observed that the amount they paid as rent was on an average one-fourth of the gross value of output.

migrant households were unlikely to return to the village and resume self-cultivation. In addition to the motivation to keep the land within the family, another factor that resulted in land being leased only to Tyagi tenants was that only Tyagi households had financial resources and means of production required for cultivating irrigated crops like sugarcane. So, even though the landowners accepted a lower share of value of output, by leasing out the land to resource rich Tyagi tenants, the landowners ensured that in absolute terms they earn higher incomes as compared to what they would have earned if the tenants cultivated wheat or paddy.

It is noteworthy that the tenants who cultivated land under such contracts themselves owned enough means of production including large amounts of land. The tenants belonged mostly to the rich peasantry and upper middle peasantry, and in addition to land, they owned substantial amount of other means of production particularly agricultural machinery. The tenancy contracts only added to the process of accumulation for such tenant households. All the tenant households in such contracts owned irrigation equipments (diesel or electric pump sets in addition to other equipments) and the average value was ₹ 32,700, 50 per cent of households owned tractor and accessories with an average value of ₹ 1,53,812, and 75 per cent of households owned draught animals with an average value of ₹ 9,792. The 25 per cent of households which did not own draught animals owned tractor and accessories instead. In addition to owning large plots of irrigated land themselves (10.4 acres on average), these tenant households leased in substantial amounts of land (7.6 acres on average) for cultivation of sugarcane.

The fixed rent contract in these cases suited both the tenants and the lessors. Under a fixed rent contract, the lessors did not have to supervise cultivation at all and as absentee landowners retained control over land. On the other hand, being rich peasants themselves, the tenants took land under a fixed rent contract for cultivation of sugarcane which facilitated in furthering the process of accumulation. It may be noted here that Dalits with little resource endowment could not get access to such tenancy contract which was clearly favourable for the tenant. Therefore, it was primarily the sufficient resource endowment of the resident Tyagi households which enabled them to get access to such favourable tenancy contract in Harevli.

Seasonal Share Tenancy Contracts between Resident Tyagi Lessors and Dalit/Muslim/OBC Tenant Households (Type 2)

This type of tenancy contract was the most widespread in the village. It not only depicted the economic relation between the landowner and the tenant, but it also clearly highlighted the position of dominance of Tyagi households in the socioeconomic hierarchy in this village. In this type of tenancy contract, resident Tyagi households belonging to landlord, rich peasant and upper middle peasant economic classes were the lessors. Tenants belonged to Dalit, Muslim and OBC households, belonging primarily to small peasant and manual labour classes, and owning no agricultural land at all or small plots of land (less than 0.5 acres). Among non-land means of production such tenant households owned simple hand implements, and they had to depend on manual labour, mainly in agriculture, for their survival.

The resident Tyagi households, being landowners in such tenancy contracts, leased out land to Dalit, Muslim and OBC households on share rent for a period of one season for paddy cultivation. The tenants could only retain one-third of the crop output, while the landowner took two-third of crop output along with the entire by-product. Landowners generally shared a part of the input cost. In most of the cases, the landowners ploughed the field with his tractor before handing over the land to the tenants. In general, the landowners shared a greater part of costs of material inputs when the land was leased out for a season to one of their long-term workers.

Clearly, the lack of ownership of means of production by Dalit, Muslim and OBC households resulted in their dependence on wage employment for survival. The tenants worked on the land cultivated by the landowners as wage labourers which could only be assured to them if they accepted the seasonal share rental contract for the cultivation of paddy. Thus, the linkage between the tenancy market and labour market was established through this relation of dependence. In other words, Tyagi households, owning the bulk of agricultural land and other non-land means of production, provided wage employment to Dalit, Muslim and OBC households conditional on their acceptance of the seasonal share rental contract. The motive for Dalit, Muslim and OBC households in accepting such onerous terms of contracts was the assurance of wage employment on the operational holdings of the landowners and food security (share from paddy cultivation). The tenants also had to provide un-paid or under-paid labour services to their landowners. The various forms of labour services involved working at the employer's house during some ceremonies, cleaning the employer's house every morning before working in the field, doing various kinds of animal husbandry work like feeding animals, milking and cleaning dung.

The lack of ownership of the means of production by Dalit, Muslim and OBC households provided the Tyagi households with the opportunity to exploit them by appropriating, on average, about 65 per cent of gross value of output. The exploitation of these tenants was extended through the inter-linkage between tenancy contracts and wage employment, which also included labour services. The situation was that of compulsive market participation for asset poor Dalit, Muslim and OBC households on extremely onerous terms for sheer survival.

Landlessness or lack of ownership of assets resulted in Dalits getting access to only such unfavourable terms of tenancy contract. It might seem at first sight that this form of tenancy contract is actually a kind of labour tying arrangement. But the point is Dalit households wanted to cultivate land as tenant (as they did not have enough land of their own), and the only way they could get access to land was to accept such onerous terms besides ensuring assured supply of labour. Dalits who did not get access to even such onerous tenancy contracts worked as wage labourers in agriculture and non-agriculture. But the preference for Dalit households was for tenancy as it simultaneously ensured their wage employment as well as food security.

Tenancy in Mahatwar

In Mahatwar, owner-cultivation was the predominant form of crop cultivation. When households were categorised in terms of cultivation status, the preponderance of owner cultivators came out sharply and it was observed that an overwhelming majority of households (82 households out of a total of 153 households) were pure owner cultivators, neither leasing in nor leasing out any land (Table 3.9).

However, tenancy did exist and it was an institution of some importance in the village particularly for the lower caste and lower class households which did not own enough productive assets for their sustenance. Fixed rental payment in kind (which implied payment by means of a specified quantity of crop per unit of land cultivated) was the most common form of tenancy contract in the village, and the rent was uniform throughout the village.[11]

Table 3.9

Characterisation of Households by Cultivation Status, Mahatwar, 2006

		Land Owning	
	Landless	*Leasing out*	*Not Leasing out*
Leasing in	Landless tenant (8)	Tenant cum lessor (0)	Owner tenant (37)
Not Leasing in			
Cultivating	na	Self cultivator cum lessor (4)	Pure owner cultivator (82)
Not Cultivating	Landless non-cultivator (21)	Pure lessor (1)	na

Note: "na" implies not applicable. The figures in brackets indicate number of households (out of 153 households in the village). The different categories are mutually exclusive and exhaustive. This scheme of categorising of households was done by Sharma and Dreze (1998). The names of the categories have been modified to suit the present purpose.

Source: Survey Data.

In Mahatwar, lessors and tenants belonged to different social groups and had very clear cut segregation in terms of economic classes. Other Caste Hindu households (rajputs and brahmins) belonging to landlord and rich peasant economic classes were the lessors, while Dalits and OBC households belonging to small peasant and manual labour economic classes were the primary lessees in the land lease transactions. It might be noted that unlike in Harevli, none of the households in Mahatwar were simultaneously tenants as well as lessors in the lease market.

Even though ownership of means of production did not determine the forms of tenancy contracts in Mahatwar, it was an important determinant influencing the extent of land leased in. In Mahatwar, the most important non-land means of production owned by the households were irrigation equipments and draught animals. Among the 126 land operating households in the village it was observed that households with higher values of irrigation equipments had larger ownership holdings.[12] But it was not the case that households with higher values of irrigation equipments

11. The village standard was 3 quintal of paddy or wheat per acre of land.
12. Ownership of irrigation equipments was contingent upon ownership of land.

leased in more land. The rental market for hiring of pump sets was well developed in the village and the owners charged uniform charges to all users. In other words, ownership of irrigation equipments was not critical in deciding the extent of land to be leased in and this was possible because of the well established market for the hiring of irrigation equipments.

Though tractors were used for ploughing in Mahatwar, draught animals were also extensively used particularly in case of wet ploughing for paddy. Given the smallness of size of operational holding, even the first dry ploughing in many cases was done using draught animals. There was no rental market for draught animals in the village because, first, hiring out only draught animals always had the possibility of being maltreated or misused which the owners definitely wanted to avoid. Second, the possibility of hiring in labour (owner of the draught animals) along with draught animals was not at all an attractive option for the hirers because in Mahatwar (given the smallness of size of operational holding) there were enough family members in the households who could do the ploughing. In the absence of any rental market for draught animals, in case of tenant households, area leased in was positively related to ownership of draught animals.

Low agricultural income, monsoon dependent nature of agriculture, smallness of size of operational holding resulted in making agriculture a non-profitable venture. This was true across all social groups and economic classes. Therefore, the dependence of asset poor Dalit and OBC households on upper caste households for agricultural wage employment did not exist. Given the standardised norms of rent payment prevalent in the village, ownership of means of production did not influence terms of tenancy contract. But among the tenants, the participation and the extent of participation in tenancy market depended on ownership of certain means of production.

Another important reason for non-existence of relation of dependence between Dalits/OBCs and upper castes was non-farm employment outside the village. A State highway connecting neighbouring towns (Rasra to Mau) passes through the village and different modes of transportation was available at frequent intervals. Therefore, scope for outside non-agricultural employment was available to the asset poor Dalits/OBCs households. In fact, boring tube-wells was an occupation in which a large number of villagers were employed. Out of 142 Dalit/OBC households in the village, 69 households had at least one member in this particular occupation either as wage labourer or as contractor. Therefore, availability of non-agricultural employment outside the village helped in breaking this relationship of dependence and bondage in the village.

Harevli, located in the agriculturally prosperous region of Western Uttar Pradesh, witnessed a relation of dependence between asset poor Dalit/Muslim/OBC households, and asset rich Tyagi households. The relation of dependence manifested itself in the form of inter-linkage between tenancy market and labour market. Such dependence was not there in the agriculturally backward region of Eastern Uttar Pradesh where Mahatwar was located. Nevertheless, for the tenant households, ownership of assets determined their extent of participation in the lease market. Thus, asset ownership had an impact in tenancy contract in both the villages even though in different ways.

Conclusion

This paper illustrates how ownership of assets influenced participation of rural households in tenancy market through a case study of two villages in Uttar Pradesh. Tenancy was an institution of major importance which provided access to land to the asset poor Dalit and OBC households in both the study villages. Tenancy was the main basis of the economy of Harevli and 69 per cent of households were involved in tenancy contracts. Both tenants and landowners belonged to all socioeconomic groups in Harevli and as such the objective and terms and conditions of tenancy contracts varied considerably depending up on the socioeconomic positions of tenants and landowners. Households owning sufficient means of production had a better bargaining power and therefore the terms and conditions of tenancy contracts were favourable to them. This was particularly true for resident Tyagi households in the village, both as tenants (in Type 1 contract) and as landowner (in Type 2 contract). In Type 1 contract the resident Tyagi households leased in land from non-resident Tyagi landowners on annual fixed rent contract payable in cash primarily for the cultivation of sugarcane. The objective was to further the process of accumulation through higher profit. In Type 2 contract, the same resident Tyagi households leased out land to asset poor Dalit, Muslim and OBC households for one season (paddy) on crop sharing basis. Through such tenancy contracts the resident Tyagi landowners obtained an assured supply of farm labourers on their own operational holdings. The asset poor Dalit, Muslim and OBC households which primarily depended on wage employment in agriculture for their livelihood had to accept such tenancy contracts for the sake of assured wage employment on the operational holdings of the landowners. In such lease contracts, access to land for subsistence consumption was the objective for Dalit, OBC and Muslim tenants belonging to small peasant and manual labour classes while control over labour power was the primary motive for the upper caste and upper class landowners. The resident Tyagi households by virtue of their asset ownership were the dominant landowners in the tenancy market and also the dominant employer in the labour market, and therefore were in a position to fully exploit the situation of lack of ownership of assets by Dalit, Muslim and OBC households.

Although owner cultivation was the dominant form of agriculture in Mahatwar, tenancy did exist where landlord and rich peasant belonging to Other Caste Hindu households leased out land, mostly on a fixed rent basis (in kind) to Dalit and OBC households, primarily belonging to lower middle peasant and small peasant classes. There were no tenants among Other Caste Hindu households and no lessors among Dalit and OBC households. Thus, tenancy in Mahatwar was characterised by a clear cut segregation of lessors and tenants in terms of social groups and economic classes.

The terms and conditions of tenancy contracts were fixed in Mahatwar and the standard norms were generally adhered to. However, ownership of certain means of production, draught animals in particular, was an important determinant regarding the extent of land leased in by the tenants.

An important distinction could be observed between the two villages in terms of relation of dependence between the asset rich (mostly upper caste) households and asset poor Dalit households. While in Harevli it was the absolute dominance of Tyagi households over the village society and economy, in Mahatwar, better connectivity and availability of non-agricultural employment outside the village helped in establishing self-reliance among Dalit/OBC households.

References

Appu, P.S. (1996). *Land Reforms in India*. Vikas Publishing House Pvt. Ltd.

Ballabh, V. and T.S. Walker (1986). "Land Fragmentation, Subdivision and Consolidation in India's Dry, Semi-Arid Tropics", *Economics Group Progress Report 76*. Hyderabad: ICRISAT.

Ballabh, V. and S. Pandey (1999). "Transitions in Rice Production Systems in Eastern India: Evidence from Two Villages in Uttar Pradesh", *Economic and Political Weekly*, March 27.

Bandyopadhyay, D. (2008). "Does Land Still Matter", Paper presented at National Seminar on *Land Reforms in Uttar Pradesh: Retrospect and Prospects*. August 12-13.

Barker, R. and R.W. Herdt (1985). *The Rice Economy of Asia*. Washington D.C.: Resources for the Future Inc.

Bharadwaj, K. (1974). *Production Conditions in Indian Agriculture: A Study Based on Farm Management Surveys*. Cambridge University Press.

———. (1994). *Accumulation, Exchange and Development: Essays on the Indian Economy*. Sage Publications.

Cheung, S.N.S. (1968). "Private Property Rights and Sharecropping", *Journal of Political Economy*, November/December.

Chopra, Kanchan (1984). "Distribution of Agricultural Assets in Punjab: Some Aspects of Inequality", *Economic and Political Weekly* 19(13).

Dreze, Jean and Haris Gazdar (1996). "Uttar Pradesh: The Burden of Inertia", in Jean Dreze and Amartya Sen (eds.), *Indian Development: Selected Regional Perspectives*. Oxford University Press.

Dreze, Jean and Naresh Sharma (1998). "Palanpur: Population, Society, Economy", in Peter Lanjouw and Nicholas Stern (eds.), *Economic Development in Palanpur Over Five Decades*. Oxford University Press.

Gill, S.S. (1989). "Changing Land Relations in Punjab and Implications for Land Reforms", *Economic and Political Weekly*, June 24.

Gupta, Dipankar (ed.) (2004). *Caste in Question*. Sage Publications.

Hanstad, Tim, Robin Nielsen and T. Haque (2008), "Improving Land Access for India's Poor", Paper presented at *National Seminar on Land Reforms in Uttar Pradesh: Retrospect and Prospects*. August 12-13.

Joshi, B.K. (1996). "Land Reforms and Agrarian Change in Uttar Pradesh Since 1947", *Social Action* 46(4).

Lanjouw, P. and N. Stern (1991). "Poverty in Pulanpur", *World Bank Economic Review* 5(1): 23-55.

Lieten, G.K. and Ravi Srivastava (1999). *Unequal Partners: Power Relations, Devolution and Development in Uttar Pradesh*. Sage Publications.

Mehrotra, Santosh (2008). "Land Relations: A 11th Plan Perspective", Paper presented at National Seminar on *Land Reforms in Uttar Pradesh: Retrospect and Prospects*. August 12-13.

Mishra, Surya, Kanta and Vikas Rawal (2002). "Agrarian Relations in Contemporary West Bengal and Tasks for the Left", in V. K. Ramachandran and Madhura Swaminathan (eds.), *Agrarian Studies*. Tulika Books.

Mohanty, B.B. (2001). "Land Distribution among Scheduled Castes and Tribes", *Economic and Political Weekly*, October 6.

Nagaraj, K. (2008). "A Note on Methods of Village Study", Paper presented at *Studying Village Economies in India: A Colloquium on Methodology*. December 21 to 24.

Neale, W.C. (1962). *Economic Change in Rural India: Land Tenure and Land Reform, 1800-1955*. New Heaven: Yale University Press.

Newbery, D.M.G. (1974). "Crop-sharing Tenancy in Agriculture: A Comment", *American Economic Review* 64(6).

Newbery, D.M.G. and J.E. Stiglitz (1979). "Sharecropping, Risk Sharing and the Importance of Imperfect Information", in J.A. Roumasset *et al.* (eds.), *Risk, Uncertainity and Agricultural Development*. New York.

Pant, C. (1983). "Tenancy and Family Resources", *Journal of Development Economics* 12: 27-39.

Patnaik, Utsa (1976). "Class Differentiation within the Peasantry: An Approach to Analysis of Indian Agriculture", *Economic and Political Weekly* : A 82-A 101. September 25.

Patnaik, Utsa (1994). "Tenancy and Accumulation", in Kaushik Basu (ed.), *Agrarian Questions*. New Delhi: Oxford University Press.

Ramachandran, V.K. (1990). *Wage Labour and Unfreedom in Agriculture: An Indian Case Study*. Oxford: Clarendon Press.

———. (1997). "On Kerala's Development Achievements", in Jean Dreze and Amartya Sen (eds.) *Indian Development: Selected Regional Perspectives*. Oxford University Press.

Rawal, Vikas (1999). "Irrigation Development in West Bengal: 1977-78 to 1995-96", Ph.D thesis submitted to Indira Gandhi Institute of Development Research, Mumbai.

———. (2001). "Irrigation Statistics in West Bengal-1", *Economic and Political Weekly*: 2537-44, July 07.

———. (2008a). "Ownership Holdings of Land in Rural India: Putting The Record Straight", *Economic and Political Weekly*, March 8.

———. (2008b). "Estimation of Rural Household Incomes in India: Selected Methodological Issues", Paper presented at *Studying Village Economies in India: A Colloquium on Methodology*. December 21 to 24.

Saha, Partha (2009). "Land Relations and Asset Holdings: A Study Based on Village-Level Evidence from Uttar Pradesh", Unpublished Ph.D. thesis submitted to Jawaharlal Nehru University, New Delhi.

Sharma, H.R. (1994). "Distribution of Landholdings in Rural India, 1953-54 to 1981-82: Implications for Land Reforms", *Economic and Political Weekly*, September 24.

Siddiqui, Kalim (1999). "New Technology and Process of Differentiation: Two Sugarcane Cultivating Villages in UP", *Economic and Political Weekly*, December 25.

Singh, A.K. (1987). *Agricultural Development and Rural Poverty*. New Delhi: Ashish Publishing House.

Singh, I. (1989). "Reverse Tenancy in Punjab Agriculture: Impact of Technological Change", *Economic and Political Weekly* 24(25).

Singh, Jagpal (1992). *Capitalism and Dependence: Agrarian Politics in Western Uttar Pradesh, 1951-1991*. Manohar Publishers.

Sharma, H.R. (2008). "Functioning of Lease Market in Rural India: Are Marginal and Small Farmers Discriminated?", Paper presented at *National Seminar on Land Reforms in Uttar Pradesh: Retrospect and Prospects*. August 12-13.

Sharma, N. and J. Dreze (1998), "Tenancy", in P. Lanjouw and N. Stern (eds.), *Economic Development in Palanpur Over Five Decades*. Oxford University Press.

Shrimali, P.D. (2003). *Agrarian Structure, Movements and Peasant Organisations in India: Uttar Pradesh*. Manak Publications Pvt. Ltd.

Singh, A.K. (1987). *Agricultural Development and Rural Poverty*. New Delhi: Ashish Publishing House.

———. (2008). "Impact of Land Reforms on Agrarian Structure and Agricultural Growth in Uttar Pradesh", Paper presented at *National Seminar on Land Reforms in Uttar Pradesh: Retrospect and Prospects*. August 12-13.

Srivastava, Ravi (2000). "Changes in Contractual Relations in Land and Labour in India", *Indian Journal of Agricultural Economics* 55(3).

———. (1989a). "Tenancy Contracts During Transition: A Study Based on Fieldwork in Uttar Pradesh", *Journal of Peasant Studies* 16.

———. (1989b). "Interlinked Modes of Exploitation in Indian Agriculture: A Case Study", *Journal of Peasant Studies* 16.

———. (1987). "Surplus Appropriation and Accumulation by Rural Households in India: A Case Study Based on Field Work in Uttar Pradesh", Ph.D Thesis submitted at the University of Cambridge.

Thorner, D. (1976). *Agrarian Prospect in India*. Delhi: Allied Publishers.

4 Problem of Farm Viability, Poverty and Agrarian Crisis

The Case of West Bengal

Rajiv Dey

Introduction

The Indian economy went along the structural adjustment path of International Monetary Fund (IMF) and introduced a series of neo-liberal economic reforms in 1991. The direct and immediate impact of this policy reforms in agriculture was, first of all, a cut in food and fertiliser subsidy and there was a rise in fertiliser price. The prices of nitrogeneous fertiliser were increased by 30 per cent and the prices of phosphatic, potassic fertilisers were doubled following their decontrol in the early 1990s (Rao, 1998). A large section of poor, small and middle farmer found it extremely difficult to bear this higher price. Particularly, the IMF directed structural adjustment policy as a part of financial liberalisation also curbed the provision of institutional credit (Bhattacharyya and Bhattacharyya, 2007). Given the fact that the price of fertiliser increased and access to credit decreased, the peasantry in large faced a hand to mouth situation. The latest "Situation Assessment Survey of Farmers 2005" of NSSO (National Sample Survey Organisation) observes that the average monthly per capita expenditure (MPCE) of a farm household has been just ₹ 503 (US$11.32) in 2003 across regions, classes and income groups, with 27 per cent of farmers not liking farming while 40 per cent felt that given the option they would prefer other career owing to poor earnings. According to NSSO 59th round report on "Indebtedness of Farmers Household" conducted in 2003 revealed that out of 89.35 million farmer households, about 43.42 million (48.6%) were indebted. This is nearly half the farm households in the country. A similar survey observed only 26 per cent of farm households were indebted in 1991.

The farmers in most of the regions in India are fed up with crop cultivation because they have not been able to recover even the cost of cultivation. The Situation Assesment Survey (SAS) carried out on various aspect of farming by the Ministry of Agriculture clearly showed that over 40 per cent of farmers are willing to get out of farming because it had become non-viable (Narayanamoorthy, 2007).

Patnaik (2007) by studying market-oriented reforms data of Indian agriculture observed that macroeconomic policies of expenditure deflation (neo-liberal economic policies guided by the Bretton Woods Institutions) is the main reason behind Agrarian Crisis.

Rao (2001) tried to work out the impact of WTO on viability of Indian agriculture and observed that the major challenge to the viability of agriculture in India is posed by the shortfalls in public investment and in the provision of agricultural services, and the failure of agricultural supplies to respond to the favourable incentive framework created by macroeconomic reforms, including trade liberalisation, in the 1990s. India was major exporter of food grain in the world but due to the unfavourable terms of trade exports have come down and finally, the price fall in the international market has significantly affected whole farming community.

Suri (2006) argues that agrarian distress is the result of the policies pursued by the governments over the years. Other factors such as changed cropping pattern due to a shift away from light crops to cash crops; liberalisation policies which prematurely pushed Indian agriculture into the global markets without a level playing field; heavy dependence on high-cost paid out inputs; growing costs of cultivation; volatility of crop output; market vagaries; lack of remunerative prices; indebtedness; neglect of agriculture by the government; decline in public investment have contributed further to agrarian crisis.

Galab and Reddy (2006), in their study examined the factors that caused crises in agriculture. They are technological, ecological, socio-cultural and policy related factors. Extensive cultivation has led to decrease in fertility and productivity. This is due to intensive use of fertilisers. Ecological factors include decreasing quality of land and water resources due to intensive chemical and fertiliser use. Socio and cultural factors include the effects of globalisation and urban culture on villages. In order to access better facilities farmers have changed their cropping pattern. Policy related factors include decrease in public investment from 4 per cent of agricultural GDP during 1980s to 1.86 during early 2000.

Mazumdar (1999), Ramachandran and Swaminathan (2001) and Shetty (2004) in their studies explored that policy fatigue with reference to agricultural credit being prime culprit of agrarian crisis in India.

This study aims to comprehend the real situation of farm viability, cause of agrarian crisis and to compare cultivators income with poverty level. This is done in three sections. First, the methodology of the study is introduced. Second section deals with the real situation of farm viability, cause of agrarian crisis and compares cultivators income with poverty level. The study is summed up with a brief conclusion in Section III.

I

Methodology

This study covered four districts of West Bengal and was carried out by the author in different time periods from July 2010 to March 2011. The districts selected purposively were Birbhum, Burdwan, Hooghly and Murshidabad. These districts have distinct development indices and different cropping patterns: Birbhum for paddy, Burdwan for paddy, mustard and potato, Hooghly for paddy, potato and mustard, and Murshidabad for paddy and jute. After selecting the districts, villages were selected according to different development indices: percentage area irrigated, percentage of literates etc., from four district village directory of Census 2001, as representative of developed samples and less developed samples. The villages which have low percentage of development indices are referred to as less developed samples whereas the villages which have high percentage of development indices are referred to as developed samples. Finally the combination of developed and less developed samples known as common samples. After this, these samples were utilised to classify the sample household according to acreage and Patnaik's labour exploitation criterion. For a better study, villages were identified as developed and less developed from the village level disaggregated data of Census 2001. The households were selected randomly (SRSWOR).The total number of households was 221.

This study have used the NSSO definition to identify farmer and farmer household. Any person who possessed some land and was engaged in agricultural activities on any part of that land during the reference year was accepted as a farmer and any household which had at least one farmer member was accepted as a farmer household.

The farm-size grouping is self-explanatory. Five categories of households were constructed:

- No land operated; that is, landless households.
- Some land operated but less than 2.5 acres.
- Operated land in excess of 2.5 acres but less than or equal to 5 acres.
- Operated land in excess of 5 acres but less than or equal to 10 acres.
- Operated land in excess of 10 acres (above 10 acres groups).

Scholars such as Sen (1966) and Thorner (1980) argue that farm sizes reflect the economic status of peasants and thus an analysis based on acreage sizes is, essentially, a class-based analysis. Patnaik (1976; 1987), on the other hand, took a more classic Marxist approach to rural class differentiation. She measured differentiation of the peasantry in terms of labour use. She developed a labour-exploitation index, or E-criterion, in order to obtain an empirical approximation to the analytical concept of the class status of the household based on labour use. The labour-exploitation index measured the farming family's use of outside labour through hiring in or employing tenants and, conversely, the dependence of the family members on working for others; both considered in relation to the extent of self-employment on their own farm. Under certain simplifying assumptions, this is identical to the surplus labour appropriated (or parted with), relative to surplus labour generated in self-employment. Patnaik argued that this index allowed a more precise study of the (class) dynamics of agrarian change. The six agrarian classes based on the labour-exploitation index are:

- Landlord: buyer of labour, not working on own land.
- Rich peasant: net buyer of labour, buying in more than spent on working own land.
- Middle peasant: net buyer of labour, but of fewer days than spent on working own land.
- Small peasant: net provider of labour, but of fewer days than spent on working own land.
- Poor peasant: net provider of labour, more days than spent on working own land.
- Landless worker: provider of labour, not working own land.

The labour-exploitation index seeks to give an empirical approximation to the analytical concept of the class status of the household. The class-status is essentially determined by the extent of the use of outside labour or to the extent the family works for others, relative to the extent of self-employment. It is identical, under certain simplifying assumptions with the surplus labour appropriated or parted with, relative to surplus labour with self- employment.

E-x/Y-{(Hi-Ho) + (Lo-Li)}/F

Where,

Hi = Labour-days hired on the operational holding of the household.

Ho = Family labour days hired out to others.

Li =Labour days worked on leased in land (whether by family or hired labour).

Lo = labour days similary worked on land leased out by the household.

F = labour days worked by household worker on the operational holding.

The index is a ratio, or a pure number, which can have positive or negative values depending on whether the household is a net employer of outside labour or is itself on balance working for others (as labourer or tenant). The range of values of E is from plus infinity to minus infinity, for at the two poles of the rural class structure, there will be diametrically opposite types of households for whom F will be zero or near zero: first, the big landlords have such a large resource endowment that they perform no manual labour themselves, but rely entirely on employing others' labour, and the landless labourers, with zero resource endowment, hence zero self employment, who are entirely dependent on working for other (Patnaik, 1976; 1987). Classes within the cultivating peasantry are identified by looking at the degree of working for others or of employing others' labour, relative to self employment. For this purpose certain limits are set upon the values of the E-ratio, which are given, in the following Table. The limits specified are to the value of E in order to classify households into a set of mutually exclusive and all-exhaustive categories (sub-categories not specified here are not ruled out). All subsequent use of class categories refers to the definitions given in the Tables taken from Patnaik (1976; 1987).

Class Category	*Defining Characteristics*	*Value of E = X/F*	*Reason*
1. Landless labourers	No self-employment; working entirely for others	($E \rightarrow -\infty$)	$F = 0$ $X < 0$ and large
2. Poor peasants (tenants and labourers with smallholdings)	Working for others exceeds self-employment	($E \leq -1$)	$F > 0$, $X < 0$ $\lvert X \rvert \geq F$
3. Small peasants	No employment of or working for others; self-employment exceeds working for others	($0 \geq E > -1$)	$F > 0$ $X \leq 0$ $\lvert X \rvert < F$
4. Middle peasants	Self-employment predominates, combined with the employment of a small amount of others' labour	($1 > E > 0$)	$F > 0$ $X > 0$, $X < F$
5. Rich peasants	Self-employment equalled by the amount of labour-power purchased	($E \geq 1$)	$F > 0$, $X > 0$ $X \geq F$
6. Landlords	No manual labour expended in the form of self-employment; large amounts of outside labour employed	($E \rightarrow \infty$)	$F = 0$, $X > 0$, and large

Source: Patnaik (1987).

The survey data represents the distribution of: (i) outputs, (ii) surplus/deficit, (iii) value-added, (iv) total income, and (v) total disposable income. There are some differences between the results obtained from the two types of grouping.

A complete list of households belonging to different occupational groups is available from the village papers lying at local *panchayat* office. Among all rural households belonging to different occupational groups, this study has taken only the household belonging to the cultivator category. From this category of households, simple random sample is drawn without replacement. The study have used MS excel, Access, SPSS for analysing the data.

Problem of Farm Viability and Agrarian Crisis

It is widely accepted that a serious crisis arises in the countryside under market-oriented reforms. The crisis is clearly marked by the continuous and widespread deceleration in agricultural growth rate. The growth of agriculture started declining since the reforms and became worse after 1995. The growth of agriculture has come down from 3.62 in 1990-91 to 1.97 by 2004-05 and the share of agriculture in the gross domestic product (GDP) has registered a steady decline from 36.4 per cent in 1982-83 to 18.5 per cent in 2006-07 (Chand *et al.*, 2007). Narayanamoorthy (2007) examined that peasantry in most of the regions in India are fed up with crop cultivation since they have not been able to recuperate even the cost of cultivation. Even the SAS revealed that over 40 per cent of farmers are willing to escape from farming since it had become non-viable.

For the depiction of farm non-viability, the study calculated some variables that project economic surpluses/deficits accruing to the peasant classes/acreage group. The variables stressed were net output and surpluses like gross value added (GVA), farm labour surplus (FLS)/deficit and farm disposable surplus (FDS)/deficit. In order to represent the distress, this study investigated: Firstly, the gross value of crop output and net output class and group wise. Secondly, surplus and deficit accruing to each class and group. Thirdly, gross output and value added, farm labour and farm disposable surplus class and group wise.

(1) Value of agricultural production = Gross output of all crops + value of by-product and livestock production.

(2) GVA of all crops = Value of agricultural production – Total material input cost.

(3) Farm labour income (FLI) = GVA of all crops – wages (kind & cash).

(4) FLS = FLI – imputed value of family labour income.

(5) Total income = Agricultural income (A) + Non-Agricultural income (B).

Where agricultural income is the sum of: (i) Farm labour income (A1); (ii) Income from agricultural wages (A2); (iii) Income from hiring out equipments, etc. (A3); (iv) Non-cultivation income from rent (A4). Therefore, A = A1 + A2 + A3 + A4.

And non-agricultural income is the sum of: (i) Income from salaries, pension,and remittances (B1); (ii) Non-cultivation income from other sources (B2). Therefore, non-agricultural Income = B = B1 + B2.

(6) Farm-disposable income (FDI) = FLI – Rents paid.

(7) Total disposable income = FDI + Non-agriculture income + Agricultural wage + Rents + Income from Hiring out.

(8) FDS or deficit = FDI – Imputed value of family labour. (Patnaik, 1987; Rakshit, 2010.)

Table 4.1 to 4.4 portray agricultural production and FLS/deficit according to economic class and farm size in the survey region.

Let us begin with Table 4.1, here the gross value of crop output per holding increased from poor peasants to rich peasants, similarly same trend is noticed for gross value of crop output per acre. This indicates that exploiting class (rich peasants) have a greater amount of crop output than the exploited class (poor peasants). Relative measures i.e., gross output (GO) per ₹ of material input cost and GVA per ₹ of material input cost are 0.56 and -0.18 respectively. In other words, for each rupee of material input cost GO obtained is more than rupee cost whereas GVA obtained is negative.

FLS or deficit is calculated by deducting from FLI the imputed value of family labour income.

FLS is presented in relative terms per holding, per acre and they are negative for all classes. The study found that it shows a uniform pattern of downward fluctuation in the exploited and self employed classes (Figure 4.1). According to Figure 4.1, all peasant class is the sufferer in terms of realisation of surplus. Similarly, the same picture is observed in Table 4.2 according to acreage analysis (Table 4.2).

Table 4.1

Class-wise Gross Output, Gross Value Added and Farm Labour Surplus

	Common Samples					
Class	*Gross Value of Crop Output Per Holding*	*Gross Value of Crop Output Per Acre*	*Gross Output Per ₹ of Material Input Cost*	*Gross Value Added Per ₹ of Material Input Cost*	*Farm Labour Surplus Per Holding*	*Farm Labour Surplus Per Acre*
Landless	0.00	0.00	0.00	0.00	3814.38	0.00
Poor	13220.08	10569.02	0.40	-0.31	-19216.33	-15362.82
Small	23925.30	18584.33	0.61	-0.04	-19547.02	-15183.43
Middle	38241.25	20288.15	0.54	-0.23	-37782.96	-20045.02
Rich	53174.56	22619.95	0.61	-0.13	-29703.78	-12635.71
Total	36948.72	20679.38	0.56	-0.18	-29525.72	-16524.89

Figure 4.1

Farm Labour Surplus/Deficit in Terms of per Holding and per Acre

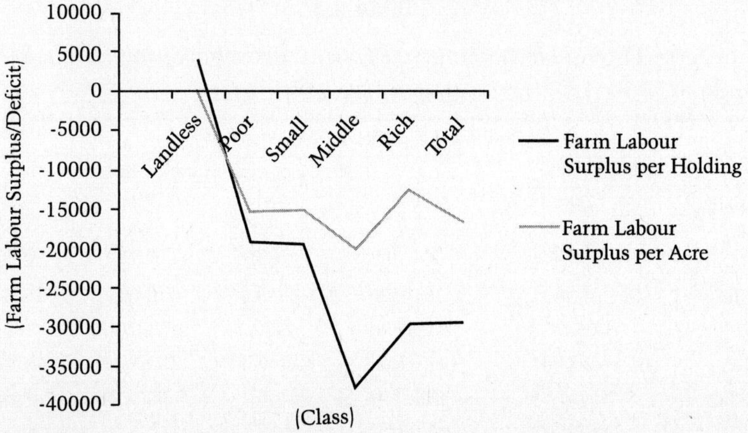

Table 4.2

Acreage-wise Gross Output, Gross Value Added and Farm Labour Surplus

	Common Samples					
Farm Size	*Gross Value of Crop Output Per Holding*	*Gross Value of Crop Output Per Acre*	*Gross Output Per ₹ of Material Input Cost*	*Gross Value Added Per ₹ of Material Input Cost*	*Farm Labour Surplus Per Holding*	*Farm Labour Surplus Per Acre*
0	0.00	0.00	0.00	0.00	3814.38	0.00
0.01-1.25	16322.64	23221.75	0.46	-0.22	-23499.64	-33432.25
1.25-2.5	37629.86	20346.06	0.62	-0.09	-25986.48	-14050.61
2.5-5.0	69178.16	19557.27	0.61	-0.17	-42541.16	-12026.76
5.0-10.0	145283.90	21277.67	0.54	-0.29	-103827.50	-15206.14
Total	36948.72	20679.38	0.56	-0.18	-29525.72	-16524.89

Now, the study came to the newer concept of surplus and deficit that is FDS or deficit. What is distinguishing about FDS or deficit in comparison to FLS is that only rent payment is additionally deducted when it is discussing about FLS. Table 4.3 depicts the FDS per acre and per holding according to economic classes. It is negative for all classes. This means that farmers are in deficit zone. They are unable to realise the disposable surplus. In addition, there is a uniform pattern of downward fluctuation in the self employed and exploited classes (Figure 4.2).

Table 4.3

Farm Disposable Income and Farm Disposable Surplus/Deficit acccording to Economic Class

Common Samples: Economic Class Wise					
Farm Disposable Income		*Farm Disposable Surplus/Deficit*			
Economic Class	*FDS per Acre*	*FDS per Holding*	*FDS per Worker*	*TDI per Acre*	*TDI per Holding*
Landless	0.00	3814.38	2653.48	0.00	15273.75
Poor	-17200.93	-21515.50	-9930.23	-2392.47	-2992.58
Small	-16925.41	-21789.63	-10663.01	2859.29	3681.02
Middle	-21665.64	-40837.68	-28667.51	2855.98	5383.26
Rich	-13308.82	-31286.13	-26697.49	3758.54	8835.52
Total	-17801.87	-31807.35	-21830.51	3619.50	6467.11

Figure 4.2

Farm Disposable Surplus/Deficit in terms of Per Holding and Per Acre

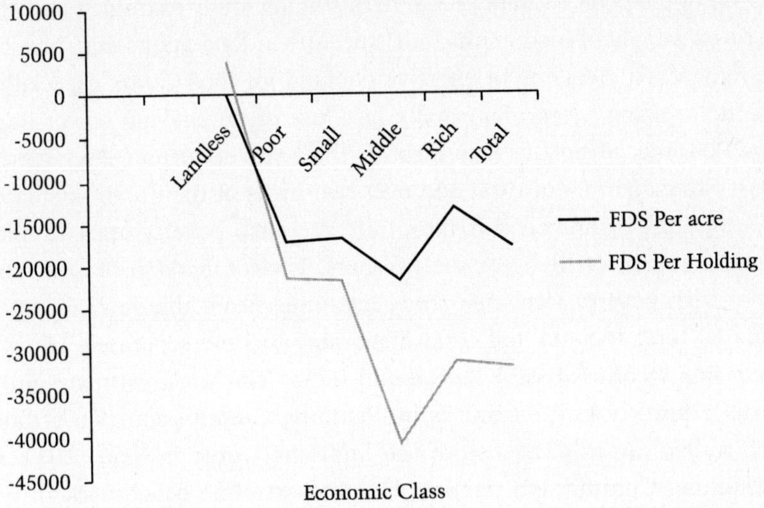

In Table 4.4 FDS per holding and per acre are negative for all acreage groups. In other words, all acreage groups are the worst sufferer in terms of realisation of surplus.

A crisis in agriculture adversely affects all the classes, farm size. The most vital incident is that it affects the rich farmers as well. The well to do farmers obtained significant degree of negative farm disposable surplus.

Table 4.4

Farm Disposable Surplus in terms of Per Holding, Per Acre and Per Worker, Total Disposable Income in terms of Per Holding and Per Acre

Common Samples: Acreage Wise					
Farm Disposable Income		*Farm Disposable Surplus/Deficit*			
Farm Size	*FDS per Acre*	*FDS per Holding*	*FDS per Worker*	*TDI per Acre*	*TDI per Holding*
0	0.00	3814.38	2653.48	0.00	15273.75
0.01-1.25	-34355.00	-24148.24	-17143.41	10512.52	7389.28
1.25-2.5	-15738.75	-29108.68	-19296.77	4648.12	8596.66
2.5-5.0	-13366.07	-47278.60	-30802.73	1637.21	5791.16
5.0-10.0	-16026.29	-109427.50	-84175.00	-3786.98	-25857.50
Total	-17801.87	-31807.35	-21830.51	3619.50	6467.11

Poverty

Poverty has intensified since reforms. Market-oriented reforms is not in favour of the poor. Patnaik (2005) in one of her study examined that there is not over supply of food grains, but a decline in food grains supply and an even more drastic decline of effective demand for food grains especially in rural India owing to an abnormally fast loss of purchasing power during 1998-2004 and poverty is very high on the basis of current data affecting at least three-quarters of rural and over two-fifths of the urban population. The study has compared cultivators income with poverty level to obtain the real standard of living of the peasants. For comparison of cultivators' income with poverty level this study has undertaken the deviation of the calculated total income and total disposable income separately from the poverty line income of West Bengal and India. The latest estimate on this per capita poverty line income is by Planning Commission. Government of India filed an affidavit before the Supreme Court in Sept. 2011 and the Planning Commission published the poverty line benchmark in rural areas in March 2012. Initially, this study has taken the first case, where the estimated poverty line per capita income for India is ₹ 781 per month and ₹ 800 for West Bengal. Secondly, the estimated poverty line per capita income for India is ₹ 673 per month and ₹ 680 for West Bengal. The study has calculated group-wise per capita income per annum by multiplying the number of members in each group with the poverty line per capita income. The complete analysis is shown in Tables 4.5a-b and 4.6a-b. For a better study the samples are divided into common, developed and less developed samples. This is performed according to economic and social criterion obtained from different development indices of the districts of West Bengal. Table 4.5a reveals economic class-wise deviation of total income/ total disposable income from poverty line. Here poverty line benchmark is ₹ 781 (rural) *vis-á-vis* Planning Commission which filed an affidavit before the Supreme Court in September 2011. Table 4.6a reveals economic class-wise deviation of total income/total disposable income from poverty line. Here poverty line benchmark is ₹ 673 (rural) as published by the Planning Commission in March 2012. Similarly, Table 4.5b and 4.6b reveals acreage analysis.

Let us first consider the case for India when the estimated poverty line per capita income is ₹ 781 per month and ₹ 800 for West Bengal. In developed samples, this study has calculated the group-wise poverty line income with respect to West Bengal's and India's per capita poverty line income. By comparing them with total income according to calculation specified before with respect to West Bengal's per capita poverty line income, this study examined all the classes to have negative deviation. The negative deviation increases with the ascending economic status. The trend is similar to FLS and FDS. The deviation from poverty line changes from less deficit to more deficit with ascending class status. The deviation between total disposable income and both type of calculation of poverty line income reveals similar trend like deviation between total income.

Now by comparing total income of household with respect to India's per capita poverty line income, the study observed all the classes to have negative deviation. The negative deviation increases with the ascending economic status. The deviation from poverty line changes from lower deficit to higher deficit with ascending class status. The deviation between total disposable income and both type of calculation of poverty line income reveals similar trend like deviation between total income.

In less developed samples, the analysis is undertaken in the same way taking the group-wise poverty line income of West Bengal and India. Then the deviation of the two with respect to total income and total disposable income is calculated. Initially, by comparing total income and West Bengal's poverty line income, the study observed that poor and small peasant classes to be in deficit though landless; the middle and rich peasant classes showed surplus and a positive deviation. They are just above the boundary and to consider them as well-off would really be an amplification.

Similarly, by considering total disposable income and comparing it with West Bengal's group-wise poverty line income, the study observed that all classes from poor to rich peasant have a negative deviation, though landless maintains higher surplus per holding. Moreover, deficit per holding decreases from poor to rich peasant. In fact, after payment of land rent, since disposable income lowers, consequently the middle and rich peasant classes reveals a deficit. In comparison to India's per capita poverty line income, the deviation of total income and total disposable income shows a similar trend like West Bengal poverty line income deviation with respect to total income and total disposable income.

It is reflected from the above analysis that middle and rich peasant classes in developed samples and poor and small peasant classes in less developed samples are the worst sufferer in the market-oriented reform. They are unable to realise the surplus value and are also the prime victim.

Secondly, consider the case when the estimated poverty line per capita income is ₹ 673 per month for India and ₹ 680 for West Bengal. In developed samples, the study has calculated the group-wise poverty line income with respect to West Bengal's and India's per capita poverty line income. By comparing them with total income according to calculation specified before with respect to West Bengal's per capita poverty line income, the study observed all the classes to have negative deviation. The negative deviation increases with the ascending economic status. The trend is similar as in FLS and FDS. The deviation from poverty line changes from less deficit to more deficit with ascending class status. The deviation between total disposable income and both type of calculation of poverty line income reveals similar trend like deviation between total income.

Now by comparing the total income of household with respect to India's per capita poverty line income, the study observed all the classes to have negative deviation. The negative deviation increases with the ascending economic status. The deviation from poverty line changes from lower deficit to higher deficit with ascending class status. The deviation between total disposable income and both type of calculation of poverty line income reveals similar trend like deviation between total income.

In less developed samples, the analysis is undertaken in the same way taking the group-wise poverty line income of West Bengal and India. Then the deviation of the two with respect to total income and total disposable income is calculated. Initially, by comparing total income and West Bengal's poverty line income, the study observed poor and small peasant classes to be in deficit though landless; middle and rich peasant classes showed surplus and a positive deviation. They are just above the boundary and to consider them as well-off would really be an amplification.

Similarly, by considering total disposable income and comparing it with West Bengal's group-wise poverty line incomes, the study observed that poor and small peasant classes are deficit though landless, middle and rich peasant classes showed surplus and a positive deviation. They are just above the boundary and to consider them as well-off would really be an amplification. In comparison to India's per capita poverty line income the deviation of total income and total disposable income shows a similar trend like West Bengal poverty line income deviation with respect to total income and total disposable income.

It is reflected from the above analysis that middle and rich peasant classes in developed samples and poor and small peasant classes in less developed samples are the worst sufferer in the market-oriented reform. They are unable to realise the surplus value and are also the prime casualty.

The rationale behind such vulnerability of middle and rich peasant classes in developed samples and poor and small peasant classes in less developed samples can be summarised as: 1) They are incurring high input costs in the respective samples. 2) Middle and rich peasant classes are unwilling to earn from wage labour and other non-farm labour particularly if they are unskilled jobs. 3) As higher classes diversified their crops in the urge to obtain a better return, they finally fail due to market volatility.

Regarding acreage analysis, similar pattern is noticed as economic class analysis. This means that higher acreage groups of developed samples and lower acreage groups of less developed samples are the key casualty.

The most vital point here is that the comparison of different poverty level income with farm labour and farm disposable income have increased the magnitude of negative difference. This reflects to the point that with the inception of market-oriented reforms, the economic status of all classes has worsened ghastly. Due to this, with no suitable alternative opportunities in the countryside they are forced to undertake other non-farm activities to supplement their farm labour and farm disposable income so as to sustain themselves in these dreaded uncertain economic environment promulgated from neo-liberal policies pursued by the Government of India since 1991.

Table 4.5a

Economic Class-wise Deviation of Total Income/Total Disposable Income from Poverty Line: Considering West Bengal and India's Rural Level Poverty Line Per Capita Income, September 2011

Economic Class	Total Income Agri+ Non Agri	Total Disposable Agri+ Non Agri	Groupwise WB's P-Line Income	Deviation WRT TI		Deviation WRT TDI		Groupwise India's P-Line Income	Deviation WRT TI		Deviation WRT TDI	
				From P. Line	Per Holding	From P Line	Per Holding		From P Line	Per Holding	From P Line	Per Holding
Common Samples												
Landless	244380.00	244380.00	585600.00	-341220.00	-21326.25	-341220.00	-21326.25	571692.00	-327312.00	-20457.00	-327312.00	-20457.00
Poor	-8321.00	-35911.00	489600.00	-497921.00	-41493.42	-525511.00	-43792.58	477972.00	-486293.00	-40524.42	-513883.00	-42823.58
Small	136243.50	84663.50	1152000.00	-1015756.50	-44163.33	-1067336.50	-46405.93	1124640.00	-988396.50	-42973.76	-1039976.50	-45216.37
Middle	894425.75	570625.75	4800000.00	-3905574.25	-36845.04	-4229374.25	-39899.76	4686000.00	-3791574.25	-35769.57	-4115374.25	-38824.29
Rich	666743.00	565473.00	2534400.00	-1867657.00	-29182.14	-1968927.00	-30764.48	2474208.00	-1807465.00	-28241.64	-1908735.00	-29823.98
Total	1933471.25	1429231.25	9561600.00	-7628128.75	-34516.42	-8132368.75	-36798.05	9334512.00	-7401040.75	-33488.87	-7905280.75	-35770.50
Developed Samples												
Landless	99400.00	99400.00	249600.00	-150200.00	-25033.33	-150200.00	-25033.33	243672.00	-144272.00	-24045.33	-144272.00	-24045.33
Poor	33268.00	-21912.00	211200.00	-177932.00	-35586.40	-233112.00	-46622.40	206184.00	-172916.00	-34583.20	-228096.00	-45619.20
Small	132407.50	132407.50	518400.00	-385992.50	-38599.25	-385992.50	-38599.25	506088.00	-373680.50	-37368.05	-373680.50	-37368.05
Middle	-92637.00	-287717.00	1920000.00	-2012637.00	-47919.93	-2207717.00	-52564.69	1874400.00	-1967037.00	-46834.21	-2162117.00	-51478.98
Rich	17065.00	56425.00	777600.00	-760535.00	-40028.16	-721175.00	-37956.58	759132.00	-742067.00	-39056.16	-702707.00	-36984.58
Total	189503.50	-21396.50	3676800.00	-3487296.50	-42528.01	-3698196.50	-45099.96	3589476.00	-3399972.50	-41463.08	-3610872.50	-44035.03
Less Developed Samples												
Landless	144980.00	144980.00	124800.00	20180.00	2018.00	20180.00	2018.00	121836.00	23144.00	2314.40	23144.00	2314.40
Poor	-13999.00	-13999.00	144000.00	-157999.00	-22571.29	-157999.00	-22571.29	140580.00	-154579.00	-22082.71	-154579.00	-22082.71
Small	55416.00	-47744.00	240000.00	-184584.00	-14198.77	-287744.00	-22134.15	234300.00	-178884.00	-13760.31	-282044.00	-21695.69
Middle	1267022.75	858342.75	883200.00	383822.75	5997.23	-24857.25	-388.39	862224.00	404798.75	6324.98	-3881.25	-60.64
Rich	635118.00	509048.00	528000.00	107118.00	2380.40	-18952.00	-421.16	515460.00	119658.00	2659.07	-6412.00	-142.49
Total	2088537.75	1450627.75	1920000.00	168537.75	1212.50	-469372.25	-3376.78	1874400.00	214137.75	1540.56	-423772.25	-3048.72

Note: In Sept. 2011, Planning Commission filed an affidavit before the Supreme Court regarding the poverty line benchmark in rural areas at ₹ 781 monthly per capita income.

Table 4.5b

Economic Class-wise Deviation of Total Income/Total Disposable Income from Poverty Line: Considering West Bengal and India's Rural Level Poverty Line Per Capita Income, March 2012

Economic Class	Total Income Agri+ Non Agri	Total Disposable Agri+ Non Agri	Groupwise WB's P-Line Income	Deviation WRT TI		Deviation WRT TDI		Groupwise India's P-Line Income	Deviation WRT TI		Deviation WRT TDI	
				From P Line	Per Holding	From P Line	Per Holding		From P Line	Per Holding	From P Line	Per Holding
Common Samples												
Landless	244380.00	244380.00	497760.00	-253380.00	-15836.25	-253380.00	-15836.25	492636.00	-248256.00	-15516.00	-248256.00	-15516.00
Poor	-8321.00	-35911.00	416160.00	-424481.00	-35373.42	-452071.00	-37672.58	411876.00	-420197.00	-35016.42	-447787.00	-37315.58
Small	136243.50	84663.50	979200.00	-842956.50	-36650.28	-894536.50	-38892.89	969120.00	-832876.50	-36212.02	-884456.50	-38454.63
Middle	894425.75	570625.75	4080000.00	-3185574.25	-30052.59	-3509374.25	-33107.30	4038000.00	-3143574.25	-29656.36	-3467374.25	-32711.08
Rich	666743.00	565473.00	2154240.00	-1487497.00	-23242.14	-1588767.00	-24824.48	2132064.00	-1465321.00	-22895.64	-1566591.00	-24477.98
Total	1933471.25	1429231.25	8127360.00	-6193888.75	-28026.65	-6698128.75	-30308.27	8043696.00	-6110224.75	-27648.08	-6614464.75	-29929.70
Developed Samples												
Landless	99400.00	99400.00	212160.00	-112760.00	-18793.33	-112760.00	-18793.33	209976.00	-110576.00	-18429.33	-110576.00	-18429.33
Poor	33268.00	-21912.00	179520.00	-146252.00	-29250.40	-201432.00	-40286.40	177672.00	-144404.00	-28880.80	-199584.00	-39916.80
Small	132407.50	132407.50	440640.00	-308232.50	-30823.25	-308232.50	-30823.25	436104.00	-303696.50	-30369.65	-303696.50	-30369.65
Middle	-92637.00	-287717.00	1632000.00	-1724637.00	-41062.79	-1919717.00	-45707.55	1615200.00	-1707837.00	-40662.79	-1902917.00	-45307.55
Rich	17065.00	56425.00	660960.00	-643895.00	-33889.21	-604535.00	-31817.63	654156.00	-637091.00	-33531.11	-597731.00	-31459.53
Total	189503.50	-21396.50	3125280.00	-2935776.50	-35802.15	-3146676.50	-38374.10	3093108.00	-2903604.50	-35409.81	-3114504.50	-37981.76
Less Developed Samples												
Landless	144980.00	144980.00	106080.00	38900.00	3890.00	38900.00	3890.00	104988.00	39992.00	3999.20	39992.00	3999.20
Poor	-13999.00	-13999.00	122400.00	-136399.00	-19485.57	-136399.00	-19485.57	121140.00	-135139.00	-19305.57	-135139.00	-19305.57
Small	55416.00	-47744.00	204000.00	-148584.00	-11429.54	-251744.00	-19364.92	201900.00	-146484.00	-11268.00	-249644.00	-19203.38
Middle	1267022.75	858342.75	750720.00	516302.75	8067.23	107622.75	1681.61	742992.00	524030.75	8187.98	115350.75	1802.36
Rich	635118.00	509048.00	448800.00	186318.00	4140.40	60248.00	1338.84	444180.00	190938.00	4243.07	64868.00	1441.51
Total	2088537.75	1450627.75	1632000.00	456537.75	3284.44	-181372.25	-1304.84	1615200.00	473337.75	3405.31	-164572.25	-1183.97

Note: In March 2012, Planning Commission published the poverty line benchmark in rural areas at ₹ 673 monthly per capita income.

Table 4.6a

Acreage-wise Deviation of Total Income/Total Disposable Income from Poverty Line Considering West Bengal and India's Rural Level Poverty Line Per Capita Income, September 2011

Farm Size	Total Income Agri+ Non Agri	Total Disposable Agri+ Non Agri	Groupwise WB's P-Line Income	Deviation WRT TI		Deviation WRT TDI		Groupwise India's P-Line Income	Deviation WRT TI		Deviation WRT TDI	
				From P Line	Per Holding	From P Line	Per Holding		From P Line	Per Holding	From P Line	Per Holding
						Common Samples						
0.00	244380.00	244380.00	585600.00	-341220.00	-21326.25	-341220.00	-21326.25	571692.00	-327312.00	-20457.00	-327312.00	-20457.00
0.01-1.25	747523.50	687203.50	3830400.00	-3082876.50	-33149.21	-3143196.50	-33797.81	3739428.00	-2991904.50	-32171.02	-3052224.50	-32819.62
1.25-2.5	691412.75	507202.75	2640000.00	-1948587.25	-33026.90	-2132797.25	-36149.11	2577300.00	-1885887.25	-31964.19	-2070097.25	-35086.39
2.5-5.0	452730.00	249020.00	2044800.00	-1592070.00	-37024.88	-1795780.00	-41762.33	1996236.00	-1543506.00	-35895.49	-1747216.00	-40632.93
5.0-10.0	-202575.00	-258575.00	460800.00	-663375.00	-66337.50	-719375.00	-71937.50	449856.00	-652431.00	-65243.10	-708431.00	-70843.10
Total	1933471.25	1429231.25	9561600.00	-7628128.75	-34516.42	-8132368.75	-36798.05	9334512.00	-7401040.75	-33488.87	-7905280.75	-35770.50
						Developed Samples						
0	99400.00	99400.00	249600.00	-150200.00	-25033.33	-150200.00	-25033.33	243672.00	-144272.00	-24045.33	-144272.00	-24045.33
0.01-1.25	197655.50	154595.50	1334400.00	-1136744.50	-34446.80	-1179804.50	-35751.65	1302708.00	-1105052.50	-33486.44	-1148112.50	-34791.29
1.25-2.5	457507.00	352427.00	1209600.00	-752093.00	-28926.65	-857173.00	-32968.19	1180872.00	-723365.00	-27821.73	-828445.00	-31863.27
2.5-5.0	-156162.00	-238922.00	758400.00	-914562.00	-60970.80	-997322.00	-66488.13	740388.00	-896550.00	-59770.00	-979310.00	-65287.33
5.0-10.0	-408897.00	-388897.00	124800.00	-533697.00	-266848.50	-513697.00	-256848.50	121836.00	-530733.00	-265366.50	-510733.00	-255366.50
Total	189503.50	-21396.50	3676800.00	-3487296.50	-42528.01	-3698196.50	-45099.96	3589476.00	-3399972.50	-41463.08	-3610872.50	-44035.03
						Less Developed Samples						
0	144980.00	144980.00	124800.00	20180.00	2018.00	20180.00	2018.00	121836.00	23144.00	2314.40	23144.00	2314.40
0.01-1.25	560278.00	536928.00	825600.00	-265322.00	-4349.54	-288672.00	-4732.33	805992.00	-245714.00	-4028.10	-269064.00	-4410.89
1.25-2.5	410595.75	150455.75	451200.00	-40604.25	-1268.88	-300744.25	-9398.26	440484.00	-29888.25	-934.01	-290028.25	-9063.38
2.5-5.0	745262.00	487942.00	412800.00	332462.00	11873.64	75142.00	2683.64	402996.00	342266.00	12223.79	84946.00	3033.79
5.0-10.0	227422.00	130322.00	105600.00	121822.00	15227.75	24722.00	3090.25	103092.00	124330.00	15541.25	27230.00	3403.75
Total	2088537.75	1450627.75	1920000.00	168537.75	1212.50	-469372.25	-3376.78	1874400.00	214137.75	1540.56	-423772.25	-3048.72

Note: In September 2011, Planning Commission filed an affidavit before the Supreme Court regarding the poverty line benchmark in rural areas at ₹ 781 monthly per capita income.

Table 4.6b

Acreage-wise Deviation of Total Income/Total Disposable Income from Poverty Line Considering West Bengal and India's Rural Level Poverty Line Per Capita Income, March 2012

Farm Size	Total Income Agri+ Non Agri	Total Disposable Agri+ Non Agri	Groupwise WB's P-Line Income	Deviation WRT TI		Deviation WRT TDI		Groupwise India's P-Line Income	Deviation WRT TI		Deviation WRT TDI	
				From P Line	Per Holding	From P Line	Per Holding		From P Line	Per Holding	From P Line	Per Holding
						Common Samples						
0	244380.00	244380.00	497760.00	-253380.00	-15836.25	-253380.00	-15836.25	492636.00	-248256.00	-15516.00	-248256.00	-15516.00
0.01-1.25	747523.50	687203.50	3255840.00	-2508316.50	-26971.15	-2568636.50	-27619.75	3222324.00	-2474800.50	-26610.76	-2535120.50	-27259.36
1.25-2.5	691412.75	507202.75	2244000.00	-1552587.25	-26315.04	-1736797.25	-29437.24	2220900.00	-1529487.25	-25923.51	-1713697.25	-29045.72
2.5-5.0	452730.00	249020.00	1738080.00	-1285350.00	-29891.86	-1489060.00	-34629.30	1720188.00	-1267458.00	-29475.77	-1471168.00	-34213.21
5.0-10.0	-202575.00	-258575.00	391680.00	-594255.00	-59425.50	-650255.00	-65025.50	387648.00	-590223.00	-59022.30	-646223.00	-64622.30
Total	1933471.25	1429231.25	8127360.00	-6193888.75	-28026.65	-6698128.75	-30308.27	8043696.00	-6110224.75	-27648.08	-6614464.75	-29929.70
						Developed Samples						
0	99400.00	99400.00	212160.00	-112760.00	-18793.33	-112760.00	-18793.33	209976.00	-110576.00	-18429.33	-110576.00	-18429.33
0.01-1.25	197655.50	154595.50	1134240.00	-936584.50	-28381.35	-979644.50	-29686.20	1122564.00	-924908.50	-28027.53	-967968.50	-29332.38
1.25-2.5	457507.00	352427.00	1028160.00	-570653.00	-21948.19	-675733.00	-25989.73	1017576.00	-560069.00	-21541.12	-665149.00	-25582.65
2.5-5.0	-156162.00	-238922.00	644640.00	-800802.00	-53386.80	-883562.00	-58904.13	638004.00	-794166.00	-52944.40	-876926.00	-58461.73
5.0-10.0	-408897.00	-388897.00	106080.00	-514977.00	-257488.50	-494977.00	-247488.50	104988.00	-513885.00	-256942.50	-493885.00	-246942.50
TOTAL	189503.50	-21396.50	3125280.00	-2935776.50	-35802.15	-3146676.50	-38374.10	3093108.00	-2903604.50	-35409.81	-3114504.50	-37981.76
						Less Developed Samples						
0	144980.00	144980.00	106080.00	38900.00	3890.00	38900.00	3890.00	104988.00	39992.00	3999.20	39992.00	3999.20
0.01-1.25	560278.00	536928.00	701760.00	-141482.00	-2319.38	-164832.00	-2702.16	694536.00	-134258.00	-2200.95	-157608.00	-2583.74
1.25-2.5	410595.75	150455.75	383520.00	27075.75	846.12	-233064.25	-7283.26	379572.00	31023.75	969.49	-229116.25	-7159.88
2.5-5.0	745262.00	487942.00	350880.00	394382.00	14085.07	137062.00	4895.07	347268.00	397994.00	14214.07	140674.00	5024.07
5.0-10.0	227422.00	130322.00	89760.00	137662.00	17207.75	40562.00	5070.25	88836.00	138586.00	17323.25	41486.00	5185.75
Total	2088537.75	1450627.75	1632000.00	456537.75	3284.44	-181372.25	-1304.84	1615200.00	473337.75	3405.31	-164572.25	-1183.97

Note: In March 2012, Planning Commission published the poverty line benchmark in rural areas at ₹ 673 monthly per capita income.

Conclusion

By summing up, the survey findings clearly reflected that the farm sector is not non-viable but it is in a very ghastly state. This study shows that farmers are in deficit zone. Peasants are unable to recover the cost of production due to inhuman rise in input cost in the market-oriented reforms. Peasants are facing losses in agriculture by being unable to recover the cost of production.

This study finds a significant degree of farm labour deficit and farm disposable deficit among all peasants and also among all acreage groups. They are unable to realise the surplus. A crisis in agriculture affects adversely all the classes and farm size. The most momentous incident is that it affects the rich farmers as well. The well to do farmers obtained significant degree of negative farm disposable surplus. Even the upper echelons in the agrarian class structure with non-agricultural income are also unable to recuperate their situation. They are in a very nasty condition.

Finally, it is interesting to note that the comparison of different poverty level income with farm labour income and farm disposable income have increased the magnitude of negative divergence. The study found that all acreage and economic class are far below the West Bengal and India's poverty line income. This clearly explores the fact that under the market-oriented reforms the economic status of all peasant groups has worsened so badly that even the non-farm activities cannot supplement their farm labour income and farm disposable income to sustain themselves in the neo-liberal policies adopted by the Government of India after 1991.

The survey findings revealed that after two decades of market-oriented reforms the economic position of all economic classes is in a very depressed state. To uplift the peasantry and to resolve the agrarian distress, the following measures need to be undertaken:

- Enhancing investment thereby strengthening the resource base of agriculture.
- Developing agricultural market and eliminating local traders and middleman.
- Introducing co-operative farming.

- Developing rural infrastructure.
- Developing agricultural research and technology.

References

Bhattacharyya and Bhattacharyya (2007). "Agrarian Impasse in West Bengal in the Liberalisation Era", *Economic and Political Weekly* XLII(52), Dec 29.

Chand, R., S.S. Raju and L.M. Pandey (2007). "Growth Crisis in Agriculture Severity and Options at National and State Levels", *Economic and Political Weekly* 42(26): 2528-33.

Galab, Shaik and Ratna Reddy (2006). "Agrarian Crisis Looking Beyond Debt Trap", *Economic and Political Weekly* 1838-42.

Majumdar, N.A. (1999). "Reviving Rural Credit", *Economic and Political Weekly* 34(25): 1577-79, June 19.

Narayanamoorthy, A. (2007). "Deceleration in Agricultural Growth, Technology Fatigue or Policy Fatigue?", *Economic and Political Weekly* 42(25): 23-29, June.

National Sample Survey Organisation (2005). *Some Aspects of Farming, Situation Assessment Survey of Farmers*, Report No 496; *Indebtedness of Farmer Households: Situation Assessment Survey of Farmers*, Report No 498; *Household Assets and Liabilities in India*, Report No 500, New Delhi: Ministry of Statistics and Programme Implementation, Government of India.

Patnaik, Utsa (1976). "Class Differentiation Within the Peasantry: An Approach to the Analysis of Indian Agriculture", *Economic and PoliticalWeekly* 11(39): A82–101.

———. (1987). *Peasant Class Differentiation: A Study in Method with Reference to Haryana*. Delhi: Oxford University Press.

———. (2005). "Theorizing Food Security and Poverty in the era of Economic Reforms", *Social Scientist* 33(7-8), Jul-Aug.

———. (2007). "New Data on the Arrested Development of Capitalism in Indian Agriculture", *Social Scientist* 35(7/8): 4-23, Jul–Aug.

Rao, C.H. (1998). "Agricultural Growth, Sustainability and Poverty Alleviation, Recent Trends and Major Issues of Reform", *Economic and Political Weekly* 33(29-30), July 18.

———. (2001). "WTO and Viability of Indian Agriculture", *Economic and Political Weekly* 3453-57, September 8.

Rakshit, Santanu (2010). "Agrarian Transition: Diversity in Nature, Notion and Observations—A Survey of Theoretical Expositions and Empirical Studies with Reference to India and West Bengal", *Research in Social Stratification and Mobility, Elsevier, Science Direct* 28(4), December.

Ramachandran, V.K. and Madhura Swaminathan (2001). *Does Informal Credit Provide Security? Rural Banking Policy in India*. Geneva: International Labour Organisation (*www.ilo.org/ses*).

Sen, A.K. (1966). "Peasants and Dualism with and without Surplus Labour", *Journal of Political Economy* 74 (5): 425-50.

Shetty, S.L. (2004). "Distributional Issues in Bank Credit: Multi-pronged Strategy for Correcting Past Neglect", *Economic and Political Weekly* 3265-69, July 17.

Suri, K.C. (2006). "Political Economy of Agrarian Distress", *Economic and Political Weekly* 31: 1523-29.

Thorner, D. (1980). *The Shaping of Modern India*. New Delhi: Allied.

Thorner, Alice (1982). "Semi Feudalism or Capitalism: Contemporary Debate on Classes and Modes of Production in India", *Economic and Political Weekly* 18 (49): 1964-70.

I would like to thank Dr Sudipta Bhattacharyya and Dr Santanu Rakshit for the useful discussion on this paper.

5 Strategy and Implications for Rainfed Organic Cotton Farmers in the Context of Agricultural Crisis

P. Prashanth, M. Jagan Mohan Reddy and N. Kishore Kumar

Introduction

The green revolution was ushered in India during sixties and it has been the cornerstone of India's agricultural achievement, transforming the country from the stage of food deficiency to self-sufficiency. During the period, the production of food grains has increased four folds from 50.82 million tonnes in 1950-51 to 246 million tonnes on 2009-10. As a result, consumption of synthetic fertilisers in India saw an astronomical increase from 70 thousand tonnes in 1950-51 to 2.6 crore tonnes in 2009-10. Initially, application of fertiliser did help in increasing food production in India but gradually crop response to application of fertiliser started getting weaker. India ranks 7th in the world in terms of total land under organic cultivation; the cultivated land under organic certification is around 1 million hectares. India also ranks 2nd in the world in terms of number of producers involved in organic farming.

In India during 2010-11 cotton was cultivated in 11.161 m ha; and in Andhra Pradesh cotton is an important cash crop and grown over an area of about 18.21 lakh ha. It is one of the leading states in cotton cultivation with the area of 1.776 m ha after Maharashtra and Gujarat. It is also the 3rd ranked state in productivity with 507 kg/ha after Gujarat (659 kg/ha) and Punjab (513 kg/ha). Indian certified organic cotton production is 15 per cent of the total world production and it is about 37 per cent of the Asian production. The overall productivity is very low as compared to US. It may be due to the many vagarious faced by farmers while cultivating cotton.

Cotton is attacked by a number of insect, pests, diseases, nematodes and weeds right from the time of sowing till picking/harvesting the crop. Yield losses due to these pests ranges from 15-25 per cent in spite of extensive use of pesticides. In India cotton is attacked by 162 insects, of which only about dozen are very serious causing 50-60 per cent loss. There are a number of problems encountered by the cotton growers in adoption of the standardised organic cotton practices. Keeping the above in mind, the present study aims to identify some of the constraints affecting the rate of adoption of organic farming in cotton in Karimnagar district of Andhra Pradesh.

Majority of the cotton growers in the district are inclined to apply organic farming in cotton cultivation. The exorbitant amounts of chemicals are used to arrest the pest population in the crop. The conventional cotton drastically effected by sucking pest, lost of soil fertility, low price of market caused many farmers suicides in the district in the 21st century in spite of adopting Bt cotton cultivation. In recent years farmers are gradually oriented to cultivate the cotton crop on organic mode realising the adverse effects of inorganic farming.

Current Trading of Indian Organic Products in the Export Market

India exported 86 items in last year (2010-11) with the total volume of 6987 MT. The export realisation was around 157.22 million US$ registering a 33 per cent growth over the previous year. Traders imported Indian organic products into the selected export markets (Germany, Switzerland, UK, Holland, US and Japan). Imports were directed mainly to Germany, the Netherlands and Switzerland. German and Dutch traders imported mainly spices, fruits, tea and nuts. Traders from the Netherlands imported the highest amount of sesame while German traders imported the highest amount of tea. Swiss traders were involved mostly with the import of tea and cotton, and very small quantities of organic spices.

The Indian organic products that are traded in the selected market are shown in Table 5.1. The values shown are totals for each product across the different countries. Sesame is one product that is imported in higher volumes, followed by organic fruits (such as frozen mango puree and pineapple juice). In the third place comes the well-known Indian spices, such as black and white pepper, ginger etc. Out of the tea varieties, higher amounts of black tea than green tea are imported.

Table 5.1

India Organic Products Traded in Selected Market

Product Categories	*Export Volume (Mt)*	*% Share*
Oil Crops (exept Sesame)	17966	25.73
Cotton & Textiles	17363	24.86
Processed Food	8752	12.53
Basmati Rice	5243	7.51
Tea	2928	4.19
Sesame	2409	3.45
Honey	2409	3.45
Rice	1634	2.34
Dry Fruits	1472	2.11
Cereals	1348	1.93
Spices-Condiments	1174	1.68
Medicinal & Herbal Plants/Products	627	0.90
Coffee	320	0.46
Vegetables	167	0.24
Aromatic Oil	39	0.06

Source: Govt of India, APEDA, 2011.

Materials and Methods

The Telangana region of Andhra Pradesh was selected purposively for the study; district Karimanagar of Telangana region was selected randomly. *Ex post facto* research design was followed. A sample of 60 organic and 60 conventional cotton practicing farmers from six villages of two *mandals* of the district were selected randomly. Constraints in the present study mean that impediments faced by the farmers for the adoption of standardised organic practices in cotton cultivation. Farmers were asked to elicite their constraints identified for the study. The frequency and percentages were calculated for each constraint expressed by the farmers. All the constraints were ranked based on the percentage of farmers.

Table 5.2

Profile Characteristics of Selected Respondents

No.	Category	Organic (n=60)		Conventional (n=60)	
		Frequency	Percentage	Frequency	Percentage
1. Age	Young age (up to 35)	22	36.66	12	20.00
	Middle age (36-45)	32	53.33	29	48.33
	Old age (>55years)	6	10.00	19	31.66
2. Level of education	Illiterate	5	8.33	15	25.00
	Primary school	25	41.66	19	31.66
	Upper primary school	13	21.66	11	18.33
	High school	7	11.66	6	10.00
	Intermediate	3	5.00	5	8.33
	Under graduate	2	3.33	3	5.00
	Post graduate	5	8.33	1	1.66
3. Farming experience	Low	18	30	15	25.00
	Medium	27	45	30	50.00
	High	15	25	15	25.00
4. Farm size	Marginal (1ha)	21	35.00	12	20.00
	Small (1.1-2ha)	22	36.66	23	38.33
	Large (>2ha)	17	28.33	25	41.66
5. Annual income	Low	21	35	15	25.00
	Medium	29	48.33	19	31.66
	High	10	16.66	20	43.33
6. Herd size	Small	23	38.33	31	51.66
	Medium	24	40.00	23	38.33
	Big	13	21.66	7	11.66
7. Organic inputs utilisation pattern	Low	16	26.66	32	53.33
	Medium	15	25.00	19	31.66
	High	29	48.33	9	15.00
8. Training received	Low	7	11.66	36	60.00
	Medium	30	50.00	15	25.00
	High	23	38.33	9	15.00
9. Decision-making behaviour	Low	21	35	16	26.66
	Medium	29	48.33	33	55.00
	High	10	16.66	11	18.33
10. Extension contact	Low	15	25.00	19	31.66
	Medium	26	43.33	27	45.00
	High	19	31.66	14	23.33

Results and Discussion

Majority of the organic cotton farmers were middle aged (53.33%) with primary school education (41.66%), medium farming experience (45.00%) having small farm size (36.66%), medium annual income (48.33%), medium herd size (40.00%), high organic inputs utilisation pattern (48.33%), medium level of training received (50.00%) with medium decision-making behaviour (48.33%) and medium level of extension contact (43.33%).

Majority of the conventional cotton farmers were middle aged (48.33%) with primary school education (31.66%), medium farming experience (50.00%) having large farm size (41.66%), high annual income (43.33%), small herd size (51.66%), low organic inputs utilisation pattern (53.33%), low level of training received (60.00%) with medium decision-making behaviour (55.00%) and medium extension contact (45.00%).

Table 5.3

Distribution of Organic and Conventional Cotton Farmers Based on their Extent of Adoption of Organic Cotton Practices

Category	*Organic (n=60)*			*Conventional (n=60)*		
	Low (51-61)	*Medium (62-72)*	*High (73-83)*	*Low (12-22)*	*Medium (23-33)*	*High (34-44)*
Frequency	12	17	31	34	18	8
Percentage	20.00	28.33	51.66	56.66	30.00	13.33

Majority (51.66%) of the organic cotton farmers had high extent of adoption of organic cotton practices, whereas it was low (56.66%) in case of conventional cotton farmers. Further, the results revealed that significant difference was observed between the mean extent of adoption of organic and conventional farmers about organic practices in cotton crop.

Organic cotton farmers had high extent of adoption on conversion and certification practices, whereas the conventional cotton farmers had high extent of adoption on land preparation, biomass development, sowing and weed management and low level of knowledge on conversion and certification practices

Constraints Expressed by the Organic Cotton Farmers in Organic Cotton Farming

The technological constraints elicited by the organic cotton farmers are shown in Table 5.4. A great majority (76.35%) of the farmers perceived the problem of complex and costly certification process for organic cotton cultivation and hence it was ranked I among the technological constraints. Two-third (66.19%) of farmers perceived lack of skill regarding the preparation, production and usage of certain organic inputs and was assigned rank II.

Table 5.4

Technological Constraints in Organic Cotton Farming Expressed by the Organic Cotton Farmers

S.No.	*Technological Constraints*	*Percentage*	*Rank*
1.	Lack of knowledge regarding different methods of pest control under organic farming	60.95	V
2.	Lack of skill regarding the preparation and production of certain organic inputs	66.19	II
3.	Lack of methods for easy detection of the pest economic threshold levels	48.00	VII
4.	Complex and costly certification process	76.35	I
5.	Lack of knowledge about the maintenance of records and conversion plan details	64.00	III
6.	Lack of knowledge on identifying the harmful and beneficial insects and to conserve natural predators	55.00	VI
7.	Lack of standardised, recommended low cost technologies for organic cotton farming.	62.00	IV

Farmers also expressed problems like lack of knowledge about the maintenance of records and conversion plan details (64.00%), lack of standardised, recommended low cost technologies for organic cotton farming (62.00%), lack of knowledge regarding different methods of pest control under organic farming (60.95%), lack of knowledge on identifying the harmful and beneficial insects and to conserve natural predators (55.00%), lack of methods for easy detection of the pest economic threshold level (48.00). These problems have been ranked as III, IV, V, VI and VII, respectively.

It can also be understood from Table 5.4 that the certification process of organic cotton was complex, laborious and needs to be simplified for acceptability by the farmers. Proficiency is needed in the process of production of certain plant and animal based organic inputs like *neem* cake, vermi compost etc., which are critical in organic farming. The conversion plans requires analytical and comprehensive skills which may not be expected among the farming community. Organic cotton farming in India is still at infant stage and primitive in nature. Hence, standardised technologies have not reached the farmers. These are the prominent technological constraints elicited by the sampled farmers of the study, and they are in conformity with the findings of Jain and Bhattacharya (2000) and Saxena and Singh (2000).

Table 5.5

Socioeconomic Constraints in Organic Cotton Farming Expressed by the Organic Cotton Farmers

S.No.	*Socioeconomical Problems*	*Percentage*	*Rank*
1.	Lack of community approach	85.70	I
2.	Lack of organisational and cooperative efforts of farmers	22.00	VIII
3.	Increased cost of cultivation due to addition of organic technologies	42.86	VII
4.	Reduction in the yields due to adoption of various organic inputs	82.00	II
5.	High rate of interest on credit	20.00	IX
6.	High cost of organic inputs	50.00	VI
7.	Lack of technical and input support during conversion period	80.00	IV
8.	Low profitability	75.00	V
9.	Pressure from family members not to volunteer for organic cotton farming	81.00	III

The socioeconomic constraints elicited by the organic cotton farmers are shown in Table 5.5. Among the socioeconomical constraints, majority of the respondents (85.70%) had expressed the constraint on lack of community approach for organic cotton cultivation and hence it was ranked I. The constraint, reduction in yields due to adoption of various organic inputs was perceived by 82.00 per cent farmers and was assigned rank II. The farmers also expressed problems like pressure from family members to not volunteer for organic cotton farming (81.00%), lack of technical and input support during conversion period (80.00%), low profitability (75.00%), high cost of organic inputs (50.00%), increased cost of cultivation due to addition of organic technologies (42.86%), lack of organisational and cooperative efforts of farmers (22.00%), high rate of interest of credit (20.00%). These problems have been ranked as III, IV, V, VI, VII, VIII and IX, respectively.

Under Indian farming situations farmers vary in terms of their land size, socioeconomic aspects etc., hence a spirit of cooperation and co-ordination may not be expected among the farmers. The stabilisation of yields under organic farming conditions will take a significant period of time; these are the testing times for the tolerance of the farmers. Farmers may not accept low yields irrespective of whatever be the goodness of the technology. The low yields and low net income levels psychologically force the family members not to grow cotton crop under organic conditions. No public, private or non-governmental organisation is supporting the farmer in any form during this arduous period of converting his/her land. These socioeconomic constraints are in line with constraints expressed by Chote and Borkar (2001) and Trivedi *et al.* (2007).

The supply, service and marketing constraints elicited by the organic cotton farmers are shown in Table 5.6. With regard to the supply, service and marketing constraints, majority (78.00%) of farmers faced constraint of weak marketing system in organic cotton trade and hence it was ranked I. More than three-fourth (76.00%) of farmers perceived lack of subsidy on bio-agents, bio-pesticides, traps and spraying equipments and was assigned rank II. The farmers also expressed problems like non-availability of certain organic inputs (bio-pesticides, bio-agents and traps) (72.00%), at times the produce of conventional cotton is pricing more in the market (70.00%), strong fluctuation of prices in the local and world markets (54.00%), lack of sufficient warehouses to store the cotton produced under organic farming (42.00%) and the market intelligence on organic cotton farming is very poor among the scientists, departmental officials and farmers (39.00%.) These problems have been ranked as III, IV, V, VI and VII, respectively.

The scope of marketing the organic cotton is still marginal. Both the marketing infrastructure and market intelligence are not robust to meet the demand of marketing the organic cotton produce. The subsidy given for various organic inputs is highly irrational and not in line with the purchasing capacity of the farmers. Most of the times organic cotton farmers are forced to sell their produce at distressed conditions with minimum prices compared to with inorganic cotton. These constraints on supply, service and marketing are in parallel with the constraints expressed by Noorjehan (2004) and Patel *et al.* (2006).

Table 5.6

Supply, Service and Marketing Constraints in Organic Cotton Farming Expressed by the Organic Cotton Farmers

S.No.	*Supply, Service and Marketing Problems*	*Percentage*	*Rank*
1.	Non-availability of critical in time organic inputs (bio-pesticides, bio-agents and traps)	72.00	III
2.	At times the produce of inorganic cotton is priced more in the market.	70.00	IV
3.	The market intelligence on organic cotton farming is very poor among the scientists, departmental officials and farmers	39.00	VII
4.	Lack of sufficient warehouses to store the cotton produced under organic farming.	42.00	VI
5.	Strong fluctuation of prices in the local and world markets on organic cotton produce.	54.00	V
6.	Lack of subsidy on bio-agents, bio-pesticides, traps and spraying equipments used in organic cotton farming	76.00	II
7.	Weak marketing system in organic cotton trade	78.00	I

The transfer of technology constraints elicited by the organic cotton farmers are shown in Table 5.7. Among the transfer of technology, majority (70.00%) of farmers faced constraint of organic recommendations not communicated in an understandable form and hence it was ranked first. More than half (54.00 %) of farmers perceived lack of regular visits by extension personnel to guide the farmers on organic cotton farming and was assigned rank II. The farmers also expressed problems like lack of coordinate efforts by all the concerned cotton related departments for promotion of organic cotton farming (45.00%) and minimal usage of various forms/channels of communication to disseminate the information on organic cotton farming (42.00%). These problems have been ranked as III and IV, respectively.

Table 5.7

Transfer of Technology Constraints in Organic Cotton Farming Expressed by the Organic Cotton Farmers

S.No.	Transfer of Technology Constraints	Percentage	Rank
1.	Lack of co-ordinated effort by all the concerned cotton related departments for promotion of organic cotton farming	45.00	III
2.	Organic recommendations not communicated in an understandable form	70.00	I
3.	Minimal usage of various forms/channels of communication to disseminate the information on organic cotton farming	42.00	IV
4.	Lack of regular visits by extension worker to give guidance to farmers on organic cotton farming	54.00	II

Precision is not maintained in recommending the organic based technologies (plant and animal based) in terms of their form, dosage and time of application. Seldom these recommendations are far from the understanding of the farmers. Extension machinery is not geared up thoroughly to advice the farmers at critical stages of crop growth. There is no unanimity among the departments or organisations concerned with the promotion of organic cotton farming and these are not working in tandem. These transfer of technology constraints were quoted similarly by Subhash Chander and Singh (2003) and Prasad (2010).

Table 5.8

Situational Constraints in Organic Cotton Farming Expressed by the Organic Cotton Farmers

S.No.	Situational Constraints	Percentage	Rank
1.	Transportation of organic and poultry manures is costly and laborious	60.00	III
2.	Small herd size (source of organic inputs)	85.00	I
3.	Non-availability of biologically treated seed in the market	80.00	II
4.	Lack of serious and continuous focus by the subsequent governments in promotion of organic cotton farming	56.00	IV
5.	Lack of access to ginning facility	30.00	V

The situational constraints elicited by the organic cotton farmers are shown in Table 5.8. With regard to the situational constraints, majority (85.00%) of farmers faced problem of small herd size (source of organic input) and hence it was ranked first. Eighty per cent of farmers perceived non-availability of biologically treated seed in the market as a major constraint and was thus assigned rank II. The farmers also expressed problems like transportation of organic and poultry manures is costly and laborious (60.00%), lack of serious continuous focus by the subsequent governments in promotion of organic cotton farming (56.00%), lack of access to ginning facility (30.00%). These problems were ranked as III, IV and V, respectively.

Due to pervading of farm mechanisation into various farm operations, the usage of cattle wealth is coming down drastically in the rural areas. Biological inputs play a vital role in organic farming; there is a great dearth of biologically treated seed in the market. The availabity of poultry and organic manures at their village is not sufficient and thus the farmers are resorting to import from far places which is leading to increase in their investment. These situational constraints are in agreement with the findings of Yadav *et al.* (2007) and Sivanarayana *et al.* (2008).

The organic cotton farmers should invariably follow the certification practices and procedures. This allows them to understand the intricacies involved in this process. However, the farmers expressed that the certification process involved in organic cotton cultivation is very complex and costlier. Lack of community approach both at pre and post harvest operations in cotton crop let to continuous low yields in organic cotton farming. Seldom the prices of organic cotton produce are wavering and there is no price stability in the local market, there by the farmers are forced to go for distressed sales. This might be the reason why the organic cotton farmers expressed that existing market system is very weak. The farmers also elicited the present organic recommendations being not communicated to them in an understandable form as one after constraint. This might be due to giving information through wrong channel to the targeted audience. Small herd size was one of the major constraints expressed. The farmers were not able to generate enough amount of organic inputs due to absence of good size of cattle, hence the constraints of small herd size.

Constraints Expressed by the Conventional Cotton Farmers in Organic Cotton Farming

The constraints elicited by the conventional cotton farmers are shown in Table 5.9. A great majority (82.76%) of the farmers perceived the problem of complex and costly certification process for organic cotton cultivation as a major problem and hence it was ranked I among the all constraints. Eighty per cent of farmers felt that organic cotton practices are labour intensive and it was assigned rank II.

The farmers also expressed problems like lack of specific package of practices (79.00%), low yields and not suitable for irrigated conditions (78.00%), high incidence of pest and disease problems (77.68%), higher weed infestation (76.00%), cost of cultivation is very high (73.00%), lack of mechanisms for bringing about awareness among the farmers regarding organic cotton farming (68.00%), poor availability of organic inputs (45.00%). These problems have been ranked as III, IV, V, VI and VII, VIII and IX, respectively. The constraints on certification, intensity of labour and lack of specific package of practices were elicited due to the fact that these conventional practices are not much habituated to certify their produce or are not inclined to maintain the rule and regulations of certification bodies. They felt organic farming requires more labour to arrest the weed population, it requires extra care while harvesting, storing etc. They are very much confused with the loosely structured scientific recommendations of organic farming in cotton.

Table 5.9

Constraints Expressed by the Conventional Cotton Farmers in Organic Cotton Farming

S.No.	*Constraints*	*Percentage*	*Rank*
1.	Labour intensive	80.00	II
2.	Cost of cultivation is very high	73.00	VII
3.	Lack of specific package of practices	79.00	III
4.	Low yields and not suitable for irrigated conditions	78.00	IV
5.	Certification is a complex and costly process	82.76	I
6.	Poor availability of organic inputs	45.00	IX
7.	Higher weed infestation	76.00	VI
8.	High incidence of pest and disease problems	77.68	V
9.	Lack of mechanisms for making aware the farmers about organic cotton farming	68.00	VIII

In tune with organic cotton farmers, conventional cotton farmers also agreed that the certification process of organic cotton is laborious expensive, and involves rigorous procedures. Labour requirement is more for various operations of organic cotton farming especially during weeding picking, roughing and post-harvest operations. The organic cotton farming as such doesn't have any established standardised and proven technologies which are universally applicable. In most of the cases, loosely structured package is given for organic farming. Farmers under irrigated conditions expects more yields as they are endowed with good natural resources which make them apathetic towards organic farming. The strict non-pesticidal usage during all operations of the crop growth makes the crop susceptible to various kinds of pests and diseases which are normally controlled by the usage of pesticides. The absence of using the herbicides makes the organic field more infested with the weeds. These constraints expressed by conventional cotton growers are in support with findings of Prasad (2010).

State Government's Policy on Organic Farming

Andhra Pradesh government, through Agriculture Department, encourages farmers who adopt organic farming by providing them with vermin compost units, vermin hatcheries, green manures, bio-fertilisers in the form of subsidy. State Horticulture Department encourages organic horticultural crops like mango, chilli, ginger, cashew nut and vegetables under the National Horticultural Mission. It also adoted cluster approach and provided subsidy of ₹ 9 lakh to the farmers for trainings and certification, and subsidy of ₹ 15000 to each farmer for manures i.e., 1st year ₹ 7000, 2nd year ₹ 4000 and 3rd year ₹ 4000. Andhra Pradesh government helped farmers in certification with the help of Vedic Certification Agency (Hyderabad) and Organic Certification Agency (Bangalore), and exporting mango to foreign countries through Mother India Farm, Bangalore.

Through Standards of National Programme for Organic Production (NPOP), certification and inspection services were offered in conformity with ISO Guide 65. The procedure for certification involves the following steps.

1) Request for Certification

Any operator who wishes to enter a certification programme, shall send a request for application form to the following address: Plot no 54,

Ushodaya Enclave, Mythri Nagar, Miyapur, Hyderabad. *email: voca_org@ yahoo.com*

2) Submission of Filled in Application Form

The operator should fill in the application form sent to him. The filled in form should be submitted back either by mail or by hard copy to the above mentioned address. He/she should submit all the relevant documents asked for in the application form.

3) Cost Estimation/Payment of Fees

Once the filled in application form is received at certified agency, a fee estimate shall be sent to the operator. The operator should pay 50 per cent of fees in advance, either in the form of demand draft in favour of "certification agency" or a cheque. The remaining 50 per cent should be paid after the inspection report is ready.

4) Signing of Inspection Contract

The operator should sign a contract which asks him/her to follow applicable organic standards. The contract should be signed by him/her to commit themselves to organic production. As per the contract, the operator should provide access to all production sites relevant to inspections record keeping. He should pay for the sampling and reserve analysis.

5) Inspection and Submission of Inspection Report

As per the inspection plan prepared by the evaluator, certified agency's inspector shall be assigned to conduct the inspection. He/she shall notify the operator at least 10 days before the visit. The inspector shall visit all sites relevant in production. He/she shall take samples for analysis if required. He/she shall check the documents, processing units, storage places etc. before the end of his visit. The inspector shall summarise his/ her findings with the operator. Any deviations or non-conformities with respect to the standards will be discussed with the operator. The inspector shall prepare an inspection report based on his/her findings and shall submit them in the certification agency office.

6) Evaluation of Inspection Report

The evaluator of Certification Agency will check the inspection report to conform that all aspects have been covered during the inspection. If any short comings are there, the inspector will be asked to give an explanation and if required the operator will be contacted for additional information

7) Issue of Certification Decision

The certification committee will receive the evaluated inspection report. A comparsion of the report is made with the requirements. The committee shall decide the final certification and the required certificate shall be issued (scope and transaction). The sanctions that are to be given to the operator are also decided by the certification committee. The scope certificate shall be renewed annually.

8) Issue of Transaction Certificate

The operator who wishes to market his organic produce shall apply to certified agency requesting for a transaction certificate. Transaction certificates are issued for every transaction made by the operator.

9) Appeals

If the operator is not satisfied with the certification decisions, he/she may appeal to VOCA in writing within 30 days of the notification of certification decision. The appeal committee shall take the final decision on the case.

10) Use of India Organic and Certified Agencyorgani Logo

Once the operator receives the certification decision, he/she shall request to VOCA for the use of India Organic logo and Vedic Organic logo, in a prescribed format. The logo shall be given to the operator as a soft copy.

11) Label Approval

The operator shall market his/her products only after the label is approved. Labels that are to be used by the operator shall be sent for approval. Certification agency shall approve the label as per the standards and send the decision to the operator.

General information—Conversion period: Any conventional farm that has been using chemical fertilisers and pesticides shall have a conversion period of three years before an organic certificate is issued. Any farm where organic management is being practiced for more than three years, organic certificate shall be issued at the end of first year of registration. However, a third party declaration needs to be submitted along with the application. The third party shall not have any personal relation with the operator.

Development of a Strategy Based on Findings of the Study to Promote Organic Cotton Crop

A strategy specifying the interventions to be taken up the farmers, extension officers, administrators, research organisations and marketing officials was developed to promote the organic farming in cotton crop by overcoming the bottlenecks and hurdles in the current organic cotton cultivation. Following are the interventions suggested.

Interventions Suggested for the Farmers for Promotion of Organic Cotton Crop

- Farmers should basically understand the pros and cons and benefits of organic cotton farming to the soil, environment and mankind as a whole.
- Farmers should understand and implement the spirit of collective approach and cooperative organic farming in cotton crop to realise maximum dividends.
- Each individual farmer should maintain enough amount of cattle population to generate and recycle the organic inputs in his/her farm.
- The farmers should strive to get maximum formal education to be able to understand the surrounding knowledge pertaining to organic farming and its applicability in the cotton crop.
- The farmers should volunteer to undergo training programmes conducted by various organisations on aspects related to organic cotton.

- The farmers should understand that the profits in organic farming are not decided by the production of the produce, rather it is decided by the quality of the produce.

Interventions Suggested for the Extension Officers for Promotion of Organic Cotton Crop

- More number of channels (print, audio, audio-video, visual form, folk songs, skits, dramas, puppet shows) should be put into use to simultaneously reach various categories of farmers at a greater geographical area in a short time.
- Need based training programmes should be planned, designed and implemented on the aspects of quarantine laws, the utility of organic inputs, certification procedure, conversion plan etc. The periodicity and duration of the training programmes should be decided based on the needs and interests of the farmers and complexity of the topic.
- The extension officials should update their market intelligence on organic cotton and they should be accessible to farmers at critical periods of crop cultivation.
- The extension officials should focus on enhancing decision-making behaviour of individual farmers by providing valid and reliable information for quick promotion of organic farming in a given geographical area.

Interventions Suggested for the Administrators/ Policy Makers for Promotion of Organic Cotton Crop

- The administrators at the helm of affairs should have broad vision and mission with regard to promotion of organic farming in cotton cultivation.
- A cooperative mechanism should be established between the government, private sector and NGOs at national level to percolate the spirit of organic farming at grossroot level.
- The role of NPM, Polambadi (FFS) and privileges (contingency) to rainfed cotton farmers should be clearly delineated in the promotional strategy of organic cotton farming.

- The programmes and policies should be designed in such a fashion that they should inculcate a spirit of collective approach and cooperative farming among the organic cotton farmers to generate meaningful profits/results.
- The formal education levels among the farmers should be improved to support them to get the information on organic cotton cultivation from various sources.
- The much needed impetus may be given to the organic farming, the subsidy should be given at the right time in right quantity to the critical organic inputs.
- The government should search for ways and means to increase the cattle population in rural areas. A specialised programme should be designed in line with cattle improvement programme to help the farmers to conserve, maintain and increase the number of cattle population.
- Institutional credit should be provided at less interest rate as an incentive for practicing organic cotton farming.
- To promote organic farming, the organic cotton crop should be covered under crop insurance scheme with the initiative being taken by the government.

Interventions Suggested for Research Organisations for Promotion of Organic Cotton Crop

1. The research should focus more on evolving non-pesticidal management strategies and measures to control pests and diseases in cotton crop.
2. The present integrated pest management (IPM) strategy to control various pests and diseases should be refined, fine-tuned by specifying the resources which are available at local level.
3. Needful research should be conducted on post-harvest handling of organic cotton i.e., judging the quality of fibre, in terms of strength, length, moisture percentage, colour, these are the parameters which decide the price of the organic cotton at international market.

- Research should be carried out to identify the alternatives for high cost organic inputs to enrich the soil fertility as the biological resources are declining day by day.
- Research should be focused on standardising the scientific procedures for preparation and manufacture of various organic inputs at local level for preparation of enough quantities of vermicompost, trichoderma and PSB (phosphorus soluble bacteria) etc.
- The recommendations for organic cotton cultivation at various practices should be standardised.
- A variety alternative to Bt cotton and tolerant to sucking pests and which could be able to withstand adverse weather conditions should be evolved.

Interventions Suggested for Marketing Officials for Promotion of Organic Cotton Crop

- A transparent, regulated and well-structured marketing system should be designed for effective marketing of organic cotton produce.
- The market intelligence on organic cotton should be well maintained and available at right time to all the stakeholders like scientists, extension personnel, marketing officials and farmers.
- The price disparity should be shown between the produce of organic and inorganic cotton.
- The fluctuation of price of organic cotton should be minimised to avoid distressed sale of organic cotton by the farmers.
- Encouraging participative group certification system.

Implications of the Study

- More number of agricultural information materials (farm literature) should be published by the department of agriculture in the form of leaflets, pamphlets, simple booklets on organic aspects of cotton farming as good size of respondents had formal education.

- Majority of the organic cotton respondents are of medium age, this indicates that the government should encourage the youth to take up farming as their profession in general and organic farming in particular.
- Majority of the organic cotton respondents were habituated to apply the organic inputs in their cotton cultivation. In this endeavour the administrators should think of keeping these inputs readily available in terms of quantity and quality at farmer's level. In contrast to this, the utility of organic inputs by the conventional cotton farmers is very less. Hence, awareness must be created among the farmers by the officials of DOA regarding the benefits and role of organic inputs in improving the soil fertility, quality of produce and maintaining the ecological balance.
- The training received by the organic and conventional cotton farmers is medium and low respectively. In order to update the knowledge level of these two categories of famers and also to increase the rate of adoption of organic farming practices, the public, private and NGOs should organise more need based location specific and specialised training programmes for the cotton farmers.
- The extension contacts of both the organic and conventional cotton farmers are medium. Hence, public service extension personnel especially the officials of DOA in the ranks A.E.O., A.O. and A.D.A should be accessible to the farmers and provide round the clock guidance to the farmers.
- The lowest rank attributed by the organic cotton respondents on plant protection indicates that they don't have sufficient knowledge on these aspects. Hence, the scientists of agricultural university, officials of DOA and private personnel should bestow special attention on evaluating new ways and means to control pests and diseases, and refining the existing IPM practices.
- The least adoption of harvesting and post-harvest management practices shows that there is every need to educate the organic cotton growers by the officials of department of marketing and

agriculture on the importance and value of post-harvest practices in enhancing the price of the cotton produce at local, national and international markets.

- The conventional cotton farmers have less knowledge and poor adoption of conversion and certification practices. The first step to bring the cotton farmers into organic farming is orienting them on conversion and certification requirements of organic farming. Hence, the organisations concerned with the promotion of organic farming in cotton cultivation should focus on enlightening the farmers to understand and apply the conversion and certification requirements.
- Majority of the respondents felt that the process/procedure followed to certify the quality of organic cotton produce is very complex and expensive. Hence, the officials of APEDA and other recognised certification agencies should cut short the procedure and design a participative group mechanism.
- The marketing system as expressed by the respondents to market the organic cotton produce is unstructured, irregular and involves lot of interventions by middle men. Hence, the department of marketing should come out with a structured and well-regulated marketing system to fetch good remunerative price for the organic cotton produce.
- The study revealed that, community approach is lacking among the respondents in organic cotton cultivation. In order to encourage the collective and cooperative farming among the cotton farmers, the government should design schemes to lend financial credit at less rate of interest to the groups formed to take up organic farming collectively with the additional benefit of creating development fund for the community and insurance coverage.

Conclusion

India has the technical skills and a sophisticated farming community capable of combining indigenous knowledge of organic farming practices with recent scientific advances to overcome challenges like agrarian crisis, food security, agricultural productivity and marketing. *De facto* transformation of organic areas with organic products can minimise the farmers suicides and increase the profits of farming. The study brought to surface the valid, reliable and pertinent constraints expressed by both the organic and conventional cotton farmers on organic cotton farming. The implications suggested for various stakeholders and the strategy drawn out can empower the organic cotton farmers to reduce the agricultural crisis in rainfed areas.

References

Chote, G.D. and M.M. Borkar (2001). "Constraints Faced by Farmers in Adoption of Bio Fertilizers", *J. Maharashtra Agric. Univ.* 19: 264-68.

Jain, R.K. and P. Bhattacharya (2000). "Farmer's Involvement in Bio-fertilizer Demonstration and Promotion Campaign", *J. Maharashtra Agric. Univ.* 19: 130-31.

Noorjehan, A.K. (2004). "Organic Farming in Tamil Nadu: A Multidimensional Analysis". PhD Thesis. Coimbatore: Tamil Nadu Agricultural University.

Patel, Yogesh, R.K. Choudhary and V.K. Garg (2006). "Adoption of Cotton IPM Technology in Eastern Nirmar of Madhya Pradesh: A Survey", *J. Cotton Res. and Dev.* 20: 286-88.

Prasad, M.S. (2010). "Adoption of Recommended Package of Practices of Dry Land Crops in Relation to Socio-economic and Personal Characteristics of Farmers", *Indian J. Dry Land Agric. Res & Dev.* 17: 39-43.

Saxena, K.K. and R.L. Singh (2000). "Adoption of Organic Farming Practices by Farmers of Malwa Region", *J. Maharashtra Agric. Univ.* 19: 53-55.

Sivanarayana, G., M. Ramadevi and P. Venkataramaiah (2008). "Awareness and Adoption of Cotton (Gossipium hirsutum L.): Integrated Pest Management Practices by the Farmers of Warangal District in Andhra Pradesh", *J. Res. ANGRAU* 36: 33-40.

Subhash Chander and S.P. Singh (2003). "Constraints in Adoption of Integrated Pest Management Practices in Cotton", *Indian Journal of Extension Education* 39: 41-49.

Trivedi, M.K., N.B. Chauhan and Umesh (2007). "Training Needs of Cotton Growers in Bt. Cotton Cultivation", *Journal of Agricultural Extension Management* 15: 117-27.

Yadav, Dharmendra, M.M. Patel and R.K. Choudhary (2007). "IPM in Cotton Adoption and Constraints", *J. Cotton Res. and Dev.* 21: 230-34.

6 Managing Drought Risk and Vulnerability

Evidence from Western Odisha

Mrutyunjay Swain

Introduction

Drought is considered by many to be the most complex but least understood disaster, affecting more people than any other hazard (Hagman, 1984). More than half of the world population is susceptible to drought every year (Kogan, 1997). It is one of the major threats to people's livelihood and community socioeconomic development. Being a slow onset insidious disaster, it results in extensive damage to crops and hydrological imbalances that impinge on different livelihood activities directly or indirectly. The rural households and their income generation activities exhibit a great deal of sensitivity to different drought risk factors. Their income, expenditure and savings and access to water resources are significantly affected. They endure several hardships on account of consumption shortfall and health related problems (Swain, 2010). Obviously, they adopt different strategies to reduce the level of drought risk and vulnerability. However, different livelihood groups are affected quite differently depending on magnitude of drought impacts. Crop production under rainfed condition and agricultural labour works are more severely affected compared to other livelihood activities. Due to decrease in agricultural output, more number of poor rural households rush to nearby forests for collecting timber and non-timber forest produces. Thus, forest resources are more intensively exploited during the drought year. The increase in pressure on forest resources during the drought year raises the level of vulnerability of those forest dwellers who solely depend on the forest produces.

This study analyses the risk management strategies adopted by the major livelihood groups in Bolangir district of western Odisha. The role played by community level insurance mechanism and institutional support system in mitigating the drought risk and vulnerability has been analysed. The vital role played by the targeted public distribution system (TPDS), community-based water resource management and other institutional mechanisms in withstanding drought by rural households have been critically examined. The evolution and efficacy of drought management policies and development schemes have also been analysed.

As far as the drought vulnerability in the study region is concerned, drought is found to be a recurring and single most insidious phenomenon in the region. The recurrent drought phenomenon in the district is mostly responsible for its 'chronic backwardness' (Pattnaik, 1998). The increasing frequency of occurrence of this hazard is one of the major factors behind the rising level of drought vulnerability in the region, which is mainly due to a widening of the variability of rainfall from season to season, rather than as a result of deficiency of annual rainfall (Sainath, 1996; Swain, 2006). One of the prime reasons for increasing drought frequency and vulnerability in Bolangir is the neglect of the traditional water-harvesting structures (WHSs) (Nayak, 2004; Roy *et al.*, 2004). Hence, the neglect of traditional WHSs, misuse and over-exploitation of natural resources like forests and minerals, and the depreciation of biodiversity in the region are major causes of the rising levels of drought vulnerability in the region. Thus, it is largely human induced process that seems to be exacerbating the vulnerability to drought in the region.

Data and Methodology

A multi-stage sampling method has been followed for the present study. At first stage, Bolangir district was selected as the district was found to be the most vulnerable to drought among the districts in Odisha (Roy *et al.*, 2004). At subsequent stages, 139 households were selected from three study villages: Samara from Saintala block (most vulnerable), Mundomahul from Patnagarh block (moderately vulnerable) and Bijepur from Titlagarh block (least vulnerable). The degree of drought vulnerability in these blocks has been calculated in the form of composite drought vulnerability index (CDVI) on the basis of ranks or weights attached to 19 key drought vulnerability factors out of which 6 are biophysical factors (i.e., drought probability, intensity, long-term rainfall variability, water holding capacity of soil, land slope and groundwater table) and 13 are socioeconomic factors covering poverty, education, irrigation, major crop production, land use pattern and some important institutional factors. The sample households (HHs) covered 12 major livelihood groups, which include large farmer (average size of operational area of more than 10 acres), medium farmer (5.01-10 acres), small farmer (2.51-5 acres), marginal farmer (up to 2.5 acres), agricultural labourer, non-agricultural labourer, forest resource dependant, rural artisan, businessman, service holder, livestock rearer, and others covering fishing community, stone merchants and tailors. The drought year 2002 and the normal year 2003 were the reference years for the study and the primary data was collected in 2003. The recall method was followed for primary data collection.

The relative degree of drought vulnerability and risk for livelihood groups has been estimated on the basis of CDVI which has been constructed using indexing and vulnerability profile method with 19 key drought vulnerability factors as stated earlier. Similarly, data on different biophysical and socioeconomic indicators of drought adaptability, physical exposure and drought risk have been used to generate aggregate indices such as composite drought adaptability index (CDAI), physical exposure index (PhyExpo) and drought risk index (DRI) for each livelihood group.

To measure crop diversification as an adaptation strategy, the Herfindahl Index (Ht)[1] was estimated as follows:

$$H_t = \sum_{i=1}^{n} Pi^2$$

Where,

$$P_i = \frac{Ai}{\sum_{i=1}^{n} Ai}$$

1. The Herfindahl Index (Ht) is defined as the sum of squares of all proportions and is a measure of concentration. The value of Herfindahl index varies from zero to one. It takes the value one when there is complete specialisation and value zero when there is perfect diversification, i.e., it has inverse relationship with diversification.

Pi = Proportion of ith crop

Ai = Area under ith crop

$\sum_{i=1}^{n} Ai$ = Total cropped area

Major Risk Management Strategies Adopted by Rural Households

Before analysing the major risk management strategies adopted by the rural households, it is imperative to shed some light on the level of their vulnerability to drought in the study area. The analysis of drought vulnerability of different livelihood groups reveals that rural artisans, livestock rearers, pretty traders, service holders and forest resource dependent come under low drought risk zone since they are not heavily dependant on agriculture as their major source of livelihood. It may be noted that the small farmers are most vulnerable with CDVI value of 0.70, followed by the agricultural labourers and marginal farmers with CDVI values of 0.69 and 0.68, respectively (Table 6.1). The livelihood groups with lower vulnerability are the large farmers (0.29) followed by service holders (0.30), rural artisans (0.44) and local business group (0.46). Since most of the forest reource dependent were having less agricultural land, they have been relatively less affected by drought.

The major causes of vulnerability of cultivator households in the study district were very low irrigation coverage and highly erratic rainfall pattern. One of the most disappointing features of rainfall pattern in Bolangir district is that the average annual rainfall is gradually declining over the years. The annual normal rainfall for the period 1901-1950 in Bolangir was 1443.5 mm that has declined to 1230 mm in the period 1951-2000 and further to 1211 mm during the period 1986-2003. The decline in annual rainfall from 1st half normal (1901-1950) to the 2nd half normal (1950-1991) is 14.8 per cent that has further declined by 16.1 per cent from the first half normal to last 18 year (1986-2003) normal. While the annual rainfall in Orissa as a whole has depicted increasing trend (Figure 6.2), the same in the case of most vulnerable Saintala block of Bolangir district exhibits a declining trend as revealed by the log rainfall graph within the period of 1986-2003 (Figure 6.1). Notably, the total average annual rainfall

Table 6.1

Relative Degree of Drought Vulnerability and Risk for Livelihood Groups

Livelihood Groups (LG)	*CDVI (Weighted with equal weights to indicators)*	*CDAI (Weighted with unequal weights)*	*PhyExpo*	*DRI* (PhyExpo CDVI)/ CDAI*	*Relative Degree of Drought Risk (Ranked by DRI)*
Rural Artisan	0.44	0.23#	0.00#	0.000#	Low Drought Risk Zone
Livestock Rearer	0.52	0.39	0.00#	0.000#	
Businessman	0.46	0.49	0.01	0.006	
Forest Resources Dependent (FRD)	0.55	0.45	0.03	0.037	
Non-agrl Labourer	0.53	0.32	0.03	0.050	
Service Holder	0.30	0.74**	0.45	0.133	
Large Farmer (LF)	0.29#	0.52	0.85	0.478	Moderate Drought Risk Zone
Others	0.68	0.31	0.27	0.506	
All LGs	0.56	0.42	0.42	0.512	
Medium Farmer (MDF)	0.59	0.54	0.90	0.865	
All Cultivators	0.57	0.45	0.93	1.073	High Drought Risk Zone
Agricultural Labourer	0.69	0.29	0.51	1.151	
Small Farmer (SF)	0.70**	0.42	0.97	1.455	
Marginal Farmer (MF)	0.68	0.33	1.00**	1.885**	

Notes: 1. *Livelihood groups are classified into different drought risk zones by following method:

If DRI (drought risk index) < 0.4, it implies low level of drought risk; If DRI $>= 1.0$, it implies high level of drought risk; and If DRI < 1.0, it implies moderate level of drought risk.

2. ** The figures are the highest of respective indices.

The figures are the lowest of respective indices.

Source: Computation from field data.

of Bolangir district is not a cause of concern since it is even higher than the state and national average. The decline in the long-term average annual rainfall is a serious problem, since it indicates gradual desertification of the area. Not only the long-term average rainfall is declining but also the erratic rainfall pattern is a significant threat to the agrarian economy of Bolangir.

Figure 6.1

Rainfall Variability in Bolangir (Database 1986-2003)

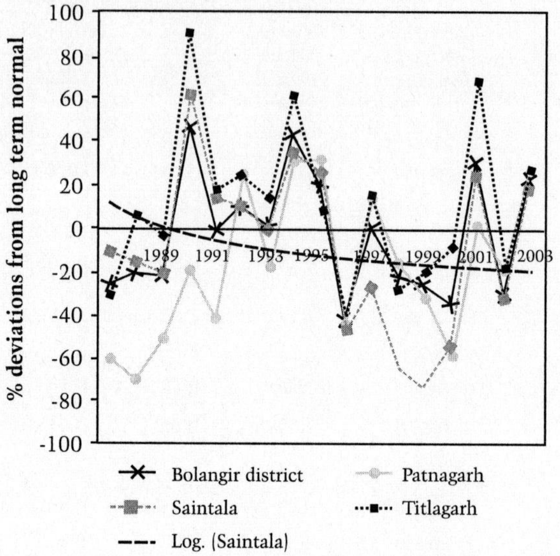

Figure 6.2

Rainfall Variability in Orissa (Database 1961-2003)

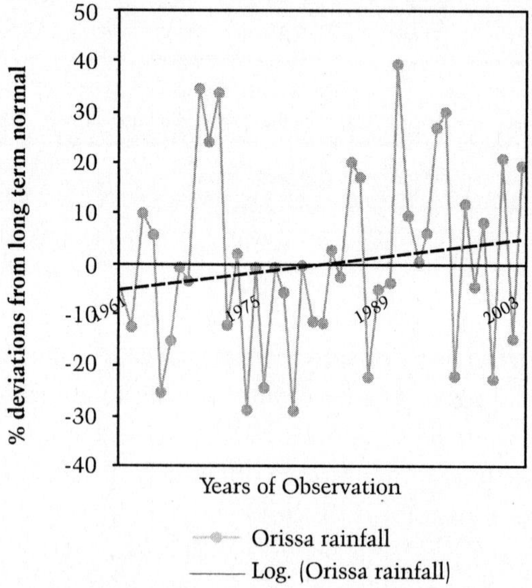

As far as the adaptation strategies of the major livelihood groups in the study region is concerned, it was found that there were certain common adaptation strategies which all these livelihood groups resorted to. Some of them were borrowings from institutional and non-institutional sources of credit, reducing the consumption, disposal of assets and livestock, increased dependence on common property resources, diversification of activities through migration or local non-agricultural employment etc. Besides, some livelihood groups also followed divergent adaptation strategies. For example, cultivator households adopted some strategies like curtailing cost of cultivation, diversifying cropping pattern, crop insurance and creating/ renovating WHSs etc. Some of these major risk management strategies adopted by the rural households have been discussed in this section.

Diversification of Income Sources

Diversification of income sources was the major risk management strategy adopted by households for managing drought risk. This strategy was found to be operational at two levels: short-term strategies for risk adjustment and long-term risk mitigation/reduction or adaptation. However, the extent of long-term adaptational income diversification is found to be much lower than the level of short-term adjustmental income diversification.

Short Term Income Diversification

Though the number of income sources on which various households are found to depend in the drought year did not vary much in relation to normal year, the extent of reliance on different sources has varied a lot. In the normal year, the share of agriculture in gross household income is found to be maximum (27.8%) considering the entire livelihood groups together (Table 6.2). The income from labour works, livestock, artisan works, local businesses and forest resources significantly contributed to the total household income, expenditure and savings. As mentioned earlier in the chapter, total number of income sources of 139 selected households accounts for 231.7 per cent of number of primary income sources that provide enough flexibility to rural households to earn the subsistence income in the period of climatic hazards including drought. In the drought year, the share of agriculture in total household income has drastically dropped to 10 per cent and the income from agriculture has drastically declined by 70.3 per cent compared to normal year since 75.6 per cent of net sown area (NSA) of sample farmers was rainfed. The share of income from all farm activities including agriculture, cattle rearing, fishing, poultry/duckery and orchards has fallen from 31.9 per cent in the normal year to 14.3 per cent in the drought year.

Table 6.2

Extent of Income Diversification by Various Livelihood Groups

Sources	*All Cultivator HHs*			*All Non-cultivator HHs*			*All HHs*		
	NY	*DY*	*% Change*	*NY*	*DY*	*% Change*	*NY*	*DY*	*% Change*
1	*2*	*3*	*4*	*5*	*6*	*7*	*8*	*9*	*10*
(a) Farm activities									
Crop farming	30137.8 (39.5)	8015.7 (14.0)	-73.4	4841.3 (14.4)	1913.7 (6.1)	-60.5	13274 (27.8)	3947.7 (10.0)	-70.3
Fishing*	719.5 (0.9)	487.1 (0.8)	-32.3	798.9 (2.4)	664.6 (2.1)	-16.8	772.4 (1.6)	605.5 (1.5)	-21.6
Cattle rearing*	638.8 (0.8)	607.9 (1.1)	-4.8	877.3 (2.6)	781.0 (2.5)	-11.0	797.8 (1.7)	723.3 (1.8)	-9.3
Poultry/Duckery*	396.3 (0.5)	406.9 (0.7)	2.7	279.2 (0.8)	269.7 (0.9)	-3.4	318.2 (0.7)	315.4 (0.8)	-0.9
Orchards*	134.9 (0.2)	123.3 (0.2)	-8.6	51.9 (0.2)	57.0 (0.2)	9.8	79.6 (0.2)	79.1 (0.2)	-0.6
Sub Total-a	32027.3 (42.0)	9640.8 (16.8)	-69.9	6848.6 (20.4)	3686.1 (11.8)	-46.2	15242 (31.9)	5671.0 (14.3)	-62.8
(b) Non-farm activities									
Wage earning#	1336.8 (1.8)	833.8 (1.5)	-37.6	1425.9 (4.2)	1214.0 (3.9)	-14.9	1396.2 (2.9)	1087.3 (2.8)	-22.1
Service/Pension	11701.2 (15.3)	11500 (20.1)	-1.7	12081 (36.0)	12081 (38.8)	0.0	11954.1 (25.0)	11886.9 (30.1)	-0.6
Business/Trading	8905.8 (11.7)	9573.1 (16.7)	7.5	1833.3 (5.5)	1448.7 (4.7)	-21.0	4190.8 (8.8)	4156.9 (10.5)	-0.8
Artisan works	139.2 (0.2)	135.6 (0.2)	-2.6	764.6 (2.3)	886.7 (2.8)	16.0	556.1 (1.2)	636.4 (1.6)	14.4
Hiring out assets/ Agrl implements	720.4 (0.9)	702.0 (1.2)	-2.6	58.1 (0.2)	66.1 (0.2)	13.7	278.9 (0.6)	278.1 (0.7)	-0.3
Forest produces	2083.0 (2.7)	2485.9 (4.3)	19.3	2005.8 (6.0)	2133.5 (6.8)	6.4	2031.5 (4.2)	2250.9 (5.7)	10.8
Sub Total-b	24886.3 (32.6)	25230 (44.0)	1.4	18168 (54.1)	17830 (57.2)	-1.9	20408 (42.7)	20296 (51.3)	-0.5

contd...

...contd...

Sources	*All Cultivator HHs*			*All Non-cultivator HHs*			*All HHs*		
	NY	*DY*	*% Change*	*NY*	*DY*	*% Change*	*NY*	*DY*	*% Change*
1	*2*	*3*	*4*	*5*	*6*	*7*	*8*	*9*	*10*
(c) Other activities (mainly for dealing with risk)									
Migration	351.3 (0.5)	594.5 (1.0)	69.3	920.9 (2.7)	994.1 (3.2)	7.9	731.0 (1.5)	860.9 (2.2)	17.8
Borrowing from formal sources	10335.4 (13.5)	9292.3 (16.2)	-10.1	2568.4 (7.6)	2537.0 (8.1)	-1.2	5132.3 (10.7)	4722.0 (11.9)	-8.0
Borrowing from informal sources	8012.5 (10.5)	7709.7 (13.4)	-3.8	5057.6 (15.1)	5396.2 (17.3)	6.7	6042.6 (12.6)	5886.4 (14.9)	-2.6
Sale/Mortgage of assets/Implements**	256.4 (0.3)	1335.4 (2.3)	420.9	0.0 0.0	0.0 0.0		85.5 (0.2)	445.1 (1.1)	420.9
Crop Insurance	471.4 (0.6)	3017.5 (5.3)	540.2	0.0 0.0	344.5 (1.1)		157.1 (0.3)	1235.5 (3.1)	686.3
Food for work	0.0 0.0	518.6 (0.9)		0.0 0.0	314.0 (1.0)		0.0 0.0	382.2 (1.0)	
Begging	0.0 0.0	0.0 0.0		27.5 (0.1)	48.6 (0.2)	76.6	18.3 (0.0)	32.4 (0.1)	76.6
Sub Total-c	19426.9 (25.4)	22468 (39.2)	15.7	8574.4 (25.5)	9634.3 (30.9)	12.4	12167 (25.4)	13565 (34.3)	11.5
Gross HH Income	76341 (100.0)	57339 (100.0)	-24.9	33591 (100.0)	31150 (100.0)	-7.3	47816 (100.0)	39532 (100.0)	-17.3
Gross HH Expenditure	59734 (78.2)	46920 (81.8)	-21.5	19409 (57.8)	14763 (47.4)	-23.9	34877 (72.9)	26941 (68.2)	-22.8
Annual Net Savings	16606	10419	-37.3	14182	16387	15.5	12939	12591	-2.7
(21.8)	(18.2)		(42.2)	(52.6)		(27.1)	(31.8)		

Notes: 1. Figures in parentheses show percentage of gross household income.

2. # does not include labour work under Food for Work Programme.

* Figures shows the monetary value of products mainly used for self consumption except some livelihood groups like livestock rearer, artisan those who sell a part of their products.

** Major implements mortgaged are tractors, pump sets, sprayers etc.

3. NY stands for normal year and DY stands for drought year.

Source: Computed from field data.

While the fall in income from agriculture was substantial, the fall in income from other farm activities was very marginal. Specifically in the case of cultivator households, the fall in income from farm activities is more pronounced compared to non-cultivator households. The income share from all farm activities has decreased from 42 per cent in the normal year to 16.8 per cent in the drought year in case of cultivator households; but in case of non-cultivator households, the extent of this decline is from 20.4 per cent in normal year to 11.8 per cent in the drought year (Figures 6.3 and 6.4). The non-cultivator households such as service holders, livestock rearers, artisans and forest resource dependent were not fully independent of agriculture as a source of livelihood. However, their extent of dependence on agriculture was less compared to that of farm households. The amount of the income from different kinds of farm activities has declined by 69.9 per cent in case of cultivator households among which crop farming has been severely affected with reduction in revenue of 73.4 per cent followed by fishing with revenue reduction of 32.3 per cent.

Figure 6.3

Composition of Annual Household Gross Income (All Cultivations)

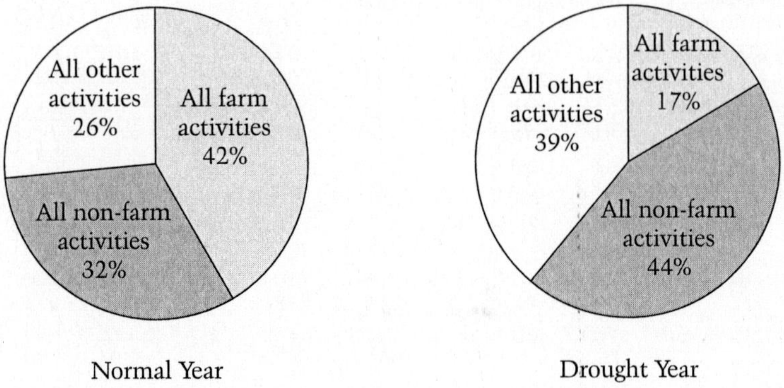

Crop insurance, non-farm activities like artisan works, sale and consumption of forest produces, informal borrowings and migration were the major income smoothing sources for majority of livelihood groups in the drought year. Non-farm activities contributed about 32 per cent, while other activities like borrowings from formal and informal sources, migration and sale or mortgage of assets/implements contributed about 26 per cent of the gross household income in the normal year. The share of non-farm and other activities has increased from 32 per cent and 26 per cent in the normal year to 44 per cent and 39 per cent in the drought year respectively. The share of crop insurance was only 0.3 per cent of total household income in the normal year, that has increased to 3.1 per cent in the drought year. The sale or mortgaging of assets and implements contributed to only 0.2 per cent.

The dependence on some specific income sources like crop insurance, food for work programme (FFW) and begging has dramatically increased during the drought year. The income from crop insurance by cultivator households has contributed to only 0.6 per cent of total household income while this share has increased to 5.3 per cent in the drought year. The income from FFW was nil in the normal year for both cultivator and non-cultivator households while the income from this source has been around 1 per cent of their total household income. The begging as a livelihood source is found to be adopted by some non-cultivator households (landless) in both normal and drought year and income from this source has increased by 76.6 per cent. The decreased income from agricultural labour works has forced them to rely on this self-effacing strategy of survival.

Figure 6.4

Composition of Annual Household Gross Income (All Non-Cultivations)

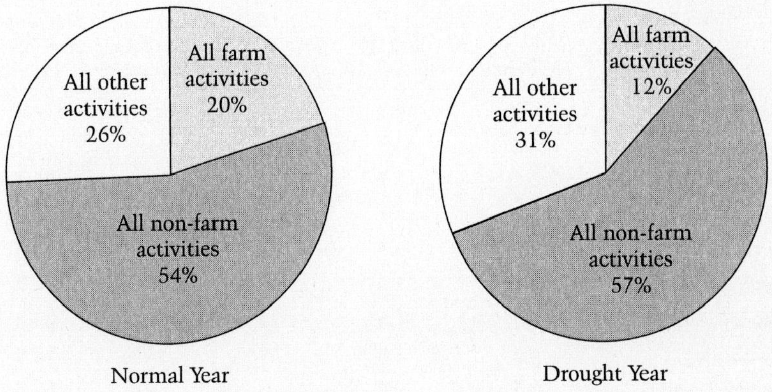

Long Term Income Diversification

Faced with frequent droughts year after year, people have learned to survive by diversifying the income sources in favour of non-farm activities. Considering fairly a longer period of time, it reveals that majority of people were depending mainly on agriculture, labour works and forest resources and other common property resources two decades ago. By now, a major diversification has occurred in favour of business, trading and services though the extent of returns they are deriving from the diversified income sources is not enough for their successful adaptation with drought risks. So the extent of migration has also significantly increased as a part of their effort to managing the scarcity situation. It may be noted that about 69.1

per cent of sample households were depending on agriculture in 1982-1983 while the proportions of households depending on agricultural labour, common property resources (CPRs), livestock and migration were 51.8 per cent, 48.9 per cent, 13.7 per cent and 10.1 per cent respectively (Table 6.3). The average number of livelihood sources for sample households was 4.9 in 1982-83. But by present (2002-03), the average number of livelihood sources has substantially increased to 9. Nearly 66 per cent of households are still depending on agriculture and the proportions of households depending on agricultural labour, CPRs, livestock and migration were 18.7 per cent, 38.8 per cent, 15.8 per cent and 15.1 per cent respectively.

Table 6.3

Adaptive Income Diversification

Sl. No.	*Livelihood Source*	*2002-03*		*1982-83**		*% Change in 2002-03 over 1982-83*
		No. of Households	*% of All Households*	*No. of Households*	*% of All Households*	
1	Agriculture	92	66.2	96	69.1	-4.2
2	Agricultural labour	73	52.5	72	51.8	1.4
3	Non-agricultural labour	26	18.7	14	10.1	85.7
4	Migration	21	15.1	14	10.1	50.0
5	Artisan	9	6.5	4	2.9	125.0
6	Petty trade	31	22.3	12	8.6	158.3
7	Money lending	16	11.5	21	15.1	-23.8
8	Livestock	22	15.8	19	13.7	15.8
9	Service/Pension	13	9.4	5	3.6	160.0
10	CPRs	54	38.8	68	48.9	-20.6
11	Transfer/Remittance	16	11.5	11	7.9	45.5
12	Others	11	7.9	2	1.4	450.0
	All HHs (Repeated)	384.0	276.3	338	243.2	13.6
	All HHs (Actual)	139	100.0	139	100.0	

Note: *The primary data for 1982-83 has been collected from the farmers using the recall method.
Source: Field level survey.

Notably, the dependence on petty trades, transfers/remittance, service/pension and migration has increased by the year 2002-03. The proportion of households depending on petty trades increased to 22.3 per cent in 2002-03 from 8.6 per cent in 1982-83 while the proportion of households depending on transfers and remittance increased to 11.5 per cent in 2002-03 from 7.9 per cent in 1982-83. The proportion of households with migrated family members increased to 15.1 per cent in 2002-03 from 10.1 per cent in 1982-83. On the other hand, the reliance on agricultural labour, CPRs and livestock relatively decreased. The money lending business also declined by 23.8 per cent during this period.

Asset Depletion-Replenishment Process

Asset holding position considerably influences the extent of drought coping capacity of rural households via its influence on the pattern of household income and expenditure and credit worthiness. It was observed that some of the households sold their assets and depleted their financial resources or savings in the drought year. While some poor households were forced to reduce their asset holding for adjusting with drought; some other households depleted their financial resources for lending at exorbitant rate. They tried to exploit the scarcity situation at village level with their past savings. On an average in the normal year, the value of total assets with a large farmer was about ₹ 6,64,308 and that with a marginal farmer was ₹ 40,857 which was just 6.15 per cent of total assets' value of a large farmer. This depicts the prevalence of significant level of inequality in asset holdings among the households in the study area that was found to be little affected in the drought year. In the drought year, the total asset value with a large farmer was ₹ 6,57,864 while that with a marginal farmer was ₹ 40,964 which was about 6.22 per cent of a large farmer's asset value.

On the whole, the asset holding of cultivator households dropped marginally in the drought year. They had to forego agricultural lands, agricultural implements and livestock by 0.47 per cent, 0.42 per cent and 1.95 per cent respectively, for adjusting with drought induced shortages. However, there were some increases in consumer durables of large and medium farmers. There were some transfers of agricultural lands and agricultural implements from medium and small farmers to large farmers since the agricultural lands and implements of large farmers increased by 0.11 per cent and 0.91 per cent respectively, in the drought year while the same of medium farmers decreased by 0.64 per cent and 0.5 per cent in the drought year.

Resorting to Formal and Informal Sources of Credit

Resorting to formal and non-formal sources of credit is an important strategy for managing drought risks. Rural households require borrowing from different formal and informal sources of credit to compensate the drought induced shortfall in income. The purpose of resorting to credit may be short-term adjustment or long-term adaptation. The short-term adaptation credit is usually meant for consumption while medium-term and long-term credit may be used for strengthening the income diversification process or to build soil and water conservation structures for increasing the irrigation coverage that would lessen the drought impacts.

It may be noted from Table 6.4 that the borrowing from informal sources increased while the borrowing from formal sources decreased in the drought year. Overall total borrowing per household decreased marginally by 5.1 per cent. The share of informal borrowing in total borrowing in the normal year was 54.1 per cent, that increased to 55.5 per cent in drought year though the magnitude of borrowing from this source declined from ₹ 6,043 in normal year to ₹ 5,886 in drought year. The share of borrowing from institutional sources drastically decreased from 57.5 per cent in the normal year to 22.6 per cent in the drought year in the case of cultivator households though the extent of the decline in the case of non-cultivator households is little less. As regards the dependence of households on informal sources of borrowing, the share of borrowings from large farmers and money lenders increased in the drought year while the share of borrowings from relatives declined from 14.3 per cent in the normal year to 11.3 per cent in the drought year. Thus, the strength of informal insurance tends to break down in the face of drought.

Interestingly, most of the poor households could not utilise the money in the intended activities for which they had borrowed during both normal and drought year. Overall, 59.7 per cent credit was originally taken by cultivator households for productive activities including capital investments in the normal year; but only 31.2 per cent credit could be utilised for the purpose. Rest 26.5 per cent was diverted to current consumption such as expenditure on food and non-food materials, health related problems and other socio-cultural obligations. On the other hand, in the drought year about 47.8 per cent credit was originally taken by cultivator households

Table 6.4

Drought Adjustment with Borrowing

Lending Agencies	*All Cultivator Households*			*All Non-cultivator Households*			*All Households*		
	NY	*DY*	*% Change*	*NY*	*DY*	*% Change*	*NY*	*DY*	*% Change*
1	*2*	*3*	*4*	*5*	*6*	*7*	*8*	*9*	*10*
Commercial Banks/ RRBs	7385.8 (53.3)	5824.3 (51.5)	-21.1	1294.3 (17.0)	844.7 (10.6)	-34.7	3324.8 (29.8)	2504.6 (23.6)	-24.7
Co-operatives	225.4 (1.6)	161.3 (1.4)	-28.4	192.5 (2.5)	450.2 (5.7)	133.9	352.2 (3.2)	611.6 (5.8)	73.7
Under Govt. Programmes	356.5 (2.6)	126.0 (1.1)	-64.7	1081.6 (14.2)	1242.1 (15.7)	14.8	1455.4 (13.0)	1605.8 (15.1)	10.3
Village Organisations*	1139.9 (8.2)	1245.5 (11.0)	9.3	966.0 (12.7)	1247.3 (15.7)	29.1	1063.3 (9.5)	1232.2 (11.6)	15.9
Money lenders	847.3 (6.1)	852.5 (7.5)	0.6	1083.7 (14.2)	1350.0 (17.0)	24.6	1435.9 (12.8)	1502.9 (14.2)	4.7
Traders/Shop Keepers	931.6 (6.7)	2202.0 (19.5)	136.4	637.9 (8.4)	675.6 (8.5)	5.9	902.0 (8.1)	822.3 (7.8)	-8.8
Large Farmers	1334.5 (9.6)	2962.5 (26.2)	122.0	914.4 (12.0)	1120.1 (14.1)	22.5	1047.1 (9.4)	1128.1 (10.6)	7.7
Relatives	1628.7 (11.8)	1498.5 (13.2)	-8.0	1455.7 (19.1)	1003.2 (12.6)	-31.1	1594.3 (14.3)	1200.8 (11.3)	-24.7
Borrowing from Institutional Sources	7967.8 (57.5)	2559.1 (22.6)	-67.9	2568.4 (33.7)	2537.0 (32.0)	-1.2	5132.3 (45.9)	4722.0 (44.5)	-8.0
Borrowing from Private Sources**	5881.9 (42.5)	8761.0 (77.4)	48.9	5057.6 (66.3)	5396.2 (68.0)	6.7	6042.6 (54.1)	5886.4 (55.5)	-2.6
Borrowing from all Sources	13849.7 (100.0)	11320.1 (100.0)	-18.3	7626.0 (100.0)	7933.2 (100.0)	4.0	11174.9 (100.0)	10608.4 (100.0)	-5.1

Notes: * Village organisations include SHGs, Village Cultural Committee (Kirtan Mandali), Youth Club (Yubaka Sangha) and Village Development Committee.

** The private informal sources of credit include the village level organisations, local money lenders, traders, large farmers and relatives.

Source: Field level survey.

for productive activities including capital investments out of which, only 24.7 per cent credit could be utilised for the intended purpose. Remaining amount was diverted for current consumption. Such diversion tended to be larger with fall in size of land holdings during both normal and drought year, i.e., the less the holding size, the more the extent of poverty, the more the size of deficit household budget and the more the diversion.

Demand Adjustment

It was revealed that the agricultural income which was the major source for sustenance in the study region has drastically declined in the drought year. This had consequential effects on consumption of goods and services by the sample households. Both consumption expenditure and investment expenditure of sample households have dropped as a result of drought situation that we have discussed in the first part of this section. Due to sharp shortfall in income, the poor households were forced to reduce daily intake of foods. Their extent of adjustment with food deficiency will be discussed in the second part of this section. The poor households heavily relied on subsidised food materials provided through the targeted public distribution system (TPDS). The extent by which TPDS has helped the poor households in their consumption smoothing is analysed in the third part of this section.

Consumption expenditure declined by 18 per cent and 19.2 per cent respectively, for cultivator households and non-cultivator households in the drought year. The expenditure on food grains increased from 37 per cent in the normal year to 40.8 per cent in the drought year; while the expenditure on vegetables increased from 6.4 per cent in the normal year to 7.2 per cent in the drought year (Figure 6.5). On the other hand, the proportion of expenditure fell on purchase of consumer durables, clothes, grocery items, meat/fish/milk, electricity and medical treatment. The consumption share on grocery items decreased to 15.9 per cent of total consumption expenditure compared to 16.2 per cent during the normal year. The consumption on clothes and non-vegetarian food items declined from 8 per cent and 7.5 per cent in the normal year to 6.6 per cent and 7 per cent respectively.

An analysis of composition of investment expenditure reveals that the expenditure share on major livelihood activity like agriculture was 60.1 per cent in the normal year and 68.4 per cent in the drought year. While interest payment/loan repayment was the second largest expenditure from the investment account in normal year, lending at village level has been the second largest investment expenditure in drought year. This shows that the dependence on informal sources of borrowings considerably increased

Figure 6.5

Effects on Composition of Annual Household Gross Consumption Expenditure (All HHs)

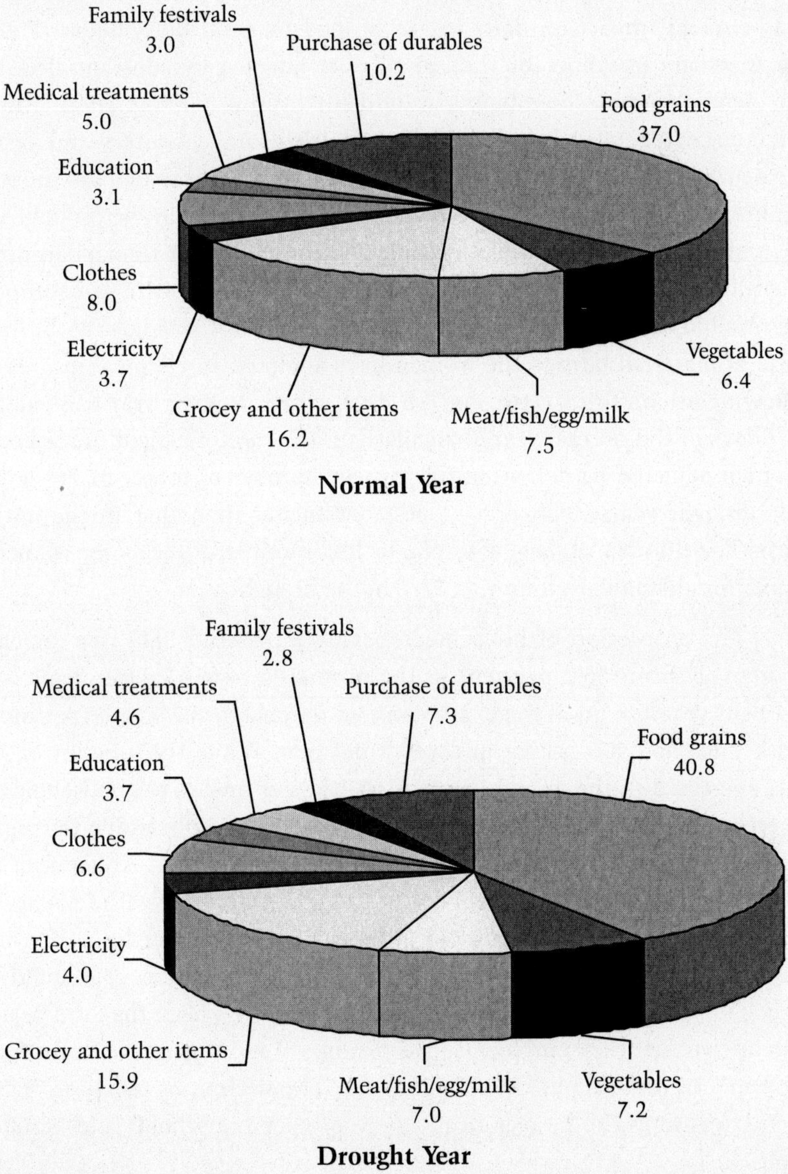

in the drought year compared to the normal year. Though there was no major shift in composition of investment expenditure, total investment expenditure has declined by 27 per cent in case of cultivator households.

Reduction in income from major income sources like agriculture has a direct impact on daily intake of food by rural households. Though all members of vulnerable households are affected by substantial decline in income, female members are hard hit by this phenomenon. Among different category of livelihood groups, agricultural labourers and 'others' category of households experienced significant level of consumption shortfall of 25.8 per cent and 25.7 per cent respectively in the drought year. Specifically women of these vulnerable groups due to their subordinate position in the family were most affected in the form of consumption shortfall. They were the last to eat and had the least to adjust with. In the case of marginal farmer, the women have accepted single meal in a day by allowing others to take two for 8.6 days in the drought year compared to 5.7 days in the normal year. Particularly in the case of agricultural labourer, women adjusted by accepting two or less number of meals in 76 days in the drought year which is 54.2 per cent higher than that in the normal year. The number of days with one or less number of meals for women of agricultural labourer group was 37.7 in the drought year.

The proportion of households having purchased PDS rice increased marginally from 66.2 per cent in the normal year to 68.3 per cent in the drought year. On an average, a household purchased 8.2 kg rice per month with price of ₹ 4.6 per kg in the normal year. While the amount of PDS rice increased in the drought year to 10.7 kg per month per household, the average price dropped to ₹ 2.2 per kg. This was particularly due to running of the subsided programmes under TPDS like Antodaya, Annapurna and some other special programmes in the drought year along with regular PDS rice of 16 kg per BPL household at a price of ₹ 4.75 per kg during both normal and drought year. Due to wide spread poverty, people could not purchase their full quota of subsidised food materials from the local dealers. On an average, they purchased PDS rice in 9.9 months in the normal year and in 11.1 months in the drought year. The decline in purchase of PDS items was found to be significant for very poor households and better off households.

Distress Out-migration

Migration is necessarily a pre-emptive move; it is the survival instinct that drives humans to seek better prospects and a better life for themselves and their families (Sundari, 2005). But the 'involuntary' distress out-migration that is commonly seen to be adopted by households in the water stress areas may not lead to better life for the migrants. Out-migration is a major socioeconomic problem in the study region. It has been an age-old seasonal process induced by agricultural instability and chronic poverty and operationalised by a chain of labour contractors. Short-term seasonal migration is more prevalent than the long-term permanent migration which is basically due to unstable income from agriculture.

Out of 830 population belonging to 139 sample households, 154 persons migrated in the drought year compared to 80 in the normal year which means that about 92.5 per cent increase in migration was observed in the drought year. While about 9.6 per cent of total population migrated in the normal year when there was no crop failure, about 18.5 per cent of total population migrated in the drought year. Among different livelihood groups, forest resource dependants, agricultural and non-agricultural labourers, marginal and small farmers migrated in larger proportion during the drought year compared to other livelihood groups. About 36.8 per cent of forest resource dependants, 35.9 per cent of non-agricultural labourers, 31.0 per cent marginal farmers and 25.7 per cent small farmers migrated in the drought year. Since the nature of migration is seasonal, none of the migrants were found to move with cattle. The duration of migration ranged from around 30 days to more than 6 months. Out of 139 sample households, 52 people migrated for more than 6 months in the drought year compared to 29 people in the normal year, i.e., 79 per cent increase in the drought year. It may be noted that more people from agricultural labourer and marginal farmer categories of households migrated to work in brick kilns. About 35 per cent of out-migrants were engaged in brick work and another 35 per cent worked in the agriculture sector as agricultural labourers, and another 24 per cent worked in factories. It may be observed that the average duration of migration increased by 51 per cent to 6.8 months in the drought year from 4.5 months in the normal year. The expenditure on migration induced health related hazards also significantly increased in the drought year compared to the normal year.

Drawing upon Common Pool Resources

Drawing upon CPRs is a major risk smoothing strategy for poor households belonging to livelihood groups such as agricultural and non-agricultural labourers, small and marginal farmers, livestock rearers and rural artisans. Some studies reveal that poor households heavily depend on CPRs due to reduced income from agriculture and labour works in the water scarce periods (Jodha, 1991; Agarwal, 2000; Chen, 1991).

Our study reveals that the poor households intensively used the forest resources and village ponds for generating additional resources in the drought year. They depleted more of different kinds of non-timber forest produces. Increase in exhaustion of forest resources from trees, bushes and grasses in the drought year was about 25.8 per cent, 24.6 per cent and 17.3 per cent respectively. *Char* (Buchanania lanjan), *Kendu* leaf (Diospyros melanoxylon), *Neem* (Azadirachta indica), Tamarind (Tamarindus Indica), *Mahul* and *Sal* (Shorea robusta) were some of the species of commercial importance those were heavily depleted due to drought induced scarcity situation. On an average, a forest resource dependant earned ₹ 24,345 in the drought year either by consuming or using for their own household or selling of the non-timber forest produces (NTFPs). Plucking of *Kendu* leaves and selling them in bundles was seen to be a major seasonal activity of forest resource dependants.

Fishing in the community ponds by village youths was another income smoothing process at community level. However, in the severe drought year 2002, the pisciculture business was significantly affected. The delayed application of fish seeds due to unavailability of water in the early phase of monsoon also delayed the process. Harvesting was administered at immature stage, prior to 2 months of actual period and full grown stage, because of water shortage in the ponds. Accordingly, harvest fell by 25 per cent and sales were reduced by 30 per cent. The demand also fell by 10 per cent due to low level of purchasing power of the local people. Among other drought induced risk factors, the incidence of theft and death of fish occurred frequently in the drought year. Due to drought condition, more people attempted to steal fishes from pond to compensate the income shortfall.

Specific Risk Management by Cultivator Households

In addition to above mentioned strategies, the cultivator households have resorted to some risk management strategies those were exclusively used against risk in crop production. Among these risk management measures, six measures were found to be prominently adopted by the farmers in the region. They were: (i) crop management including curtailment in cost of cultivation and salvage operation, (ii) crop diversification, (iii) tenancy contract and production contract, (iv) management of irrigable water resources, (v) soil and water conservation measures, and (vi) crop insurance.

The quantity of most of the inputs was reduced in the drought year by farmers. Since paddy was the major crop grown in the region, the case of both rainfed and irrigated paddy has been discussed. Since the irrigation amount was not sufficient, the crop growth with supplementary irrigation was also affected. So the cost of cultivation decreased in the case of irrigated paddy too. In case of rainfed paddy, the reduction in the amount of inputs was by 13.2 per cent basically on account of fertiliser and labour. On the other hand, the average cost of cultivation was considerably more in the case of irrigated paddy compared to rainfed paddy but the amount of inputs for irrigated paddy declined in the drought year by 11.3 per cent. Surprisingly, the use of pesticide increased in the drought year in both the cases. This was basically due to increase in the extent of drought induced pest attack and farmers' effort to save the remaining crop.

In the drought year, certain measures were also adopted for extracting whatever little the adversely affected crops could offer. Table 6.5 shows the loss minimising measures adopted by the affected farmers. Knowing the fact that getting the matured crop from some plots is not possible to harvest, farmers collected only byproducts from about 32 per cent of affected cultivated plots. They harvested mature crop from only 22.3 per cent of agricultural plots in the drought year compared to 95.5 plots in the normal year.

Table 6.5

Risk/Loss Minimising Crop Practices

(% of plots)

Sl. No.	*Loss Minimizing Measures*	*Drought Year*	*Normal Year*
1	Collected weeded material as fodder	29.2	5.0
2	Harvested premature crops	16.0	
3	Harvested crop byproduct only	32.1	1.5
4	Harvested mature crop	22.3	95.5
5	Weeding done more than once	26.2	88.0
6	Post-sowing operations abandoned	22.4	

Source: Field survey.

The total cropped area during normal year was 6.02 acre per household that considerably declined to 4.59 acres in the drought year resulting in a decline in gross cropped area (GCA) by 24.7 per cent. Surprisingly, the extent of reduction is more during *Rabi* season compared to *Kharif* season. While the total area under cereals decreased by 17.9 per cent, the area under pulses dropped by 28 per cent. The decline in area under oilseeds happens to be the maximum at 54.1 per cent in the drought year. The total area under vegetables also fell by 37.5 per cent. Overall, the extent of short-run diversification was not successful due to lack of necessary infrastructures and incentives in the study region though the farmers have made a notable progress in their effort of long-term diversification of cropping pattern as a part of their risk reduction strategy.

The Herfindhal Index value, as a measure of crop diversification, is much larger (0.59) during the drought year compared to the normal year (0.46) which confirms the fact that farmers have failed to diversify in the drought year and have depended more on non-farm sources of livelihood for their sustenance. The Herfindahl Index for Bolangir has gradually declined from 0.44 in 1970-71 to 0.27 in 1990-91 and has increased thereafter. Though the farmers have successfully progressed in their endeavour to diversify the cropping pattern during 1970s and 1980s, they have apparently failed in their efforts to manage risk through crop diversification during 1990s due to the frequent occurrence of droughts and the presence of significantly a large proportion of arable area not under assured irrigation.

The major sources of irrigation in the study region were micro-level water resources like dug wells, tanks and cross bunds. It is worth mentioning that the major, medium and minor irrigation systems have not contributed a single drop of irrigation water to the sample households. It may be noted from Table 6.6 that the percentage of irrigated area under dug wells to gross irrigated area has increased from 35.6 per cent in normal year to 42.1 per cent in drought year. So the dug well is the only reliable source of irrigation in the study region.

Besides dug wells, tank/cross bund and check dams are found to be effective in the study area where the contribution of these sources to total irrigated area was about 34.2 per cent in the normal year and 27 per cent during the drought year. Overall, the irrigated area as a percentage of net sown area has declined from 29.9 per cent in the normal year to 24.4 per cent in the drought year. Most of the water harvesting structures (WHSs) and soil and water conservation measures in the watershed areas of the study area performed satisfactorily. Mainly these were effective during *Kharif* season only though some of these WHSs were capable of providing supplementary irrigation during the *Rabi* season. In some cases, the bed of WHSs has gradually been silted that requires renovation. The usual methods of irrigating from these WHSs were pumping and traditional WLD (water lifting devices). However, the use of pumps for irrigating from WHSs is very limited because of problem in power supply. Besides the water harvesting structures under watersheds, a number of additional WHSs previously existed around the study villages. However, the mechanism to regularly monitoring the condition of these structures and renovating these structures when required was too weak.

The tenancy contract, particularly share-cropping is the most popular form of risk sharing arrangement prevailing in the study region. The total number of owner cultivators as tenants surprisingly increased by 26.7 per cent. The percentage of tenanted cultivators was 21.7 per cent in the normal year that has increased to 26.1 per cent in the drought year. Except large farmers, all other cultivator groups cultivated more tenanted lands mainly leased out by large farmers. Thus, the large farmers have shifted a part of the drought risk to the other categories of farm households.

Table 6.6

Irrigation Coverage by Source and Farmer Category (Area in Acre per Household)

Sl. No.	Farmer Category	Dug well	Tank/Cross bund	Nalla	River Lift	Surface Lift	Check Dam	Canal (Major/medium /minor)	Total Area Irrigated	% of NSA Irrigated
						Normal Year				
1	LF	0.86 (34.5)	0.52 (21.0)	0.37 (14.7)	0.26 (10.4)	0.15 (6.0)	0.33 (13.3)	0.00 (0.0)	2.48 (100.0)	24.29
2	MDF	0.67 (33.4)	0.45 (22.4)	0.31 (15.5)	0.00 (0.0)	0.29 (14.3)	0.29 (14.3)	0.00 (0.0)	2.01 (100.0)	34.98
3	SF	0.56 (37.6)	0.22 (14.8)	0.19 (12.8)	0.03 (2.0)	0.16 (10.7)	0.33 (22.1)	0.00 (0.0)	1.49 (100.0)	37.53
4	MF	0.11 (65.0)	0.01 (6.1)	0.02 (10.4)	0.01 (6.1)	0.00 (0.0)	0.02 (12.3)	0.00 (0.0)	0.16 (100.0)	14.95
	Total	0.53 (35.6)	0.28 (18.8)	0.20 (13.4)	0.10 (6.7)	0.15 (10.1)	0.23 (15.4)	0.00 (0.0)	1.49 (100.0)	29.88
						Drought Year				
1	LF	0.65 (30.7)	0.41 (19.2)	0.29 (13.6)	0.25 (11.5)	0.22 (10.1)	0.32 (14.8)	0.00 (0.0)	2.13 (100.0)	21.06
2	MDF	0.55 (33.3)	0.38 (23.0)	0.23 (13.8)	0.00 (0.0)	0.30 (18.2)	0.19 (11.6)	0.00 (0.0)	1.65 (100.0)	28.64
3	SF	0.94 (69.1)	0.08 (5.9)	0.22 (16.2)	0.00 (0.0)	0.12 (8.8)	0.00 (0.0)	0.00 (0.0)	1.36 (100.0)	24.20
4	MF	0.08 (30.8)	0.06 (23.1)	0.00 (0.0)	0.09 (34.8)	0.00 (0.0)	0.03 (11.4)	0.00 (0.0)	0.26 (100.0)	24.16
Total	0.53	0.22 (42.1)	0.17 (17.5)	0.07 (13.5)	0.15 (5.6)	0.12 (11.9)	0.00 (9.5)	1.26 (0.0)	24.43 (100.0)	

Note: Figures in parentheses show percentages of total.

Source: Field survey.

Effectiveness of Institutional Mechanism

The policy framework, institutional sources of credit, the TPDS and village level institutions play pivotal role in strengthening the risk management strategies of rural households. It has been analysed in the preceding sections that the institutional sources of credit apparently failed in fulfilling the credit needs of vulnerable households. Mainly large farmers and elite groups could avail more of institutional credit with subsidised interest rates, whereas the small and marginal farmers had to resort to private money lenders with exorbitant interest rates. As far as the TPDS is concerned, the PDS items were made available in required quantity in the sample villages but people could not purchase their full quota of subsidised food materials from the local dealers due to lack of purchasing power induced by wide spread poverty.

The community level mechanism to fight with drought was stronger in earlier periods compared to the present in the study area. About 30 years back, paddy collection group known as Jagannath Dhana Fund was operational in the village. The group was providing loan in terms of paddy, transacting 1 *khandi*[2] against 1.5 *khandi*, thereby making a profit of 0.5 *khandi* on 1 *khandi* paddy invested. It was helping the poor farmers in the time of distress that they were repaying after harvesting of their crops. The organisation invests the profit paddy on developmental works like construction of temple and strengthening rural infrastructure. Such kind of disciplined and well-managed organisations were gradually weakened due to increased reliance on government run PDS. As a matter of fact, such kind of institutions are non-existent in the study villages at present.

The shortage of water in the drought year has resulted in a number of intra-village and inter-village conflicts. Since water for irrigation was very scarce, particularly in the drought year, the frequent conflicts among water users were obvious outcome. Again, the inter-village conflicts are found to be more in the study area compared to intra-village conflicts. This shows the lack of cohesion among the village level institutions in the study area.

2. I *khandi* is equivalent to 20 kg.

Crop insurance is considered as one of the effective institutional mechanism to compensate the farmers for their losses due to events which are quite unpredictable and cannot be prevented. However, the extent of crop insurance coverage in the study area is found to be very low. Only 3.8 per cent of total farmers in Bolangir were covered in the drought year 2002. Among the study blocks, the coverage was as low as 1.67 per cent in Titlagarh and 3.34 per cent and 4.23 per cent in Patnagarh and Saintala respectively. Among different categories of sample farmers, large farmers were insured to the larger extent compared to all other category of farmers. About 12.1 per cent of large farmers were covered under crop insurance as against 2.6 per cent of marginal farmers and 2.4 per cent of small farmers. On the whole, about 4.9 per cent farmers were insured under crop insurance scheme.

Evolution and Efficacy of Drought Management Policies and Programmes

Mitigating drought risks requires a multi-pronged attack with a participatory approach that requires institutional support and encouragement. Government has an important role to play by encouraging the community-based disaster mitigation measures such as development and renovation of community based WHSs, developing common property resource base, strengthening rural socioeconomic infrastructures like education, health and financial institutions etc., and supporting the long-term income and crop diversification process. There have been marked improvements over the years in government's approach to mitigating drought both in terms of policy formulation and action (Samal, 2003). However, the steps taken so far are not enough for mitigating the drought risks. A gradual improvement to the approach of famine and drought management was noticeable during the 19th century; and every major drought/famine brought about some qualitative improvement to the relief approach during pre-independence period. However, these development programmes have not been implemented properly and they were found to reach a very small proportion of the target groups.

Though the frequency of droughts during 18th and 19th centuries was less compared to 20th century, the intensity of scarcity and famine conditions was more severe (Bhatta, 1997; Swain 2010). It is imperative to analyse how the qualitative improvements in drought management policy have occurred with occurrence of every major drought. The historical data and literature suggest that the occurrence of great famine of 1866 in Orissa along with parts of Bengal, Bihar and Madras led the British government to appoint the first Famine Commission in 1880, which suggested for providing of employment to the affected persons on the public works, extension of irrigation and improved methods of agriculture, improvement in communication and establishment of famine insurance fund to meet the expenditure on relief works. The 2nd Famine Commission (1898), mainly recommended suspension and diminution of land revenue and payment of wages to the persons engaged in the public works subject to a minimum and a maximum daily wage. The 3rd Famine Commission (1989) emphasised the need of having a well-designed relief plan in advance of drought including the mitigation programmes. The cultivation of fodder crops, grant of loan, opening of cattle camps and relief to aged and destitute were recommended by the commission (Samal *et al.*, 2003). The severe drought of 1965-66 contributed to building up of reliable PDS to ensure food security of drought affected people. The periodic occurrence of drought during 1972, 1974 and 1976 forced the government to focus on the need for evolving massive employment generation programme with a view to enhance the purchasing power of the people instead of providing subsidies and free ration to the affected population. This resulted in beginning of Food for Work (FFW) programme in 1977. The drought of 1979 forced the government to emphasise the need for creating durable community assets for enabling the people of the affected area to withstand future droughts with greater resilience. This gave rise to National Rural Employment Programme (NREP) and Integrated Rural Development Programme (IRDP) in 1980. The consecutive drought of 1980, 1981 and 1982 resulted in start of the Rural Landless Employment Guarantee Programme (RLEGP) in 1983. Under NREP and RLEGP programmes, food grains were given to workers as a part of the wage component of the programme that aimed at providing food security to rural poor during drought situations. These two programmes were later merged to form Jawahar Rojagar Yojana (JRY) in 1989. During drought period of 1987-88, the self-employment programmes namely, IRDP, Development of Women and Children in Rural Area

(DWCRA), Training of Rural Youth for Self Employment (TRYSEM) were in operation which aimed at improving the economic condition of BPL households by arranging productive economic ventures for them through a mix of bank credit and government subsidy (GoI, 2003). In spite of implementation of a number of programmes during the 1980s, the study region faced frequent droughts throughout 1980s and 1990s.

Due to growing emphasis on long-term programme with participatory approach, a Long Term Action Plan (LTAP) for the three undivided districts of Kalahandi-Balangir-Koraput (KBK) was prepared in 1993 in consultation with Govt. of India. The LTAP was conceptualised for a period of seven years from 1995-96 to 2001-02 with two objectives in view: (a) drought and distress proofing, and (b) poverty alleviation and development saturation. After a review in 1997-98, the state government prepared a Revised Long Term Action Plan (RLTAP) which is under implementation. The Mahatma Gandhi National Rural Employment Guarantee Act (MNREGA) which has been implemented since 2006 has a potential to provide huge safety nets to vulnerable rural households.

Though the government programmes for reducing drought risk have been refined over the years and the institutional support system has been strengthened with the experiences of successive droughts, the institutional performance in allocating resources and implementing the developmental programmes are observed to be influenced by a number of factors. The nature of centre-state relationship and influence of pressure groups have played a key role in the sanction and implementation of the development programmes. On the other hand, the extent of coverage and performance of all these developmental programmes in the study area have not been very encouraging. As Table 6.7 shows, only 2.84 per cent of sample households were benefited by SGSY which is a combination of three earlier schemes (IRDP, TRYSEM and JRY).

Table 6.7

Impact of Drought Mitigation Programmes on Sample Households

Sl. No.	Study Villages	% of Sample Households Benefited by:								Benefited through SHGs
		DPAP	IRDP	TRYSEM	JRY	SGSY	IAY	EAS	All Programmes	
1	Samara	3.33	1.52	1.55	0.00	2.60	5.48	0.85	12.20	12.86
2	Mundo-mahul	3.89	1.31	1.09	0.00	2.14	6.64	1.10	11.29	10.85
3	Bijepur	1.24	2.10	1.96	0.70	4.06	6.80	1.87	15.43	9.17
4	All	2.78	1.58	1.51	0.22	2.84	6.11	1.17	12.13	10.29

Source: Field survey.

The proportion of people benefited by TRYSEM (Training Rural Youth for Self-Employment) was as low as 1.5 per cent of sample households. Only 0.22 per cent households have been benefited by the JRY programme. The average proportion of people benefited by different programmes was about 12.1 per cent among sample households. While about 6.11 per cent of households have been benefited by the Indira Awas Yojana (IAY), the people benefited under DPAP, which was a major programme in the drought prone area, was very low. Though the programme was undertaken with a watershed approach, there has been no significant increase in irrigation through water harvesting in the region. However, farmers have realised the importance of exploiting the groundwater resources and have relied on water harvesting structures as their major source of irrigation. The watershed programmes under different schemes including DPAP were not also very successful in increasing the cropping intensity or bringing about the sustained changes in cropping pattern. Though the programme has already been saturated in one of the study village, most of the sample households did not have any knowledge of the programme. Obviously, the extent of peoples' participation was very low. Notably, the extent of self help group (SHG) based activities has not been well developed though there has been a perceptible progress. On the whole, many developmental programmes have been implemented in the drought prone study region but the activities under these programmes and benefits accrued have not been sustained due to lack of long-term vision for these programmes. There is a need of livelihood focused interventions with high priority to peoples' participation for their sustainability.

Conclusions

There is a need for strengthening the coping capacity for effectively dealing with drought risk that seems to be rising in the region along with the process of climate change. There are many areas where coping capacity can be strengthened with effective policy interventions. Increasing irrigation coverage has to be given due importance. There is huge scope for increasing irrigation in the district through developing micro level water resources. The traditional tanks (locally known as *Kata*, *Bandha*, *Chahala* etc.) have proved to be extremely useful not only in normal years but also in water scarce years. Small size WHSs are advantageous over medium and large irrigation projects in Bolangir due to its uneven and hilly topography and other socioeconomic and biophysical factors. So instead of going for big dams those have already consumed lot of time and resources but not yet completed, efforts should be made to increase the irrigation coverage through WHSs such as dug wells, check dams, tanks and renovate the existing defunct WHSs. Though WHSs are quite feasible in the region, poor economic standards of majority of farmers, insufficient power availability, political negligence and weak institutional set up are the major hindrances for their sustainability. The financial constraints may be eased by encouraging community mobilisation of resources, provision of incentives and effective institutional development. Continuous and concerted efforts should be made to increase farmers' participation and their adaptation which is essential for sustainability of WHSs. The problem of low voltage is required to be looked at and the delivery of electricity at farmers' field needs to be assured so as to encourage the use of groundwater which has been underutilised in the region.

The lack of proper marketing facilities coupled with the problem of credit availability from institutional sources and shortage of power supply have forced many prospective farmers to avoid cultivating highly remunerative cash crops like sugarcane and cotton. Thus, much emphasis is required to be given to infrastructure development in the region. Infrastructure development is the first step for accelerating growth and livelihood promotion and diversification in the region.

The institutional support system is required to be strengthened for improving the socioeconomic conditions of drought afflicted people in the region. The crop insurance coverage is only 4.1 per cent that needs to be increased for reducing the level of drought risk of farmer community. There is a need of harmonious relationship between the centre and state government so as to facilitate implantation of development programmes in the region effectively.

References

Agarwal, Anil (2000). "Catch the Rain", *The Indian Express*, April 26.

Bhatta, Brajabandhu (1997). *The Natural Calamities in Orissa in the 19th Century*. New Delhi: Commonwealth Publishers.

Chen, M. (1991). *Coping with Seasonality and Drought*. New Delhi: Sage Publications.

GoI (2003). "Drought Situation in the Country", *Forty Fifth Report of the Standing Committee on Agriculture*. New Delhi: Department of Agriculture and Cooperation.

Hagman, G. (1984). *Prevention Better than Cure: Report on Human and Natural Disasters in the Third World*. Stockholm: Swedish Red Cross.

Jodha, N.S. (1991). "Drought Management: Farmers' Strategies and their Policy Implications", *EPW* 26(39): A-98-A-104, September 28.

Kogan, F.N. (1997). "Global Drought Watch from Space", *Buletin of American Meteorological Society* 78(4): 621–36.

Nayak, L. (2004). "Indigenous Water Harvesting Methods in Bolangir: A Case Study (in Oriya)", *Pani Samastanka Paini* 3(1). Agragamme, Bhubaneswar.

Pattnaik, C.R. (1998). "Why Drought in Bolangir is A Man-made Phenomenon", *Exchange* 22, September.

Pradhan, M.K. (1994). "State of Orissa's Environment: A Citizen's Report". Bhubaneswar: Council of Professional Social Workers (CPSW), pp.159-87.

Roy, B.C., S. Selvarajan and C. Mohanty (2004). "Vulnerability and Adaptation to Climate Change in Orissa: A Livelihood Analysis", *NCAP Working Paper* no.6, New Delhi: National Centre for Agricultural Economics and Policy Research.

Sainath, P. (1996). *Everybody Loves A Good Drought*. Penguin Books India Limited Publication. pp.315-370.

Samal, K.C. (2003). "Coping Strategy of Weaker Sections (Dalits and Women) in Post Super Cyclone Orissa (Case study of Ersama Block)", *Mimeo*. Bhubaneswar: NKC Centre for Development Studies (NCDS).

Sundari, S. (2005). "Migration as a Livelihood Strategy: A Gender Perspective", *EPW* 40(22&23): 2295-303.

Swain, M. (2006). "Sustainable Development of Rainwater Harvesting in Drought Prone Sub Humid Region: Determinants and Viability", *Social Change* 36 (4): 84-100.

———. (2010). *Drought Vulnerability and Livelihoods: Impacts and Adaptation Strategies in Western Orissa, India*. Saarbrucken (Germany): VDM Verlag.

Technological Dimensions in Agriculture

7

Appropriate Technology *versus* Adopted Technology

A Case Study of Farmer's Choice of Technologies from Sericulture Sector in India

G.K. Rajesh

Introduction

One 'way out' of agrarian crises is evolving appropriate technologies and ensuring their diffusion among practitioners of agriculture. During and ever since the green revolution, a large number of agricultural technologies came into being, with varying levels of appropriateness to the farm conditions. Owing to the location specific nature of agriculture, the appropriateness of any technology change with the location where it is supposed to be used. Thus, it is necessary to innovate such technologies to suit the location specific requirements before attempting to diffuse them in the society. The case of Bivoltine (BV) hybrid silkworm in Indian sericulture is a classic example of a promising technology (by scientific standards) that failed to diffuse among the farmers. As evolving a hybrid is very expensive, it is important to look at the reasons behind its sub-optimal diffusion, in order to evolve suitable policy prescriptions for rectification. This paper analyses the reasons behind slow diffusion of 'BV hybrid technology' among sericulture farmers in the light of an empirical study conducted in Mandya district of Karnataka.

Theoretical and Conceptual Framework

The process of technological change is understood in terms of the Schumpeterian trilogy of invention, innovation and diffusion (Rogers, 1962; Stoneman and Diederen, 1994). Diffusion research, in the past century had analysed the manner in which innovations are adopted or rejected temporally and spatially by participants in a social system. A major goal of diffusion research in agriculture has been to identify factors, which contribute to the variations in adoption behaviour of farmers. Once these factors are known they can be manipulated to expedite the diffusion rate among the potential adopters.

The spread of new agricultural technologies have received attention of researchers because they can raise the income of smallholders (Ruttan 1977; Barham *et al.*, 1995), generate broad and equitable benefits to society (Lipton and Longhurst, 1989; De Franco and Ricardo, 1993), and lower pressure on renewable natural resources (Almeida and Campari, 1994). The importance of diffusion of technology in agriculture has been realised by economists since 1950s. After Griliche's (1957) landmark study of diffusion of hybrid corn, researchers of adoption have focused on the role of economic variables (principally prices) in the diffusion of new technologies. Since then a large body of work surfaced inquiring the nature and causes of differential diffusion rates of various agricultural technologies. This study is embedded in the theoretical frame work of 'economics of technology diffusion'.

Indian Sericulture Industry: Important, Issues

India is world's second largest silk producer. It is also the largest consumer and importer of silk and silk goods (UN Comtrade Data 2007). Sericulture is important to Indian economy as a cottage industry spread over 53814 villages employing nearly 56 lakh people (Central Silk Board Database, 2007). As a labour intensive activity practiced throughout the year it is identified as a means for rural employment generation and as a remedy for seasonal unemployment (Jayaram *et al.*, 1998). The other merits of sericulture as an agro-industry are: its short gestation period to establish potential for regular returns to the farmers, reelers and weavers, environment friendly production and processing technologies, potential for farm diversification, cash flow from rich to the poor, sustainability as a rural-based activity involving family labour and women and high value addition to the end products with potential export markets (Benchamin and Giridhar, 2005). The Indian sericulture industry is currently faced with the problems of stagnation in production, low productivity, poor quality of produce, high cost of production and competition from cheap raw silk imports.

The sericulture industry is built upon two living organisms: an insect namely silkworm and its food plant namely mulberry.[1] Thus, the quality and quantity of raw silk output are primarily dependent on the genetic potential of mulberry and the silkworm breeds. Almost 95 per cent of silk produced in India is from traditional low yielding indigenous multivoltine silkworm varieties or crossbreeds[2] which are relatively poor yielders (CSB Database 2007). The cocoons produced by them are unsuitable for reeling in sophisticated reeling machines and the raw silk produced from which is characterised by lower filament length and lesser tensile strength leading to breakages making it unfit for high speed power loom weaving (Kumaresan *et al.*, 2002). Thus, the power loom industry is heavily dependent on imported Chinese raw silk which is of superior quality (Vasumathi, 2000; Thomas *et al.*, 2005a). The indigenous raw silk is largely consumed by the handloom sector and partly by the power loom sector as weft[3] (Vasumathi, 2000). The import price of raw silk has been lower than the domestic raw silk, as the cost of production of Indian silk is high (Kumaresan, 2002). Moreover as shown by Naik and Babu (1993), the price of imported Chinese raw silk is dependent on the prevailing prices of Indian raw silk, though the causative nature of indigenous raw silk price has not been clearly elucidated. This has affected the indigenous raw silk prices and in turn the domestic cocoon prices (Tikku, 1999). This could probably be one of the reasons for large scale uprooting of mulberry plantations which resulted in considerable labour displacement in the farm sector (Central Silk Board Database, 2007). There is a growing demand-supply gap of raw silk in the domestic-industry. Naik and Babu (1993) estimated that the

---

1. Though there are four species of silkworms, the mulberry silkworm—Bombyx mori—accounts for the lion share of silk production. During 2010-11, the percentage share of mulberry and non-mulberry silks in India were as follows: Mulberry: 80.7 per cent, *Eri*: 13.13 per cent, *Tasar*: 5.53 per cent and *Muga*: 0.58 per cent (CSB, 2011). This study is exclusively on mulberry silk.

2. Crossbreed (CB) is a hybrid between Pure Mysore (an indigenous multivoltine race known for its hardiness), and NB4D2, a bivoltine breed developed in India. CB is comparatively easy to rear but yield relatively poor quality silk.

3. Weft is the yarn running breadth wise in the fabric, the mechanical tension on which is lower as compared to that on the warp, which run length wise. As warp is stretched under tension, if the yarn is not strong enough it is liable to break, rendering the weaving process difficult. The tension on warp is more in power looms as compared to handlooms.

total high quality silk production in India could meet at the most 60 per cent of the estimated demand.

A solution to the qualitative and quantitative problems of Indian silk industry is popularisation of high yielding silkworm hybrids that can also yield better quality silk. The BV silkworm races prevalent in the temperate countries are characterised by high productivity (800-1250 kg cocoons/hectare of mulberry) and high quality silk as compared to multivoltine races of tropical countries (160-440 kg cocoon/ha. of mulberry) (Jayaswal *et al.*, 2001). The comparative performance of BV hybrid *vis-á-vis* cross breed furnished in table 7.1 clearly establishes the superiority of BV hybrid. The superiority of BV hybrids is not confined to the quality of silk they produce. The hybrid yields significantly higher quantities of cocoons and thus supposed to be more remunerative to farmers. The comparative data furnished in Table 7.2 is illustrative of this.

Table 7.1

Comparative Performance of BV Hybrids and CBs

Hybrid	*Cross Breed (PM X NB4D2)*	*BV Hybrid (CSR2 X CSR4)*
Colour	Yellow	White
Silk quality[1]	B	2A to 4A
Renditta[2]	8	6
Filament length[3]	750 m.	1150 m.
Yield per 40,000 larvae	50 kg.	70 kg.
Survival %	70%	53%
Cocoon price per kg. (₹)	100-150	180-240

Note: 1. The international quality standards prescribe grading of raw silk from A to D, A being the higher quality. Above A grade a further classification in the ascending order of 2A, 3A etc., is done.

2. Renditta is the measure that indicates the quantity of cocoons required to produce 1 kg of raw silk, for the crossbreed it is above 8. This means that an average of 8 kg cocoons are required to produce 1 kg of raw silk. On the other hand, the new BV hybrids have the Renditta less than 6. Hence, the silk production can be improved by 30 per cent by merely switching over to BV raw silk production.

3. Filament length is the length of the continuous filament that could be recovered from the cocoon.

Source: Dandin *et al.* (2005).

Considering this the Tropical Sericultural Technology was developed in India during 1970s and a National Sericulture Project (NSP) was launched in 1990 with World Bank support (World Bank, 1997). The major thrust of these projects was development of BV silkworm hybrids and appropriate agronomic practices for rearing them and development of sophisticated technology for processing cocoon and silk. Against the 1000 tonnes per annum target of BV hybrid cocoon production under NSP, only 400 tonnes was realised (World Bank, 1997).

Table 7.2

Comparison of BV and CB Cocoon Production (per annum)

Items	*BV Hybrids*		*Cross Breeds*	
	₹/acre	%	*₹/acre*	%
Leaf cost	27864.49	32.13	18825.83	27.31
Silkworm seed	1782.44	2.06	3766.69	5.46
Disinfectants and materials	7301.97	8.42	3755.73	5.45
Labour	13566.9	15.64	11975.61	17.37
Depreciation on fixed capital	32972.45	38.02	25485.78	36.97
Other costs	3232.12	3.73	5122.92	7.43
Total cost	86720.37	100	68932.56	100
Revenue	123519.92		86175.33	
Net return per annum	36799.55		17242.77	
B:C ratio	1.42		1.25	

Source: Kumaresan, 2002.

Statement of the Problem

The problem addressed by the present study is very low diffusion rate experienced by a very promising technology namely 'BV hybrid silkworm' among the sericulture farmers in India and the objective of the study is to find out the factors influencing the decision of sericulture farmers of Karnataka on adoption of BV hybrid silkworm.

The efforts to popularise BV hybrids met with limited success at the adoption level (Ramakrishnan, 2001; Kumaresan, 2002). It is seen that at present percentage share of BV silk in India's total raw silk production is only 7 per cent, the remaining 93 per cent being produced from traditional inferior breeds and CBs (CSB, 2011). This indicate that only around 7 per cent of the farmers have adopted BV hybrids in the country and the remaining are with conventional CBs or other inferior breeds, the silk produced out of which is of low quality suitable for handlooms only.

Figure 7.1 illustrates the diffusion of BV hybrid in major silk producing states, for the period from 1990 to 2012. Except Tamil Nadu and Maharashtra, no other states have shown any encouraging trend in diffusion.

This reality starkly contrasts with the claims of the scientific community and policy makers who prescribe 'BV hybrid' as a panacea for all ills of the sericulture sector and project it as a solution for poor yields and poor quality of produce obtained by Indian sericulture farmers. This is a very serious issue. Huge investments made towards developing suitable BV hybrids, developing appropriate agronomic practices and extension efforts have not resulted in matching diffusion of the BV hybrid in the country. It is established that without producing BV silk in sufficient quantities, India cannot hold its ground in the domestic silk market, let alone compete in the global market. Considering the fact that the domestic sericulture and silk industry is undergoing a 'struggle for existence' in the post liberalisation era, facing tough competition from cheap imports of raw silk and silk products mainly from China (DGAD, 2005), the issue of slow diffusion of BV hybrid silkworm assumes importance.

Figure 7.1

Diffusion of BV Hybrid in Various States in India 1990 to 2012

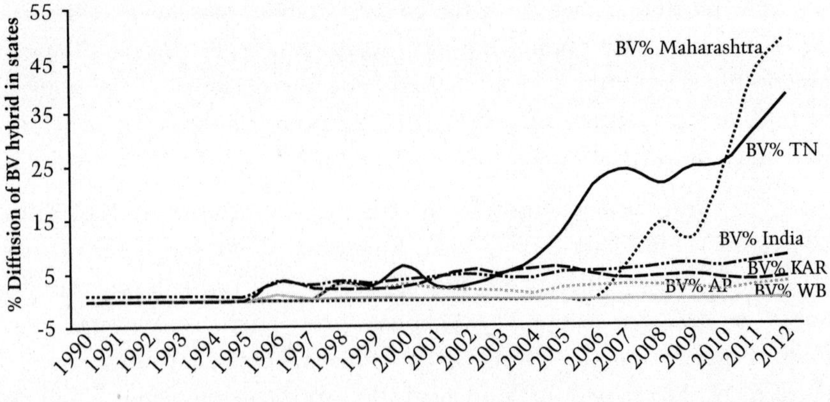

Source: CSB Database 2012.

This issue has not been subjected to much systematic economic investigation.[4] The available, limited numbers of studies on this topic were conducted by specialists in agriculture extension. Their studies dealt either with the problem of differential acceptance as a function of status, role and motivation or with the problem of communication of innovations. Therefore, the present study approaches the problem from the perspective of economics of technology diffusion.

Methodology and Source of Data

Both primary and secondary data were used for the study. The study was primarily based on the micro data generated from a sample survey conducted in Sreerangapattanam *taluka*, Mandya district, Karnataka. Out of the five sericultural ranges in the *taluka* two ranges (comprising 22 villages and 665 farmers) were selected by purposive sampling method considering large number of farmers and contiguity of farms for easiness of data collection. This consideration is important as the element of slow diffusion due to information asymmetry will be minimum in a such a sample. While information asymmetry could be an important variable in extension studies, too much of information asymmetry could be a bias in the current study where economic variables are considered. The isolated and far located farmers could be outliers in this respect. While the case of such farmers and the determinants of their adoption decision are equally important, this demand a study of wider scale, which is outside the scope of the current study.

From the entire list of farmers of these ranges 71 farmers were selected at random. The data was collected through direct interview method by using a pre-tested schedule. The data was analysed and a 'Probit Regression model' was constructed.

4. The various reports and studies available are mostly departmental studies undertaken by central and various state governments. Most of the economic investigations undertaken by western scholars were restricted to the period up to 1930 may be the period up to when there was active western interest in silk. Sinha (1989) reported that "....within a substantial body of literature on silk production systematic information on the socioeconomic dimensions of the activity is lacking."

The categorical dependent variable namely 'BV hybrid adoption decision, took either value 1 (YES) or 0 (NO). Since the majority respondents have reared both BV hybrid and CB silkworm batches at some point of time during the year, a bench mark of 50 per cent was used to categorise them. According to this if BV hybrid forms at least 50 per cent of a farmer's total annual silkworm seed consumption, he is designated as a BV hybrid adopter and if the adoption percentage is less than 50 per cent he is designated as a non-adopter.

The relationship between selected explanatory variables and the BV hybrid adoption decision was analysed by estimating a probit regression model. The probability that farmer 'i' adopts BV hybrid is modeled as a function facet of explanatory variable X as given below:

$$P(Y_i = 1 \mid X_i) = \Phi(X_i'b)$$

Where Xi contain variables related to farmer 'i'. Φ is the cumulative standard normal probability distribution function.

The 16 explanatory variables used are as follows. Their short forms furnished in parenthesis.

1. Age of the farmer ('Age'): The age of the farmer is in years as reported by the farmer.
2. Social participation of the farmer ('Soc part'): In order to measure the social participation of farmers the respondents were asked whether they participate in any of the following seven events *viz.*, political parties, *panchayat*, farmers club, private/voluntary organisations, school parent-teacher association, resident's associations or any other organisations during the previous year. One mark was assigned to each participation and the total count was taken as an index.
3. Farm Size ('Land Tot') represents the total land area owned by the farmer and is used to capture the effect of farm size on adoption decision.
4. Ratio of annual intake of silkworm eggs to mulberry acreage ('Ra egg-mulb'), which is the ratio of total silkworm eggs or Chawky

reared worms[5] (both BV and CB) purchased by the farmer during the previous year to his mulberry acreage. This variable is taken as a measure of the farmer's infrastructure and entrepreneurial capacity.

5. Ratio of mulberry acreage to total land holding ('Ra Mulb-Tot') is used as a proxy for the relative importance attached by the farmer to sericulture. When the relative importance of sericulture is more, it is expected that the farmer's dependency on it for his livelihood is more and it is more probable that the farmer adopts, the new technology.

6. Ratio of total annual cocoon production to the mulberry acreage [Ra coc-mulb] is the ratio of the total quantity of cocoon produced by the farmer (both BV hybrid and CB) during the previous year to the total mulberry area under his disposal. This is used as a measure of productivity which in turn is determined by his ability to harvest successful crops. Better productivity leads to increase in net returns and might influence BV hybrid adoption decision.

7. Difference in annual average yield between CB and BV hybrid [Diff av yld CB-BV] is calculated by subtracting the yield per 100 layings[6] obtained by the farmer for BV hybrid from that of CB, during the previous year. It is expected that this variable will have a negative influence on the BV adoption decision, when the increase in this value is indicative of better performance of CB[7] as compared to BV.

8. Difference between the maximum and minimum price received by the farmer for CB and BV cocoon [CB Pmax-min] is arrived at

---

5. The term 'chawky' refers to the young silkworm up to the age of about 12 days after hatching. Separate rearing of these young worms by exclusive Chawky Rearing Centers (CRC) is a recommended practice to ensure proper care and healthy upbringing of worms which will be sold to the farmer by the CRC instead of eggs directly. This not only gives a better crop but also saves time of farmer, permitting him to take more number of crops round the year.

6. Layings means egg masses. 100 layings (or corresponding chawky worms) gives 40,000 individual eggs or worms. The term average yield means cocoon produced in kg out of 100 layings.

7. When we take the difference as "average yield of CB minus average yield of BV", a positive value means that average yield of CB is higher than that of BV. That means increase in this value implies lesser profitability of BV.

by subtracting the minimum price received by the farmer for BV cocoons from the maximum price for CB during the previous year. This variable is hypothesised to have a negative impact on the BV adoption decision

9. The number of mass contact programmes attended by the farmer [Mass contact 0/1] indicates whether the farmer has participated in any mass contact programmes such as sericulture krishimela, exhibition, field day etc., which may positively influence his BV adoption decision by reducing uncertainty and his perceived risk.
10. Sericulture experience [Exp]: Number of years of total sericulture experience, is used as a proxy variable for uncertainty which is expected to impact the farmers BV adoption decision positively.
11. Education level (Edn): This was recorded in number of years of formal education received by the farmer starting from '0' for the uneducated. It is assumed that with increasing levels of education the tendency to adopt BV hybrid increase.
12. Labour availability (Fam lab): Labour availability to the farmer is captured in the variable as the family labour available in the household. It is assumed that availability of family labour will encourage farmer to adopt BV hybrid.
13. Credit [Credit 0/1]: Availing credit facility by the farmer is expected to encourage adoption of BV hybrids among farmers.
14. Number of extension contacts by government agency (Extn cont): It is expected that the extension contacts from the government agencies will have a positive influence on BV hybrid adoption by farmers.
15. Extension by the private CRC (Extn pvt CRC): The extension contacts from the private Chawky Rearing Centre are expected to have a negative impact on BV adoption decision of the farmers.
16. Government subsidy (Subsidy 0/1) might positively influence adoption of BV hybrid.

Results and Discussion

The Probit model was estimated using maximum likelihood method. The results are reported in Table 7.3. As there are 16 explanatory variables, placing all of them in a single regression would reduce the degrees of freedom drastically. Therefore, the estimation is done using five regression models each containing seven variables at a time. The chi square values reported under each equation tests the null hypothesis that all coefficients in the model except the constant term are zero. Given the degrees of freedom it is found that the null hypothesis is rejected in all the five estimated equations. Thus, each of these different combinations of explanatory variables in the probit model jointly has significant effect. The regression results are discussed along with the summary statistics drawn from the actual survey data.

Table 7.4 gives the percentage of respondents belonging to various age groups among adopters and non-adopters. It can be seen that in the whole sample, 76 per cent of the farmers fall within the age group of 30 to 50 years. Among adopters 87.1 per cent belong to this age group whereas among non-adopters it forms only 67.55 per cent. Out of the total sample 18.3 per cent belong to age group 51 years and above and 5.63 per cent belong to below 30 years categories. The percentage of farmers with age 51 years and above among adopters is only 9.7 per cent whereas that among non-adopters is 25 per cent. Similarly percentage of younger farmers (below 30 years) is lower among adopters.

Model 1 (Table 7.3) gives the relationship between age of the farmer and BV adoption. Age of the farmer was found to have a significant negative impact on the BV adoption decision of the farmer. The result is in conformity with the theoretical postulate and are in agreement with the findings of Shetty (1966) and Radhakrishnamenon *et al.* (1980) that the middle age group show greater tendency to adopt recommended farming practices. The older farmers are too traditional and security conscious, do not take the risk of adopting the innovation. Another reason for this may be that being in a traditional sericultural area; more age also means more experience. The older farmers (with the exemption of the too traditional and risk averting ones) by virtue of their long experience might be able to harvest better crops using the highly productive but more susceptible BV hybrids. Thus, age appears to be a decisive factor in adoption of BV hybrids.

Table 7.3

Probit Regression Results

Variable	*Model -1*	*Model -2*	*Model -3*	*Model -4*	*Model -5*
Constant	-.1316925 [-0.13]	-2.118839** [-2.62]	-.9958572 [-1.49]	-.1669868 [-0.26]	.8771456 [0.83]
Age	-.0613273* [-2.39]				
Edn	-.126698* [-2.40]	-.146846* [-2.53]			
Soc part			.0672721 [0.36]	.1294353 [0.62]	
Exp	.0462769* [2.17]	.0514218* [2.09]	.0348412 [1.46]	.069614** [2.65]	.030957 [1.08]
Fam lab				-.2817048* [-2.42]	-.4219281** [-2.56]
Land tot	.2320935* [2.31]	.0246611 [0.24]			
Ra Mulb-tot	2.104048** [2.64]	2.051582* [2.26]	1.734529* [2.31]		
Extn cont			.0323301 [0.34]		
CB Pmax-min			-.0220159** [-2.99]		.0045317 [0.37]
Ra egg-mulb				.001877* [2.06]	.004035** [3.26]
Ra coc-mulb				-.0043808** [-3.11]	-.0073815** [-3.25]
Diff av yld CB-BV					-.0486666** [-2.96]
Credit 0/1	.8866021* [1.94]				
Extn pvt CRC 0/1		-2.234289** [-2.88]	-1.687972* [-2.42]	-2.046914* [-2.32]	
Mass contact 0/1	1.030694* [2.28]	1.645243** [2.91]			
Subsidy 0/1		1.746456** [2.98]	1.31716** [2.67]	2.138132** [3.41]	2.20197** [2.83]
LR chisquare [Prob chisquare]	29.12 [0.0001]	39.58 [0.0000]	37.49 [0.0000]	40.73 [0.0000]	63.790 [0.000]
Pseudo R2	0.2993	0.4069	0.3853	0.4187	0.6557
No. of observations	71	71	71	71	71

Note: * Significant at 5 per cent.

** Significant at 1 per cent.

Figures in brackets are z values except for LR chisquare where P values are reported.

Table 7.4

Age-wise Grouping of Adopters and Non-Adopters of BV Hybrid

Age of Farmer	*% of Each Age Group Among Adoption Categories*		
	Non Adopters	*Adopters*	*Total*
Below 30 years	3 (7.5 %)	1 (3.2 %)	4 (5.63%)
30 to 40 years	14 (35 %)	13 (41.9 %)	27 (38%)
41 to 50 years	13 (32.5 %)	14 (45.2 %)	27 (38%)
51 to 60 years	7 (17.5 %)	3 (9.7 %)	10 (14.1%)
61 and above	3 (7.5 %)	0	3 (4.2%)
Total	40 (100%)	31 (100%)	71 (100%)

Source: Primary survey data.

The variable social participation was used as a proxy for the uncertainty reducing factor which could positively influence the farmers BV adoption decision by reducing uncertainty and perceived risk by acquiring more information. From Table 7.5, it is seen that 62.1 per cent of the farmers had some social participation. Among non-adopters percentage of people without any social participation was slightly more (50%) as compared with that among adopters (45.2%). Only 15 per cent of the non-adopters had participated in more than one social event. None of the non-adopters participated in more than three events. Among adopters 54.8 per cent had some kind of social participation. 25.7 per cent of the adopters participated in more than one event and 12.8 per cent participated in more than three events.

Table 7.5

Social Participation of Farmers

	% Participation within Adoption Categories		
Marks	*Non-Adopters*	*Adopters*	*Total*
0	20(50%)	14(45.2%)	34(47.9%)
1	14(35%)	9(29%)	23(32.4%)
2	5(12.5%)	3(9.7%)	8(11.3%)
3	1(2.5%)	1(3.2%)	2(2.8%)
4	0	2(6.4%)	2(2.8%)
5	0	1(3.2%)	1(1.4%)
7	0	1(3.2%)	1(1.4%)
Total	40(100%)	31(100%)	71(100%)

Source: Primary survey data.

From the regression results, social participation of the farmer does not reveal any significant influence on BV adoption decision, as per models 3 and 4 (Table 7.3). However, participation in mass contact programmes during the previous year has a significant positive impact on BV adoption decision as per models 1 and 2. This implies not only that awareness is an important factor in determining BV adoption decision but also that the mass contact programmes are effective.

The variable 'experience' is found to impact positively on BV adoption decision of the farmer

Table 7.6

Grouping of Adopters and Non-adopters Based on Total Sericulture Experience

Sericulture Experience (Years)	*Adoption Categories*		
	Non Adopters	*Adopters*	*Total*
Less than 5 years	7 (17.5%)	1 (3.2%)	8 (11.3%)
5 to 10 years	14 (35%)	9 (29%)	23 (32.4%)
11 to 20 years	13 (32.5%)	14 (45.2%)	27 (38%)
Above 20 years	6 (15%)	7 (22.6%)	13 (18.3%)
Total	40 (100%)	31 (100%)	71 (100%)

Source: Primary survey data.

From Table 7.6 it can be seen that the largest chunk of non-adopters (67.5 %) fall within 5 to 20 years experience category and the smallest group under non-adopter category is of veterans with more than 20 years experience. Here we can see that the non-adopter category follows the pattern of the sample (given under the total column) and thus it just represents the population under experience wise stratification. But the case of adopters is different. Among adopters there is a clear majority (45.2%) which belong to the category of 11-20 years experience followed by those with 5-10 years experience (29%). That is 74.2 per cent of adopters have 5-20 years of experience. It can also be noted that while among the non-adopters the percentage of farmers with below 5 years experience and above 20 years experience are almost equal (17.5% and 15% respectively), among adopters only 3.2 per cent were of less than 5 years experience and 22.6 per cent had more than 20 years experience. The data shows an increasing trend of adoption with experience.

From the regression results, the variable 'experience' is found to impact positively on BV adoption decision of the farmer as given by models 1, 2,4 and 5 (Table 7.3). In these models 'number of years of experience' has been used as a proxy for the risk bearing capacity of the farmer. Long experience also implies that the farmer has been exposed to many silkworm breeds and can make a better judgement of the BV hybrid than a novice in the field. The experienced farmer could also be expected to have the confidence to handle BV hybrid which is prone to diseases leading to crop losses. Thus, the farmer by virtue of his experience may overcome the 'perceived risks' of crop losses and adopt BV hybrid. As both the variable used to capture perceived risk and uncertainty reduction have shown positive influence on BV adoption decision, it can be inferred that risk and uncertainty play a vital role in poor adoption and diffusion of the BV hybrid. Thus, the promotional agencies should consider this aspect also.

In order to analyse the impact of farm size on BV hybrid adoption decision the adopters and non-adopters were grouped into three size classes of land area *viz.*, small (up to 2.5 acres), medium (2.5 to 5 acres) and large (above 5 acres). This was considered under three categories of ownership namely, A) total own and lease land, B) total own land, and C) acreage under mulberry.

Table 7.7

Grouping of Adopters and Non-adopters into Size Classes of Land Area

Size Classess	*A. Total Own and Lease Land (Acre)*			*B. Total Own Land (Acre)*			*C. Land Under Mulberry Cultivation (Acre)*		
	Non-adopters	*Adopters*	*Total*	*Non-adopters*	*Adopters*	*Total*	*Non-adopters*	*Adopters*	*Total*
Small up to 2.5 acre	24 (61.54)	15 (38.46)	39 (100)	25 (62.50)	15 (37.50)	40 (100)	37 (58.73)	26 (41.27)	63 100)
Medium 2.5-5 acre	10 (58.82)	7 (41.18)	17 (100)	10 (58.82)	7 (41.18)	17 (100)	2 (33.33)	4 (66.67)	6 (100)
Big Above 5 acre	6 (40.00)	9 (60.00)	15 (100)	5 (35.71)	9 (64.29)	14 (100)	1 (50.00)	1 (50.00)	2 (100)
Total	40 (56.34)	31 (43.66)	71 (100)	40 (56.34)	31 (43.66)	71 (100)	40 (56.34)	31 (43.66)	71 (100)

Note: Figures in parenthesis give percentage of adoption category.

Source: Primary survey data.

From part A of Table 7.7, considering total own and lease land, it is understood that majority of small and medium farmers (61.54% & 58.82%) were non adopters while majority of the big farmers (60%) were BV hybrid adopters.

From part B of the Table 7.7, considering total own land area, a similar trend was observed. From part C, considering acreage under mulberry cultivation alone it is found that while majority of small farmers (58.73%) were non-adopters and majority of medium farmers (66.67%) were found to be BV hybrid adopters. The big farmers were found to be indifferent, being distributed equally between adopter and non-adopter categories. This analysis reveals that with increasing size of landholding the farmers tend to adopt BV hybrids; however the mulberry area alone doesn't determine the adoption decision. While the size of the landholding can proxy the economic status of the farmer, a farmer who devotes his entire land or a considerable portion of it for sericulture and thus left devoid of any subsistence crops will consider various economic factors such as income, relative risk of crop loss, labour availability etc., while deciding on adoption of BV hybrids. This may be because in the absence of subsistence crops the impact of potential crop losses could be fatal.

From the regression results the two proxy variables representing farm size are found to be significantly influencing the adoption decision. Size of total landholding has a significant positive influence in BV adoption as given by model 1 (Table 7.3). This implies that the farm size indirectly impacts on hybrid adoption decision through several factors namely fixed adoption costs, risk preferences, human capital, credit constraints, labour requirements, tenure arrangements etc., as found by Feder *et al.* (1985). Ratio of number of eggs reared by the farmer per year to the mulberry acreage has a highly significant positive impact on BV adoption decision, as postulated (given by models 4 and 5 in Table 7.3). This indicates that farmers with higher infrastructure facilities for rearing more number of worms per unit area, who are also able to produce more leaf per unit area, tend to adopt BV hybrid. Thus, mulberry leaf productivity and infrastructure capacity could be a major determinant of BV adoption. Both these variables indirectly represent the economic status of the farmer.

Ratio of mulberry acreage to total landholding as given by models 1, 2 and 3 (Table 7.3) has a highly significant positive influence on BV adoption decision by the farmer. This ratio indicates the relative importance attached by the farmer to sericulture among all his farming activities. It is understood from the result that if the relative importance of sericulture is more the farmer's dependency on it for his livelihood is more and it is more probable that the farmer will adopt the BV hybrid technology. Thus, farmers who cultivate a substantial portion of their land with mulberry tend to adopt BV hybrids more. This decision in spite of the potential risk involved in it is difficult to understand. For a clearer understanding of this decision-making process, a detailed investigation is to be conducted taking into account the relative share of mulberry in the total of various size classes of landholding across different income groups.

All the three variables related to profitability of the enterprise and relative profitability between CB and BV rearing are found to be significantly influencing the BV adoption decision indicating the importance of profitability in adoption decision of the farmer.

Productivity plays an important role in profitability. In sericulture the major factor determining cocoon productivity is average yield which is dependent on the ability of the farmer to take successful crops. Table 7.8. gives the average yield[8] realised for CB and BV by the farmers in the sample. Since almost all the farmers reared CB and BV at least once in the year, the information contains the experience of all the farmers in both breeds

Table 7.8

Average Yield of CB and BV

Yield per 100 Layings in kg (Annual average)	*CB*		*BV*	
	Number of Farmers	*%*	*Number of Farmers*	*%*
Below 40	0	0	7	14
40-60	16	29	36	72
Above 60	40	71	7	14
	56	100	50	100

Source: Primary survey data.

8. Average yield means cocoon yield realised from 100 layings (eggs laid by 100 silk moths equalling 40000 individual eggs). Here for the study, the average yield obtained from each silkworm crop by every farmer was collected separately for BV and CB, and the average for the year calculated for each breed for the farmer.

From Table 7.8 it can be seen that 71 per cent of the farmers reported average yield above 60 kg for CB but only 14 per cent of them reported 60 kg average yield for BV. Majority (72%) of farmers, who reared BV, reported average yield between 40-60 kg and 14 per cent of them got very poor yield of less than 40 kg. None of the farmers reported less than 40kg average yield for CB. This result shows that comparatively CB crops give better yield than BV crops. This observation is strongly in opposition with the claims in the literature on BV hybrid performance, based on field trials under supervision.

Table 7.9 gives the annual average cocoon yield of adopters and non-adopters for both BV and CB crops put together. The table indicates that a higher percentage of non-adopters get better yield of 60 kg and above and a considerable number of BV adopters get very poor yield of below 40 kg per 100 layings. This indicates the better productivity level of non-adopters.

Table 7.9

Average Yield of Adopters and Non-adopters

Yield per 100 Layings in kg of Both Breeds (Annual Average)	*Non-adopters*	*Adopters*
Below 40	6 (15%)	11 (38%)
40-60	17 (44%)	14 (48%)
Above 60	16 (41%)	4 (14%)
Total	39 (100%)	29 (100%)

Source: Primary survey data.

Table 7.10 gives an overall picture of crop failure in CB and BV. During the previous year, the total crop failure in the sample was 15 per cent (for both breeds). Seventy per cent of the total crop loss was contributed by BV hybrid rearing.

Table 7.10

Crop Failures in CB and BV

	CB	*BV*	*Total*
Layings supply	55383 (53 %)	49915 (47 %)	105298 (100%)
Failure	4595 (30 %)	10890 (70 %)	15485 (15 %)

Source: Primary survey data.

Table 7.11 compares the percentage crop failure incidence in BV and CB rearings in the sample. Crop failure is calculated by taking the percentage of layings completely failed during the year. Seventy-one per cent farmers who reared CB reported 100 per cent succees of their CB batches as against mere 28 per cent in BV. And it is understood that crop failure is more in BV batches.

Table 7.11

Percentage Crop Failures in BV and CB

Crop Failure %	*Number of Farmers*	
	BV	*CB*
No failure	20 (28%)	42 (71%)
Up to 10%	22 (31%)	0
11 to 50%	24 (34%)	14 (24%)
Above 50%	5 (7%)	3 (5%)
	71 (100%)	59 (100%)

Source: Primary survey data.

Analysis of profitability will be incomplete without an analysis of relative market prices of the two breeds of cocoon.

Table 7.12

Average Cocoon Price Per Kg in Rupees Obtained by Farmers

Price in ₹	*Percentage of Farmers*	
	BV	*CB*
Below 100	0	26 (44%)
100-125	15 (30%)	29 (49%)
125-150	20 (40%)	3 (5%)
Above 150	15 (30%)	1 (2%)
	50 (100%)	59 (100)

Source: Primary survey data.

Table 7.12 summarises the relative prices received by the farmers for BV and CB cocoons. The farmers were asked the prices they received per kg of cocoon during the previous year against each crop for both CB and BV. From this data average price for the year for each farmer is worked out separately for the two breeds and tabulated. The table shows that the minimum price for BV is ₹ 100.00 whereas 44 per cent of the farmers got less than ₹ 100.00 for CB cocoon. Similarly, a good number of farmers reported average price for BV more than ₹ 150.00 per kg. There is a clear indication that BV cocoons fetch a better price than CB cocoons.

The regression results from Table 7.3 supports the above observations related to profitability. Difference between the maximum and minimum price received by the farmer for CB and BV cocoon [CB Pmax-min] during the previous year has created a significant negative impact (at 1% level) on the BV adoption decision (model 3 in Table 7.3) and is in accordance with our hypothesis. This is because, the increase in this value is indicative of an increase in the maximum price received for CB cocoon. This means that the BV adoption decision of the farmer has a bearing on the relative (maximum) price per kg of CB cocoon. If the farmer gets a higher price for CB cocoon, his subsequent adoption decision doesn't favour BV. Difference in average yield of CB and BV hybrids reared by the farmer during the previous year has a significant negative impact on the BV adoption decision (model 5, in Table 7.3). This implies that as the average yield of CB increases in comparison with that of BV the subsequent adoption decision will not be in favour of BV. In case of the farmers rearing both the breeds, this variable gives the yield difference and hence comparative profitability of the two breeds at the farmer's level.

Ratio of total annual cocoon production to the mulberry acreage is found to have a significant negative influence on the BV hybrid adoption decision of the farmer (models 4 and 5 in Table 7.3). This ratio is a measure of the farmer's productivity which in turn is determined by his ability to harvest a successful crop and disease management. The regression result points to the fact that farmers with higher productivity of land (out of both BV and CB rearing) during the previous one year are averse to adopt BV. However, the results pertain to one single year only. Nothing definite can be said about the outcome because the productivity analysis was confined to only one year. More analysis of more number of years is necessary to establish the linkage between productivity and adoption.

Number of years of formal education received by the farmer has a significant negative impact on the adoption decision as illustrated by models 1 and 2 (Table 7.3). This is a disturbing finding since the theory of diffusion and majority of empirical studies do not support it. However, we hope that an impartial analysis of the present sericulture scenario in India will explain this. We have two breeds, the traditional CB—though low yielding is sure to yield a reasonable crop with minimal attention and inputs use; the BV hybrid is a better yielder but notorious for its proneness to disease and demand for more food and extra care. Thus, there is a tradeoff between expected profit and perceived risk in BV sericulture. The comparative advantage is determined by the relative profitability. Thus when market prices of BV cocoon are low, difference in net returns tends to be small. In such circumstance, educated farmers who keep accounts of their profit and losses tend to dis-adopt BV for CB.[9]

To further analyse this variable, the farmers are grouped according to the number of years of formal education received (Table 7.13), into five categories, *viz.*, illiterate (cannot read and write and no formal education), those who can read or write or received any formal education up to 10th standard including 10th fail, 10th standard pass, 11th to 12th and finally graduates and above.

Table 7.13

Grouping of Adopters and Non-adopters with Respect to Education

Education	*% Level of Educated in Each Adoption Category*		
	Non-adopters	*Adopters*	*Total*
Illiterate	15(37.5%)	14 (45.16%)	29 (40.8%)
Less than 10 yrs	17 (42.5%)	8 (25.8%)	25 (35.2%)
10th pass	5 (12.5%)	1 (3.2%)	6 (8.5%)
11-12 years	2 (5%)	7 (22.6%)	9 (12.7%)
Graduates	1 (2.5%)	1 (3.2%)	2 (2.8%)
Total	40 (100%)	31 (100%)	71 (100%)

Source: Primary survey data.

From Table 7.13 it can be observed that 40.8 per cent of the farmers have not received any formal education at all and are illiterate, 35.2 per cent received below 10 years schooling. The percentage of farmers who have received any formal education among BV adopters is 54.84 per cent

9. This finding is in line with the finding of Griliches (1957) that economic incentive is the primary determinant of diffusion.

as against 62.5 per cent among non-adopters. Percentage of farmers with 10th standard pass is higher (12.5%) among non-adopters than that of BV adopters (3.2%). Also 45.16 per cent of BV hybrid adopters are illiterates whereas only 37.5 per cent of non-adopters are illiterates. However, BV adopters include a higher percentage of 11-12th educated farmers (22.6%) than that of non-adopters (5%). Among the adopters the percentage of graduates is also comparatively high. The results show that 25.8 per cent of the adopters are well educated and only 7.5 per cent of the non-adopters are well educated.

Another finding which is in contradiction with the empirical regularities in agricultural technology diffusion is with respect to the availability of family labour. It is generally held that the high yielding varieties is labour demanding hence labour shortage can lead to non-adoption. Our study results indicate a significant negative relationship between family labour availability and BV hybrid adoption decision.

Our analysis indicated that the BV adoption is favoured by the higher income groups as compared to the low income group. The probit regression also revealed that farm size has a significant positive impact on BV adoption. This implies that BV sericulture is practiced with hired labour rather than family labour. Our field survey revealed that latest labour saving technology[10] adoption which was originally developed for BV hybrids is complete in the area and is fully adopted by CB rearers also. This has considerably brought down the labour requirement. However, due to the non-uniform growth and maturation of the CB worms the cocoon spinning days get extended increasing labour requirement. Thus farmers are of the opinion that CB rearing demands more labour. Thus risk averting farmers who can effectively utilise their family labour tend to adopt CB.

As hypothesised, credit availed for sericulture has a significant positive influence on BV adoption decision as given by model 1 (Table 7.3). Availing credit is thus indicative of the farmer's capacity to execute collateral security and his inclination to be innovative. The farmers who has availed

10. These are subscription to Chawky Rearing Centers, Two time feeding, Shoot rearing and use of hormones to speed up maturation of larvae reducing the period of operations by 4 to 5 days.

credit for sericulture would be able to invest on the infrastructure required for BV rearing which increases the chances of adoption of BV hybrid technology. The importance of credit as a supply side factor also reveals its policy implications.

Contrary to our expectations the number of extension contacts with the government extension agency has not produced any significant impact on BV adoption decision. However, this result is in accordance with the finding that government intervention in diffusion process rarely speed diffusion and public enterprises move slower than private owned (Hann and Mc Dowell, 1984; Rose and Joskow, 1990). This may be indicative of the inefficiency of extension system which fails to create the desired impact on the mindset of the farmer regarding BV adoption.

Extension support from private Chawky Rearing Centre supplying CB worms is found to have a significant negative impact on BV adoption decision. This is in accordance with our hypothesis. It is evident that the private CRCs do a lot of promotional efforts to popularise the CB hybrids in the production of which majority of them are involved. Personal interview with farmers also revealed that the private CRCs regularly visit the crop and give guidelines for proper crop raising. In case of a severe crop loss they even replace the batch free of cost.

Government subsidy for sericulture has a significant positive impact on BV adoption decision. This indicates that the probability of continuing BV adoption is more with farmers who have received subsidy. It is to be noted that to become eligible for subsidy the farmers need to satisfy a number of conditions stipulated by the government in which separate, scientific rearing house and HYV mulberry garden are included. It can be deduced that government subsidy is a motivating factor for BV hybrid adoption.

Summary and Conclusion

The focus of this paper has been the decision-making process of South Indian sericulture farmer in his rational choice between two available technologies *viz.*, CB silkworm and BV hybrid silkworm. While the BV hybrid is an advanced technology, an improvement over the existing CBs, the present analysis raises the question whether the improved technology is appropriate in the potential adopter's context. The results indicate that it is not. The reasons for which include economic factors.

Our analysis of the adoption process of BV hybrid silkworm provides a few empirical facts. In agreement with the empirical literature on adoption, age and education are found to impact negatively on BV adoption decision. So also farm size exerts a positive influence. Mitigation of perceived risk and uncertainty are found to be important on adoption of BV hybrid. While family labour availability discouraged BV adoption, credit encouraged it. The importance of profitability is emphasised by the results. The three variables capturing reduction in profitability of BV sericulture showed significant negative influence on its adoption; however the linkage between productivity and adoption could not be explained due to data constraints. From the supply side while subsidies appear to be an encouraging factor for BV adoption the extension efforts of private CRCs (supplying CB worms) is found to deter the choice of BV hybrid by the farmer. The extension efforts of the government agencies are found to be insignificant in this respect.

Policy Implications

The study results raise serious questions about the appropriateness of BV hybrid technology. The results of the study clearly elucidate the reasons as to why the technology has failed to diffuse among the farmers. The results stress the necessity of economic incentives on the successful diffusion of BV hybrid. The probable factors determining the economic incentive to BV hybrid rearing are crop stability, yield, and a suitable price differential with CB cocoon. It is seen that crop failures are rampant in BV which affect profitability. It is also shown by our study that the average yield realised at the farmer level is far lower to that demonstrated at laboratory level or produced under expert supervision. This indicates the necessity for evolving more hybrids resistant to diseases (robust hybrids) which do not compromise with yield and are better suited to the location specific microclimate. Here arise a need for research aimed at developing location specific and trait specific silkworm breeds. Since crop loss is an important determinant of profitability, the government should also implement a crop insurance scheme to support the farmers.

Even though our analysis indicate a price differential favouring BV cocoon, the farmers complained during the interview that the price differential of BV is narrow and sometimes prices even fall below that of CB (during rainy season when cocoon quality goes down). This is partly due to the fact that the reelers are hesitant to offer a higher price for BV cocoon because of low prices experienced by the reeling industry for their produce, which has direct bearing on diffusion of technologies in the reeling sector. The government has to take appropriate policy decisions for the technology up-gradation in the reeling sector. The government can also think of installing a price stabilisation mechanism by way of support price for BV cocoon, or a production incentive system for encouraging production of quality BV cocoon.

The government extension system is found to be unable to create any strong influence on the adoption decision of the farmers. However, the credit availability and subsidies have strong influence on BV adoption decision. The government must improve its official extension system by capacity building and organise more number of mass contact programmes demonstrating the benefits of the BV hybrid. Since potential credit constraints could undermine the diffusion of BV hybrids in the country, the government should facilitate credit for sericulturists on affordable collateral.

The government extension system has a lesson to learn from the private CRCs in the study area. The extension efforts of the private CRCs are more business oriented than service oriented. However, its externalities are positive to the farmers. The government should encourage production of BV hybrid chawky worms in the private sector by way of human resources development, financial incentives for infrastructure development for quality production and credit facilities. There is need to increase the transfer of market relevant knowledge enhancing skills of the sericultural extension system. This would require action from both demand side and supply side.

Research Gaps

The level of technology diffusion in the reeling sector could play a vital role[11] in diffusion of BV hybrid technology. Further, the determinants of diffusion may have ramifications into the processing industry (yarn processing to weaving) and the prevailing administrative and political set up, manifested through its policy initiatives. Nevertheless there is no empirical evidence available to support this hypothesis, as of now. Hence, there is a need to assess the level of technology diffusion in silk reeling and yarn processing industries and to inquire into the determinants of the same in order to generate information useful for taking suitable policy initiatives or for rectifying existing policies towards better diffusion of BV hybrid technology.

Reference

Almeida, Anna Luiza Ozorio de and Joao S. Campari (1994). "Sustainable Settlements in the Amazon", *World Bank Discussion Paper* 26. Washington, D.C.: Education and Social Policy Department.

Barham, Bradford, Michael R. Carter and Sigel, W. Wayne (1995). "Agro Export Production and Peasant Land Access: Examining the Dynamics between Adoption and Accumulation", *Journal of Development Economics* 46: 85-107.

Benchamin, K.V. and Kshama Giridhar (2005). "Sericulture Industry in India", Conference Papers III, *The 20th Congress of the International Sericultural Commission*. Central Silk Board, Bangalore. pp.158-61

CSB (2007). *Central Silk Board Database.* Ministry of Textiles, Government of India.

———. (2011). *Central Silk Board Database*, Ministry of Textiles, Government of India.

Dandin, S.B., H.K. Basavaraja and N. Suresh Kumar (2005). *Silkworm Breeds and Hybrids at Galore*. Mysore: Central Sericultural Research and Training Institute.

De Franco, Mario and Godoy Ricardo (1993). "Potato-Led Growth. The Role of Agricultural Innovations in Transforming Bolivian Agriculture: A Macroeconomic Perspective", *Journal of Development Studies* 29(3): 561-87.

---

11. The bulk of research in sericulture was conducted on the farm sector for evolving high yielding mulberry and silk worm hybrids and for developing appropriate agronomic packages for sericulture. Most of the research conducted in reeling sector was aimed at developing and popularising reeling gadgets best suitable for reeling bivoltine cocoons (Vasumathi, 2000). But diffusion of these technologies has been slow in reeling sector, which has rendered it, incapable of processing high quality bivoltine hybrid cocoons into raw silk of matching quality. Thus there is no incentive for the reeling sector to absorb the bivoltine hybrid cocoons at competent prices. This in turn creates a downward price shift for bivoltine cocoons in the market. It appears as if, the technological advancement made in the farm sector has been offset by the lack of it in the reeling sector.

Directorate General of Anti Dumping & Allied Duties (DGAD) (2005). *Initiation of Anti-Dumping Investigation Concerning Import of Silk Fabrics 20-100 ms/meter from People's Republic of China. http://commerce.nic.in/adint_silkfabric20-100_chinapr.htm*

Feder, Gershon, Richard E. Just and David Zilberman (1985). "Adoption of Agricultural Innovations in Developing Countries: A Survey", *Economic Development and Cultural Change* 33(2): 255-298.

Griliches, Zvi (1957). "Hybrid Corn: An Exploration in the Economics of Technological Change", *Econometrica* 25: 501-22.

Hannan, T. and J. McDowell (1984). "The Determinants of Technology Adoption: The Case of the Banking Firm," *Rand Journal of Economics* 15: 328-35.

Jayaram, H., V. Mallikarjuna, S. Lakshmanan, R. Ganapathi, Rao and R.G. Geetha Devi (1998). "Labour Employment Under Different Mulberry Farm Holdings: A Comparative Study", *Indian Journal of Sericulture* 37(01): 52-56.

Jayaswal, K.P., P. Rama, Ohana Rao., M.M. Ahsan and R.K. Datta (2001). "Bivoltine Silkworm Breeding Strategies in the Tropics. Global Silk Scenario-2001", *Proceedings of the International Conference on Sericulture*. Oxford and IBH Publishing Co. Calcutta. pp.222-32.

Kumaresan, P. (2002). "Quality Silk Production: Some Economic Issues", *Economic and Political Weekly:* 4019-22. September 28.

Kumaresan, P., K. Boghesha, H. Tsuchiya, N.B. Vijaya Prakash and K. Kawakami (2002). "An Economic Analysis of CSR Hybrid Cocoon Production Under PPPBST Project in Karnataka", *Advances in Sericulture Research, Proceedings of The National Conference on Strategies for Sericulture Research and Development*. November 16-18. Mysore: Central Sericultural Research and Training Institute. pp.500-04.

Lipton, Michael and Richard Longhurst (1989). *New Seeds and Poor People*. Longon: Unwin Hyman Ltd.

Ministry of Textiles (2007). *Annual Report of Ministry of Textiles*. Government of India.

Naik Gopal and K.H. Babu (1993). *Demand and Supply Prospects for High Quality Silk*. New Delhi: Oxford & IBH Publishing Co. Pvt. Ltd.

Radhakrishnamenon, K., R. Subramaniam and S. Veerasamy (1980). "Time Lag in Adoption of Poultry Farming", *Indian Journal of Extension Education* 16(1/2): 59-60.

Ramakrishnan, S.R. (2001). "Bivoltine Sericulture: A Complementary Activity", *Global Silk Scenario-2001, Proceedings of the International Conference on Sericulture*. Calcutta: Oxford and IBH Publishing Co. pp.472-77.

Rogers (1962). *Diffusion of Innovations*. New York: Free Press.

Rose N. and P. Joskow (1990). "The Diffusion of New Technologies: Evidence from the Electric Utility Industry", *Rand Journal of Economics* 21(3): 354-73.

Ruttan, Vernon (1977). "The Green Revolution: Seven Generalizations", *International Development Review* 19: 16-23.

Shetty, N.S. (1966). "Agricultural Innovations: Leaders and Laggards," *Development Digest* (April): 11-17.

Sinha, Sanjay (1989). "Development Impact of Silk Production, A Wealth of Opportunities", *Economic and Political Weekly*: 157-63. January 21.

Stoneman, P. and P. Diederen (1994). "Technology Diffusion and Public Policy," *Economic Journal* 104(425): 918-30.

Thomas, Jacob, Arun Kumar, K.S. Reddy and A. Lalith (2005a). "Quantification of Relationship Between Silk Cocoon and Silk Yarn Prices: An Application of ARIMA Model", in K.S. Arun Kumar, R.K. Datta and G. Srinivas (eds.), *Reading in Sericulture Economics Marketing and Management*. Bangalore: Santhosh Creations.

Tikku, M.K. (1999). "Tangled Threads Silk Growers and Imports", *Economic and Political Weekly*: 578, March 6-13.

Vasumathi (2000). "An Analytical Study of the Silk Reeling Operations in Karnataka". PhD. Thesis. Bangalore: Department of Management Studies Indian Institute of Science.

World Bank (1997). *Implementation Completion Report*. National Sericulture Project. Report No. 17028.

8 Evaluation of Integrated Scheme of Oilseeds, Pulses, Oil Palm and Maize

Binod K. Das*

Introduction

During last two decades several challenges have surfaced in Indian agriculture. These relate to growth of output, efficiency, equity and sustainability. The biggest challenge is to reverse the sharp decline in growth rate of agriculture sector experienced after mid-1990s. The growth rate in agriculture turned lower than the growth rate in population implying that per capita income in agriculture is declining. Per capita income of the agriculture population was around half at ₹ 5,505 of the national per capita income at ₹ 11,433 during 1980 however, it came down to around 42 per cent at ₹ 6,652 of the national per capita income at ₹ 16,020 during 2000 (PHDCCI).[1] This is considered a major factor for large-scale rural distress in different parts of the country. The second biggest challenge is to ensure sustainable use of natural resources. The third challenge is the less performing oilseeds and pulses crops. It is unfortunate that per capita availability of pulses, an important source of proteins, has registered a downward trend. India produces a quarter of the world's total production of pulses, accounting for one-third of the total global acreage under pulses.

Indians consume 30 per cent of the world's pulses, but domestic production of pulses has not kept pace with population growth. Pulses production has grown at only one per cent CAGR (compound annual

---

* This evaluation study was undertaken by Amity Humanity Foundation, Amity University. Author was part of the study team while he was at Amity University.

1. Press Release, PHD Chamber of Commerce and Industry (*http://www.phdcci.in/admin/admin_logged/banner_images/1313574247.pdf*, accessed on December 15, 2012).

growth rate) from 1951–2008 compared to a population growth of about two per cent during the same period. Per capita domestic production of pulses has dropped from 63 g in 1951 to 36 g in 2008.[2] In respect of oilseeds, the country has to import large quantities of edible oil almost every year.

Although we have achieved self-sufficiency in case of wheat and rice, the production of oilseeds and pulses has generally remained stagnant in the last five years. The gap between demand and supply of edible oils and pulses has also widened due to increase in population. Considering the fact that only about 40 per cent of area under all crops is under irrigation and oilseeds, pulses and maize being the major crops of rainfed areas, these crops require special attention at Central and State levels.

The Department of Agriculture and Cooperation (DAC) was implementing the following centrally sponsored schemes under technology mission on oil seeds, pulses, maize and oil palm.

- Oilseed Production Programme (OPP).
- National Pulses Development Project (NPDP).
- Accelerated Maize Development Programme (AMDP).
- Oil Palm Development Programme (OPDP).

The above four schemes were modified and merged into one centrally sponsored integrated scheme of Oilseeds, Pulses, Oil Palm and Maize (ISOPOM) to provide flexibility to the states in implementation of these programmes based on regionally differentiated approach, promote crop diversification and provide focused approach to the programmes. The programme was commenced during the Tenth Five Year Plan in the year 2004-05.

Special Features of ISOPOM

The ISOPOM Scheme had following special features:

- Flexibility to the states to utilise the funds for the scheme/crop of their choice.

---

2. Media Kit: *http://www.growmorepulses.com/media_kit_pulses.pdf*, accessed on December 14, 2012.

- Annual action plan to be formulated by the state governments.
- Flexibility to the states for introducing innovative measures or any special component to the extent of 10 per cent of the financial allocation.
- Involvement of the private sector by the state governments in the implementation of the ISOPOM scheme with a financial cap of 15 per cent
- Flexibility for inter-component diversion of funds up to 20 per cent for non-seed components only.
- Diversion of funds from seed component to non-seed components with the prior approval of the DAC.

States and Crops Covered Under ISOPOM

The ISOPOM scheme was being implemented in 14 states namely, Andhra Pradesh, Bihar, Chhattisgarh, Gujarat, Haryana, Karnataka, Madhya Pradesh, Maharashtra, Odisha, Punjab, Rajasthan, Tamil Nadu, Uttar Pradesh and West Bengal, for oilseeds and pulses. All the districts under oilseeds and pulses programmes continued under Tenth Plan barring Jammu and Kashmir and Himachal Pradesh. Under oil palm, programme was taken in 12 states namely, Andhra Pradesh, Karnataka, Maharashtra, West Bengal, Tamil Nadu, Gujarat, Goa, Odisha, Kerala, Assam, Tripura and Mizoram. In 2010, pulses programme was implemented in 14 states, maize in 15 states and oil palm in 8 states. With the merger of pulses component of ISOPOM with NFSM-Pulses, the pulses component of ISOPOM would be winded up.

The ISOPOM scheme received three types of input and support services in case of oilseed, pulses and maize which were: (i) improved seed component, (ii) transfer of technology, and (iii) capacity building and publicity support. The states had operational flexibilities in terms of selection of crops, districts, inter-component adjustments and even addition of new components, considering the fact that inadequate availability of seed of improved varieties of these crops continued to be a major constraint. More emphasis was to be given to these components to improve the quality seed availability. The ISOPOM scheme completed three years of its implementation from 2004-05 to 2006-07 during the Tenth Plan Period. An impact evaluation study was undertaken in late 2007 during the tenth plan period.

Data and Methodology

The study covered 24 states producing one or more ISOPOM crops. Specifically, 14 oilseed and pulses producing states, 15 maize states and 12 oil palm states were studied. Many states had taken multiple crops.

Farmers' survey was conducted in two districts of each ISOPOM state. Further, two blocks in each district and five villages in each block were selected. Thus 24 states, 42 districts, 84 blocks and 420 villages were covered under the study. Moreover, 7250 farmers were interviewed during the survey. On an average, 170 farmers in a district or 17 farmers in a village were interviewed. Districts were selected on the basis of coverage of ISOPOM crops and non-contiguous locations. The identification of blocks and villages was left to state government on pragmatic considerations and suitable local conditions. Despite efforts hardly three to four per cent female farmers could be identified for survey. This indicates farming is still a male prerogative. Similarly, many ST category farmers could not be identified. Majority (39.9%) interviewed were small farmers holding the land between 2.5 to 5 acres each. The effectiveness of ISOPOM scheme was assessed through the following steps:

- The factors, which determined the success of the ISOPOM schemes for better yield and production of oilseeds, pulses, oil palm and maize etc.
- Availability of the agriculture input, irrigation, agriculture know-how, market support and state government support etc.
- Comprehension of the messages/programme covered under ISOPOM scheme.

Results and Discussion

Oilseeds Cultivation

The major oilseeds cultivated in India consist of edible and non-edible groups. The edible group consists of groundnut, soybean, rapeseed, mustard, sunflower, sesame, safflower and Niger. Non-edible group consists of castor and linseed. Groundnut, soybean and rapeseed and mustard together contribute about 85 per cent of the country's oilseeds production. While production of oilseeds has risen over the years, it has not been able to keep pace with the domestic demand for edible oil. Therefore, there has been a persistent gap between domestic demand and output of edible oil. In order to bridge this gap, a centrally sponsored Oilseeds Production Programme (OPP) was implemented in 28 states. From 2004-05, this programme was brought under the aegis of ISOPOM with a view to enhance the production and productivity of oilseeds in the country. In last four years (2006-2010) the average oil seed production has hovered around 27.01 million MT, area cultivated 26.78 million hectare and productivity/yield 1005 kg per hectare. In a recent note, oilseed production of 219.22 lakh tonnes in the year 2010-11 is highest in the decade.[3] The year wise production is described in the following matrix.

Table 8.1

Oil Seed Production

Year	*Area (Million Ha)*	*Production (Million MT)*	*Yield (Kg/Ha)*
2006-07	26.15	24.28	916
2007-08	26.69	29.75	1115
2008-09	27.56	27.72	1006
2009-10	26.75	26.32	984
Average	26.78	27.01	1005

Source: Annual Reports of Directorate of Oilseeds. Government of India (2007). *Agricultural Statistics at a Glance 2006*. New Delhi: Department of Agriculture and Cooperation, Government of India.

3. *Economic and Political Weekly* Vol XLVII (42): 111, Statistics, October 20, 2012.

Pulses Cultivation

During the period of five years (2001-02 to 2005-06), pulses have grown in an area of 22.22 million hectare. Similarly, production of 13.18 million tonnes and productivity around 593 kg per hectare was noticed. The area under pulses mostly remained constant in said period. However, there has been improvement in productivity from 472 kg per hectare in first plan (1951-56) to 598 kg per hectare during 2005-06.[4]

There have been not many fluctuations in area, production and productivity from year 2001-02 to 2005-06. Area of cultivation, production and productivity hovered around 22 million hectare, 13 million MT and 600 kg per hectare respectively in the said period. The area under pulses is almost constant compared to other food crops like wheat and paddy. The pulses production of 7.12 million tonnes during *kharif* season of the year 2010-11 is highest in the decade. Since the year 2004-05, the productivity has increased from 577 kg per hectare in year 2004-05 to 689 kg per hectare in 2010-11.[5]

Maize Cultivation

There has been an eight-fold increase in production and nearly four-fold increase in productivity of maize since independence. In India, approximately 10 million tonnes of maize is produced annually. According to statistics of 2005-06, the cultivated area was 75.90 lakh hectares, production is 147.10 lakh MT and productivity is 1938 kg per hectare.[6]

With increase in area under the crop, the research for new hybrids is undertaken by private seed companies. This has helped farmers in two ways augmenting supply of seeds to the farmers and enhancement of seed quality. The average yield per hectare of maize is 20 quintals. However, some farmers harvest maize to the extent of 60 to 70 quintals per hectare.

---

4. Directorate of Economics and Statistics, Dept of Agriculture and Cooperation, Govt of India, 2012.
5. Integrated Scheme of Oilseeds, Pulses, Maize and Oil palm (ISOPOM): *http://www.advanceagriculturalpractice.in/w/images/a/a5/Integrated_Scheme_of_Oil_Seeds%2C_Pulses%2C_Maize_and_Oil_Palm_%28ISOPOM%29.pdf*, accessed in October 2012
6. *Annual Report of Directorate of Oilseeds Development (2011-12)*, Hyderabad.

Average productivity in Andhra Pradesh has been recorded at 40 quintals per hectare. Farmers have been encouraged to rotate maize with groundnut, red gram and sunflower crops, and intercrop the same with short duration crops such as pulses and oilseeds for better and sustainable yields. Ever increasing demand for maize from the domestic poultry industry and low cultivation costs has attracted maize cultivation. Exporters too have increased their purchases to ship maize to Australia. India exports nearly five million tonnes of maize annually. The maize production of 16.64 million tonnes in the year 2010-11 is highest in the decade.

Oil Palm Cultivation

Oil palm was initially introduced in Kerala and Andaman and Nicobar Islands. Since 1992, under the aegis of TMOP&M (Technology Mission on Oilseeds, Pulses and Maize), cultivation of oil palm was extended through the centrally sponsored scheme—OPDP (Oil Palm Development Programme) to other potential areas in the states of Andhra Pradesh, Karnataka, Tamil Nadu, Odisha, Gujarat and Goa. When the plantation was started in 1992-93, the coverage under oil palm went up rapidly. The trend continued for a few years. But, when extractors took time to set up processing units; and oil palm fruiting was ripe for harvesting, farmers were discouraged for additional area cultivation. However, after the extractors launched their units and commenced procuring oil palm, farmers were encouraged for more oil palm cultivation. During the year (2006-07), highest area (about 29000 ha) was freshly brought under the crop ever since its introduction.[7] Besides, it was observed that the uprooting of oil palm plantation had come to a halt for the first time in 2006-07. While analysing the area of cultivation, it was found that country has achieved 62.3 per cent of targeted figures. While in year 2009-10, the achievement in area of cultivation was 57.16 per cent.[8]

---

7. National Food Security Mission: Pulses Operational Guidelines. *http://www.indg.in/agriculture/rural-employment-schemes/national_food_security_mission-pulses-operational_guidelines.pdf*, accessed in October 2012
8. Approved Changes in ISOPOM for 11th Plan. *http://agricoop.nic.in/tmop&m/5Modification.pdf*, accessed in October 2012

Comparative Economics of Oil Palm Cultivation

Oil palm accrues more income as compared to other competing crops such as paddy, tapioca, sugarcane, coconut and banana. Even though the farmer has no income from oil palm in first three years, fourth year onwards, oil palm delivers good income to compensate initial investments. The net income continues increasing thereafter because of low cost of cultivation and increased yield of fresh fruit bunches (FFB); whereas in case of other crops the income remains stagnant. It is therefore vital to substitute oil palm plantation with other competing crops, if situation permits. Comparative economic analysis of these crops is analysed in Table 8.2.

Table 8.2

Estimated Profits in Major Crops for Ten Years (2003–2013)

Net Income (₹ per Hectare)

Crops	*1st Year*	*2nd Year*	*3rd Year*	*4th Year*	*5th Year*	*6th Year*	*7th Year*	*8th Year*	*9th Year*	*10th Year*
Paddy	10652	10652	10652	10652	10652	10652	10652	10652	10652	10652
Tapioca	16113	16113	16113	16113	16113	16113	16113	16113	16113	16113
Sugarcane	39087	48229	43407	39040	39087	48229	43407	39040	39087	48229
Coconut	-30836	-7050	-7050	5050	14450	25450	38598	38598	38598	38598
Banana	35259	49050	59155	35259	49050	59155	35259	49050	59155	35259
Oil Palm	-19656	-3650	12256	52750	58750	67850	76485	76485	76485	76485

Note: 1. The figures have been estimated on the basis of input cost of cultivation. Net income has been estimated after deduction of cultivation cost from gross income

2. First three years figures (2003-2006) are based on actual cultivation cost obtained from Ministry of Agriculture (GoI) and figures of remaining are estimated.

Assumptions:

1. Two paddy crops are harvested in a year
2. In case of sugarcane, three ratoons are supped to be taken after first year of plantation (1+3 = 4 years)
3. In case of banana, new plantation after every three years.
4. Coconut and oil palm plantations are expected to remain productive beyond 20 years.
5. There will not be drastic variations in productivity of crops because of agro-climatic conditions.
6. Unit price realisation, unit cost of labour and other inputs are assumed to remain constant.
7. Government subsidy and taxation policy will be unchanged.

Source: Author's estimation of figures.

Figure 8.1

Estimated Net Income for Ten Years

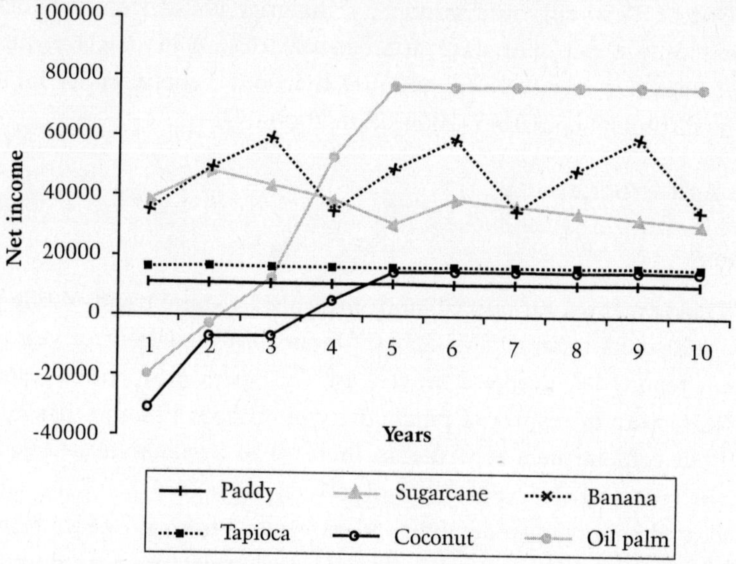

Findings from Field Data

The study had focused on small and marginal farmers depending on landholding status. Data were collected from the farmers on five major crops grown by them during last five years. This gave the information of various varieties used by the farmers and the areas sown for each crop. The survey information included cropping pattern for ISOPOM crops during last three years and reasons for fluctuations in area, production and productivity thereof, if any. The information was also collected on the cost of cultivation of major ISOPOM crops in recent years.

Altogether 7250 farmers were surveyed under the study. This included 6680 men and 570 women farmers. There was nearly eight per cent representation of woman farmers in the survey. Due care was taken to include SC/ST farmers in the survey. Out of 7250 farmers surveyed, 16 per cent belonged to SC category, 6 per cent to ST and remaining 78 per cent belonged to other category. Their representation may not be in proportion to their population but an effort was made to maximise their involvement in the study.

Out of 7250 surveyed farmers, a majority (94%) were beneficiaries whereas only 6 per cent were non-beneficiaries. A beneficiary means a farmer who has received one or more than one benefit under ISOPOM scheme during its operation (2004-05 to 2006-07).

Area and Productivity

Oilseeds

The area sown for oilseed crops presented a static trend. While 55.35 per cent farmers reported "no change" in area during last three years, 6.21 per cent reported a "decrease" in area and 7.31 per cent reported "increase" in area. Picture in respect of productivity of oilseeds is somewhat brighter. 26.19 per cent farmers reported an increase in productivity; 14 per cent farmers reported decrease in productivity and 26.6 per cent farmers viewed a constant in productivity. So as evident from the data, nearly 80 per cent farmers cultivating oilseeds reported constant in production. In Chhattisgarh majority of farmers indicated increasing trend in area for oilseed crops. Haryana led in the decreasing trend for area sown under oilseed crops.

Productivity in oilseeds presented a mixed result. Nearly, 41 per cent farmers reported a constant trend in productivity of oilseeds, 39 per cent indicated an increase in productivity. Almost 20 per cent of the farmers reported a decreasing trend in the productivity of the crop. In West Bengal, 89 per cent of farmers reported increasing trend in oilseed productivity; whereas in Haryana, Bihar and UP, declining trend in productivity was reported by a sizable proportion of farmers.

Pulses

Similar to oilseeds, the area sown under pulses represented a static trend. Twenty-one per cent farmers reported an increase in coverage, 7 per cent informed a decrease in area and 73 per cent accounted no change. Though most of the states viewed a static trend in area sown in pulses, Chhattisgarh, Maharashtra, UP and West Bengal had significant number of farmers with increased area of cultivation. Many farmers in the state of Maharashtra, UP and West Bengal reported decrease in pulses cultivation.

So far as productivity of pulses is concerned, number of farmers reporting an increase in productivity was two-and-half times higher than those reporting a decrease? So productivity in pulses demonstrated an increasing trend with 43 per cent of farmers informed about increasing productivity. Contrarily, 39 per cent farmers replied a constant trend in productivity of pulses. About 18 per cent of the farmers revealed a decreasing trend in the productivity of pulses. Karnataka, Madhya Pradesh, Chhattisgarh, Maharashtra and West Bengal were leading states in increased productivity of pulses. Bihar and Uttar Pradesh had more farmers with decreasing productivity in pulses.

Maize

Most of the farmers informed a constant trend in the area sown under maize cultivation. Nearly 74 per cent farmers indicated that their area of cultivation in maize remained constant over the years. Nearly equal number of farmers had shown increasing and decreasing trends in the area sown in maize crop. West Bengal demonstrated an increasing trend in area under maize cultivation; whereas Jammu and Kashmir confirmed a declining trend in the area of maize cultivation.

Productivity in maize had shown quite encouraging results with nearly 53 per cent farmers indicated an increasing trend and about 33 per cent farmers confirmed a constant trend. Only 14 per cent farmers had shown a declining trend in maize productivity. West Bengal, Punjab, Himachal, Jammu and Kashmir and Bihar were the leading states with increasing trend in the productivity of maize. Maharashtra and Punjab indicated a declining trend in the productivity of maize.

Oil Palm

Thirteen per cent farmers from the oil palm growing states reported an increase in area of cultivation under oil palm, while 87 per cent reported it constant over the years. Regarding the productivity, 12 per cent farmers reported an increase in productivity and majority of the farmers reported it constant. Only 3 per cent farmers reported a decrease in productivity. Farmers expected around ₹ 7-10 per kg for oil palm crops.

Reasons for Low Productivity

The farmers were asked about the reasons for the fluctuations in the area sown and production. Famers cited reasons such as erratic rainfall (72%), pests and diseases (14%), low fertiliser use, late sowing, low fertility of soil, etc., for low productivity. The state-wise production of ISOPOM crops is described in Annexure A-8.1.

Feedback on ISOPOM Components

Input Supply

Famers were provided planting materials, seed mini-kits, plant protection chemicals (PPC), weedicide and bio pesticide (WBP), plant protection equipment (PPE), nuclear polyhydrosis virus (NPV), sprinkler sets, drip irrigation sets, irrigation pipes, diesel pumps, fertilisers/bio-fertilisers etc. Nearly, 75 per cent farmers acknowledged receipt of seed mini-kits and 39 per cent of famers endorsed receipt of PPC/WBP.

Seed Usage

Overwhelming 91 per cent farmers reported use of certified seeds. Again 22 per cent informed about seed multiplication by themselves. Some farmers used both certified and uncertified seeds. In case of certified seeds, 74 per cent farmers reported its germination to be over 80 per cent.

Suggestions to Constrain

Many suggestions have emerged from the farmers, agriculture officers and technical directorates. The suggestions are as under.

Firstly, breeder seed indents may be scrutinised before final indents are placed with the suppliers by State Agriculture Departments as well as Ministry of Agriculture to ensure the following:

- Whether indent quantity is sufficient to meet the requirement of the certified seed to cover the targeted level of seed replacement ratio of the individual crops of oilseeds and pulses?

- Whether renovation of the existing old varieties is required, if farmers still demand for the old varieties?

Secondly, foundation seed and certified seed production may be taken up only on lands where assured irrigation facilities are available. Thirdly, the seeds approved by agriculture universities and ICAR institutions are more popular in the regions in which these institutions are located. A higher priority in seed distribution under ISOPOM should be accorded to aforesaid varieties on local/regional basis. Fourthly, the level of subsidy on production of foundation seeds and certified seeds was fixed in 2004-05. In the Twelfth Five Year Plan the level of subsidy should be examined; since the seed plays the most crucial role in increasing production/productivity, an upward revision is recommended. Fifthly, crash programme for quality seed production for oilseeds and pulses continued during Twelfth Plan. According to present provision, state governments were required to submit variety-wise requirement of seed production under seed programme well in advance. Subsidy was also envisaged at ₹ 500 to seed producer and handling agency taken together. Under this component no progress was reported. It is suggested that the Ministry of Agriculture may have a dialogue with the seed producing agencies, so that required quantities of quality seeds are ensured for farmers. Sixthly, sunflower seed mini-kits should be supplied only during *rabi* season and northern and eastern states during *kharif* season. Seventhly, integrated pest management (IPM) demonstrations may be clubbed with the block level demonstrations so that the farmers can see these two demonstrations at one place. Eighthly, supply of micronutrients was dropped as a component when ISOPOM was launched in 2004-05. Since oilseeds and pulses are mostly grown on marginal lands with poor irrigation facilities, role of nutrients require pilot study and modifications in guidelines thereafter, if warranted. Finally, in the past, special projects were undertaken for increasing the production of the groundnut in Gujarat and soybean in Madhya Pradesh. Similar specific projects may be experimented for oilseeds and pulses to achieve quick and desired results.

Oil Palm

Firstly, oil palm cultivation has excellent scope; oil palm growing states may be encouraged to bring more area under the crop through implementation of ISOPOM. Secondly, Mizoram, Odisha, Gujarat and

Maharashtra have huge potential to expand oil palm plantation, but still not crossed 5,000 hectare mark. They should be encouraged to undertake additional plantation. Thirdly, demand for planting material is increasing every year. Efforts are being made to meet this demand by raising seedlings in the country itself. Seedlings are also being imported from other countries to meet the demand. Efforts may be made to produce the seedlings through tissue culture, so that the sufficient planting materials can be ensured for farmers. Fourthly, there are certain oil palm areas where plantations are in fruiting stage but arrangement for procurement of FFBs (fresh fruit bunches) does not exist. The farmers having waited for long years with all their investments find themselves in a fix. In such cases, crash programme of procurement of FFBs should be undertaken. Such areas have to be temporarily tagged to a nearby processing plant and incentives given to the farmers to sell FFBs to the plants. Finally, remunerative prices to the growers of oil palm are *sine qua non* for sustainable growth in oil palm cultivation. If the prices crash to abnormal levels a mechanism for procurement at reasonable prices may have to be put in place.

Changes in ISOPOM during Eleventh Plan

Ministry of Agriculture has approved certain changes in the scheme during Eleventh Plan. First, direct funding to ICAR, State Agriculture Universities and international research institutions for need based contract research. Second, assistance has been enhanced to ₹ 1000 per quintal production of foundation and certified seed of oilseeds, pulses and maize. NAFED and IFFCO are to be eligible for assistance. Third, assistance for infrastructure development to agriculture universities and Krishi Vigyan Kendras having farm facilities for development of irrigation facilities and threshing floors at seed farms and storage space for seed of oilseeds, pulses and maize. Fourth, the other components are supply of improved farm implements, supply of micronutrients in deficient areas. Fifth, improved assistance for distribution of gypsum pyrite/liming/dolomite, supply of rhyzobium culture/phosphate solubilising bacteria, block demonstrations of polythene mulch technology in groundnut.[9]

9. Component-wise Operational Guideline for ISOPOM, 2004-05; *http://india.gov.in/allimpfrms/alldocs/9097.pdf, accessed in October 2012 k*

Restructuring of Scheme during Twelfth Five Year Plan

Oilseeds, Pulses and Maize

Firstly, a few states have recommended use of weedicide in maize crops on the pattern of dwarf/high yielding varieties of wheat. In the Twelfth Plan, demonstration on the use of weedicide in maize crops is recommended. Secondly, in oilseeds and pulses, availability of certified seeds is not adequate to meet the requirements. The deficit is met by purchase of seeds from the open market or distribution of so called truthfully labelled seeds by public agencies. The prevailing system is harmful for farmers' interest. Tightening the procedure for truthfully labelling of seed is recommended.

In view of unreliable nature of seeds available in open market, public agencies should gradually phase out distribution of truthfully labelled seeds. In case of emergencies like floods, rigorous procedure should be prescribed before public agencies for procurement of seeds from the market. Thirdly, quantity and time of import of edible oils and pulses ought to be decided carefully after gauging the shortage. Import should not affect producers. Fourthly, a subsidy limit of ₹ 800 per quintal on distribution of certified seed is inadequate in case of groundnut and gypsum. Few states have approached the Government of India on this issue. This may have to be examined while deciding the pattern of subsidy in the Twelfth Plan. Finally, support prices for oilseeds and pulses are fixed every year by the Ministry of Agriculture. NAFED may ensure procurement of oilseeds and pulses in case the prices rule below the support price.[10]

Oil Palm

Firstly, a floor price of FFBs in the Twelfth Plan is strongly recommended to assure the existing/prospective growers that their investment will not be wasted even after waiting four to five years. Considering this, each state may continue to fix price, on a quarterly basis, as per the existing mechanism. Secondly, price so fixed in each state and revised every quarter, if necessary, should be enforced rigidly by direct and indirect instruments. Thirdly, additional potential areas were identified by the committee

10. Government of India, *Annual Report (2006-07)*, Department of Agriculture and Cooperation, Ministry of Agriculture, Government of India

constituted by the Ministry of Agriculture under the chairmanship of Dr K.L. Chadha. The identified states may be persuaded to take up these areas for oil palm plantation under ISOPOM scheme. Fourthly, indigenous production of sprouts and planting material of oil palm has to go up in the light of Chadha committee report. Lastly, a national level workshop on oil palm is recommended in Andhra Pradesh or Karnataka to deliberate on the issues raised and build up consensus amongst oil palm states. Processing companies can also be represented in the workshop.

Conclusion

Recently, ISOPOM scheme has been merged with National Food Security Mission (NFSM). With the merger of pulses component of ISOPOM with NFSM-Pulses and coverage of all districts of 14 states under NFSM-Pulses, the pulses component of ISOPOM would cease to operate altogether in the country. All other states would receive financial assistance for pulse development under the Macro Management Mode of Agriculture (MMA). Moreover, an accelerated pulse production programme has been launched recently for integrated development of 60,000 pulses and oilseed villages in rainfed areas of the country. A trend analysis of agriculture production of last one decade showed that production of ISOPOM crops was highest in the year 2010-11.[11] Notwithstanding, production in last two years presented a negative trend. Hopefully, the interventions under NFSM will arrest the negative trend.

The study has opened a few research questions for further investigation. Since oilseeds and pulses are mostly grown on marginal lands with poor irrigation facilities, role of nutrients require pilot study and modifications in guidelines thereafter, if any. As some states have recommended more areas to be covered under pulses under zero tillage system, research may be undertaken on zero tillage system to identify its efficacy. Documentation of successful case studies will be useful for policy makers. Many farmers have cited erratic rainfall or no clue about the cause of low productivity of oilseeds and pulses. These need to be investigated and substantiated. Despite so much support to the farmers, the area of production of oilseeds

11. Annual Reports of Directorate of Oilseeds, Govt of India, 2005-06, 2006-07, 2007-08, 2008-09, 2009-10

and pulses is constant. These need to be researched further. The economies of oil palm in relation to other major crops both in irrigated and non-irrigated conditions must be analysed to measure profitability of the ISOPOM crops.

Notes:

a. Pulses are basically grain legumes. They occupy an important place in human nutrition due to their high protein content than cereal grains. In Indian dietary regime it occupies an important place. Since majority of Indians are vegetarians, they depend largely on grain legumes (pulses) for their dietary protein. They are also a fairly good source of vitamins like thiamine, niacin, riboflavin and iron.

b. Oil palm, a high oil-yielding crop, augurs well in the process of augmenting vegetable oil production in the country. It generates sustainable income over a period of 25-30 years thereby helping in raising socioeconomic status of the growers.

c. A detailed questionnaire was used to obtain relevant information from farmers in the ISOPOM states. The data collected pertained to personal profile, family profile, survey on four ISOPOM crops—oilseeds, pulses, maize and oil palm, feedback on ISOPOM components etc. Two districts were covered in each of ISOPOM states. Two blocks in each district with five villages in each block were selected. On an average 17 farmers were surveyed in a sampled village.

Annexure A-8.1

Evaluation of Integrated Scheme of Oilseeds, Pulses, Oil Palm and Maize

Oil Seed/Pulses/Maize: Area/Productivity, Number and Percentage of Farmers

Sl No	State	Oil seeds											
		Area						Productivity					
1	2	3	4	5	6	7	8	9	10	11	12	13	14
		Increasing	Increasing %	Decreasing	Decreasing %	Constant Number	Constant %	Increasing Number	Increasing %	Decreasing Number	Decreasing %	Constant	Constant %
1	Andhra Pradesh	2	0.4	1	0.2	160	4.0	104	5.5	22	2.2	44	2.2
2	Bihar	7	1.3	8	1.8	234	5.8	47	2.5	130	12.7	73	3.7
3	Chhattisgarh	115	21.7	5	1.1	377	9.4	209	11.0	30	2.9	244	12.2
4	Gujarat	64	12.1	127	28.1	275	6.9	112	5.9	173	16.9	180	9.0
5	Haryana	84	15.8	103	22.8	211	5.3	70	3.7	163	15.9	162	8.1
6	Himachal Pradesh	NA	NA	NA	NA	NA	NA	NA	NA	NA	NA	NA	NA
7	Jammu & Kashmir	NA	NA	NA	NA	NA	NA	NA	NA	NA	NA	NA	NA
8	Karnataka	2	0.4	0	0.0	289	7.2	218	11.5	33	3.2	43	2.2
9	Madhya Pradesh	26	4.9	5	1.1	281	7.0	202	10.6	30	2.9	58	2.9
10	Maharashtra	65	12.3	25	5.5	153	3.8	141	7.4	50	4.9	52	2.6
11	Odisha	19	3.6	20	4.4	214	5.3	70	3.7	77	7.5	108	5.4
12	Punjab	32	6.0	21	4.6	340	8.5	97	5.1	56	5.5	228	11.4
13	Rajasthan	45	8.5	96	21.2	291	7.3	185	9.7	110	10.8	135	6.8
14	Tamil Nadu	11	2.1	7	1.5	176	4.4	70	3.7	21	2.1	80	4.0
15	Uttar Pradesh	32	6.0	32	7.1	267	6.7	38	2.0	118	11.5	159	8.0
16	West Bengal	26	4.9	2	0.4	352	8.8	336	17.7	9	0.9	34	1.7
17	Assam	0	0.0	0	0.0	392	9.8	0	0.0	0	0.0	392	19.7
18	Goa	NA	NA	NA	NA	NA	NA	NA	NA	NA	NA	NA	NA
19	Kerala	NA	NA	NA	NA	NA	NA	NA	NA	NA	NA	NA	NA
20	Mizoram	NA	NA	NA	NA	NA	NA	NA	NA	NA	NA	NA	NA
21	Tripura	NA	NA	NA	NA	NA	NA	NA	NA	NA	NA	NA	NA
	Grand Total	530	100.0	452	100.0	4012	100.0	1899	100.0	1022	100.0	1992	100.0

contd...

...contd...

Sl No	*State*	*Pulses*											
		Area						*Productivity*					
1	*2*	*15 Increasing Number*	*16 Increasing %*	*17 Decreasing Number*	*18 Decreasing %*	*19 Constant Number*	*20 Constant %*	*21 Increasing Number*	*22 Increasing %*	*23 Decreasing Number*	*24 Decreasing %*	*25 Constant Number*	*26 Constant %*
1	Andhra Pradesh	1	0.1	0	0.0	149	5.8	81	5.3	26	4.0	43	3.2
2	Bihar	19	2.7	13	5.2	231	9.0	70	4.6	98	15.0	95	7.0
3	Chhattisgarh	87	12.2	15	6.0	349	13.6	205	13.5	36	5.5	212	15.5
4	Gujarat	2	0.3	8	3.2	126	4.9	6	0.4	4	0.6	127	9.3
5	Haryana	14	2.0	12	4.8	45	1.8	5	0.3	13	2.0	45	3.3
6	Himachal Pradesh	NA	NA	NA	NA	NA	NA	NA	NA	NA	NA	NA	NA
7	Jammu & Kashmir	NA	NA	NA	NA	NA	NA	NA	NA	NA	NA	NA	NA
8	Karnataka	0	0.0	1	0.4	231	9.0	169	11.2	33	5.1	40	2.9
9	Madhya Pradesh	27	3.8	5	2.0	251	9.8	171	11.3	39	6.0	63	4.6
10	Maharashtra	44	6.1	45	18.1	258	10.0	181	11.9	49	7.5	116	8.5
11	Odisha	15	2.1	3	1.2	124	4.8	34	2.2	30	4.6	78	5.7
12	Punjab	1	0.1	0	0.0	61	2.4	20	1.3	8	1.2	35	2.6
13	Rajasthan	24	3.4	33	13.3	65	2.5	50	3.3	17	2.6	56	4.1
14	Tamil Nadu	3	0.4	2	0.8	134	5.2	38	2.5	12	1.8	89	6.5
15	Uttar Pradesh	44	6.1	46	18.5	274	10.7	52	3.4	128	19.6	186	13.6
16	West Bengal	42	5.9	65	26.2	269	10.5	229	15.1	57	8.7	90	6.6
17	Assam	393	54.9	0	0.0	2	0.1	204	13.5	102	15.6	89	6.5
18	Goa	NA	NA	NA	NA	NA	NA	NA	NA	NA	NA	NA	NA
19	Kerala	NA	NA	NA	NA	NA	NA	NA	NA	NA	NA	NA	NA
20	Mizoram	NA	NA	NA	NA	NA	NA	NA	NA	NA	NA	NA	NA
21	Tripura	NA	NA	NA	NA	NA	NA	NA	NA	NA	NA	NA	NA
	Grand Total	716	100.0	248	100.0	2569	100.0	1515	100.0	652	100.0	1364	100.0

contd...

...contd...

Sl No	State	Maize											
		Area						Productivity					
1	2	27 Increasing Number	28 Increasing %	29 Decreasing Number	30 Decreasing %	31 Constant Number	32 Constant %	33 Increasing Number	34 Increasing %	35 Decreasing Number	36 Decreasing %	37 Constant Number	38 Constant %
1	Andhra Pradesh	0	0.0	0	0.0	10	0.5	7	0.5	1	0.3	2	0.2
2	Bihar	2	0.6	2	0.6	203	10.0	102	7.0	64	16.5	46	5.1
3	Chhattisgarh	35	10.1	6	1.7	88	4.3	48	3.3	11	2.8	69	7.6
4	Gujarat	17	4.9	24	6.9	87	4.3	30	2.1	42	10.8	59	6.5
5	Haryana	NA	NA	NA	NA	NA	NA	NA	NA	NA	NA	NA	NA
6	Himachal Pradesh	49	14.1	39	11.1	323	15.9	269	18.4	40	10.3	110	12.2
7	Jammu & Kashmir	14	4.0	203	58.0	209	10.3	167	11.4	12	3.1	247	27.3
8	Karnataka	2	0.6	0	0.0	76	3.7	64	4.4	2	0.5	12	1.3
9	Madhya Pradesh	0	0.0	0	0.0	7	0.3	3	0.2	2	0.5	2	0.2
10	Maharashtra	19	5.5	12	3.4	75	3.7	68	4.7	23	5.9	13	1.4
11	Odisha	9	2.6	1	0.3	78	3.8	44	3.0	18	4.6	26	2.9
12	Punjab	73	21.0	27	7.7	265	13.0	189	13.0	75	19.3	101	11.2
13	Rajasthan	18	5.2	16	4.6	212	10.4	133	9.1	58	14.9	64	7.1
14	Tamil Nadu	0	0.0	2	0.6	107	5.3	37	2.5	5	1.3	67	7.4
15	Uttar Pradesh	9	2.6	6	1.7	32	1.6	8	0.5	13	3.4	27	3.0
16	West Bengal	100	28.7	12	3.4	261	12.8	290	19.9	21	5.4	59	6.5
17	Assam	1	0.3	0	0.0	0	0.0	0	0.0	1	0.3	0	0.0
18	Goa	NA	NA	NA	NA	NA	NA	NA	NA	NA	NA	NA	NA
19	Kerala	NA	NA	NA	NA	NA	NA	NA	NA	NA	NA	NA	NA
20	Mizoram	NA	NA	NA	NA	NA	NA	NA	NA	NA	NA	NA	NA
21	Tripura	NA	NA	NA	NA	NA	NA	NA	NA	NA	NA	NA	NA
	Grand Total	348	100.0	350	100.0	2033	100.0	1459	100.0	388	100.0	904	100.0

Emerging Trends in Agriculture and Food Security

9 Agrarian Crisis and Food Security in India

S.S. Kalamkar[1]

Introduction

Agriculture is often described as the backbone of the Indian economy. On the eve of independence, the agricultural sector was the prominent sector of the Indian economy both in terms of its contribution (approximately 55%) to the gross domestic product (GDP) and in providing employment to the country's labour force (about 70%). Thus, the fortunes of a large majority of people in India were basically linked with agricultural performance (Bhalla, 2007). Besides, due to deep-rooted poverty, rapidly growing population, low agricultural productivity and resultant food and nutritional insecurity during early independence periods, country had to give high priority to make our population food secure which would in turn mean economic growth and reduce poverty. Therefore, agricultural policy focus in India across decades has been on self-sufficiency and self-reliance in food grains production. Indian agriculture has made great strides in the course of the last half a century of planned development. The grit and toil of Indian farmers has greatly contributed in transforming Indian agriculture from a moribund state at the time of independence to a resilient production system of a food secure nation. Food grains production rose from 52 million tonnes (mt) in 1951-52 to 250.42 mt in 2011-12. The share of agriculture in real GDP has fallen given its lower growth rate relative to industry and services. However, what is of concern is that growth in the agricultural sector has quite often fallen short of the Plan targets. Though contribution

1. Author thanks seminar participants and anonymous referee for their constructive comments and suggestions. Usual disclaimer applies.

of agriculture in GDP has registered a steady decline from 36.4 per cent in 1982-83 to around 13.9 per cent in 2011-12, yet this sector continues to support more than half a billion people providing employment to 52 per cent of the workforce. Farming is the principal means of livelihood for about 65 per cent of India's population (NCF, 2005). Agricultural sector also contributed about 10.5 per cent of national exports in 2010-11 (GoI, 2012a). Agricultural growth plays an important role in achieving certain national goals, such as reducing rural poverty, providing food and nutritional security, supplying raw materials to major industries such as textiles, earning foreign exchange, etc. It is proved beyond doubt that agricultural growth also significantly impacts on the reduction of rural poverty, through increased employment opportunities and wage rates in India.[2] Increased agricultural production always makes dent on the prices of agricultural commodities, which benefit millions of landless rural and urban consumers. Research also suggests that growth in agricultural sector also impacts substantially on the growth of industrial sector (Rangarajan, 1982). Thus, agricultural sector not only plays an important role in improving the growth of rural economy but also the overall growth of the economy in India and therefore, prosperity of the rural economy in India is closely linked to agriculture and allied activities.

During the last five decades, the performance of Indian agriculture has been characterised by large fluctuations; there were phases of significant growth and stagnation (Sawant, 1983; Sawant and Achuthan, 1995). Experience so far indicates that a few bright spots are often followed close on the heels by a spate of distress (Deshpande and Arora, 2010). Agriculture, the most important sector in India economy, is presently at a crossroad and passing through a period of severe crisis. Despite the beneficial effects of several programmes and policy initiatives launched by the central and the state governments, the economic conditions of the farmers in many areas have not improved much due to inadequate coverage of all regions and sections of farmers, and lack of targeting. Infact, the agricultural sector of India has been facing one of the worst crises ever since the independence.

---

2. See, Ahluwalia, 2000; Narayanamoorthy, 2001; Saleth, *et al.*, 2003; Shah and Singh, 2004; Narayanamoorthy and Deshpande, 2003; World Bank, 2006.

Although certain features of the crisis such as deceleration in growth rates, declining share of agriculture in GDP and falling public investment in agriculture, started manifesting themselves in certain parts of India during the late 1980s, the crisis assumed a serious dimension since the middle of the 1990s. While the overall growth of the economy has been in the range of seven to nine per cent since the introduction of economic reforms, the agricultural growth is almost stagnant or decelerating over last one decade or so. While the annual compound growth rate of GDP during the period 1992-93 to 2005-06 was 6 per cent, the growth rate of the agricultural sector during the corresponding period was 2.57 per cent. In contrast, the growth rate in the service sector was 7.72 per cent per annum. There was sharp decline in growth rate of yield of all crops during the period 1990-91 to 2003-04 as compared to the period 1980-81 to 1990-91. The share of public sector investment in agriculture which was 43 per cent in 1980-81 declined to 15.5 per cent in 2002-03 and then marginally increased to around 20 per cent in 2008-09 (GoI, 2012a). Another important manifestation of the crisis in agriculture was the stagnant if not deteriorating terms of trade for agriculture (GoI, 2007a).

One of the tragic manifestations of the crisis is the large number of suicides committed by the farmers in some parts of India, occurred commonly both in developed and underdeveloped regions, though the intensity is found to be higher in the rainfed regions. The Situation Assessment Surveys of the National Sample Survey Organisation (NSSO, 2005; 2005a; 2005b) reconfirmed the worsening situation of farming households which indicated that 48.6 per cent of the farmers' households in India are indebted, and about 40 per cent farmers' households in the country did not like farming because it is not profitable, risky and it lacks social status, and felt that, given a choice, they would take up some other career. The crisis has been exacerbated further by rapid environmental degradation, plateauing of the existing agricultural technology (GoI, 2007a) and declining per capita availability of food grains in the country. Though food availability and price stability were achieved through green revolution and public distribution system (PDS), the chronic food insecurity which is primary associated with poverty, still persisted in the country. While physical access to food was achieved, economic access at micro-level lagged behind indicating food and nutritional insecurity. Thus, despite substantial progress made towards agricultural and rural development since independence, there are clear signs of approaching crises which need priority attention.

With this background, an attempt has been made in this paper to study the present state of agrarian crisis and food insecurity in India. The study is based on secondary data and literature collected from the different sources such as government publications, published research reports/papers and related websites. The results and discussion are organised in three sections. Section I present dimensions and magnitude of agrarian crisis and underlying factors for same. The three main pillars of food security, i.e. availability, accessibility and absorption are discussed in Section II. The last section presents the future strategies.

I

Dimensions and Magnitude of Agrarian Crisis in India

The agricultural sector in India has been going through a painful phase since the last few years. For the last several years, particularly since the launching the economic reforms in the 1990s, Indian agriculture has been passing through a serious crisis.[3] It is not merely a crisis of deceleration of growth of agricultural production and productivity, but also increasing distress experienced by a growing proportion of the farming community which has not been able to meet their basic consumption needs from their dependence on agricultural income (Reddy and Mishra, 2010). The distress among the rural community, allegedly manifested in farmers' suicide, is commonly attributed to debt trap, crop failure and/or yield loss. The

3. 'Agricultural crisis' refers to performance of the agricultural sector in terms of changes in growth of productivity and production and the underlying factors. 'Agrarian crisis' is structural and institutional in nature, as could be seen in growing marginalisation and failure of support systems, especially as a part of the reforms agenda because of the shift in institutional emphasis from state to market. 'Agrarian crisis' and 'agricultural crisis' are not exclusive but interrelated and are nothing but two sides of the problems associated with agriculture in developing countries (Reddy and Mishra, 2010).

agrarian crisis and distress[4] of the rural population which is the outcome of accumulated eventualities and adversity, has emerged as a nationwide phenomenon and leading policy issue. Decelerating agricultural growth, stagnating yields, erosion of farm income, shrinking access and entitlement to food are the manifestations of agrarian crisis (Barah, 2011).

The Mid-Term Appraisal (MTA) for the Tenth Five Year Plan (Planning Commission, 2005) had drawn attention to the loss of dynamism in agriculture and allied sectors after the mid-1990s. In fact, during the last decade or so Indian agriculture has faced a number of severe challenges, superimposed on the long-term demographics (Planning Commission, 2008). Recent trends that have raised concern regarding food security, farmers' income and poverty are: (i) slowdown in growth; (ii) widening economic disparities between irrigated and rainfed areas; (iii) increased vulnerability to world commodity price volatility following trade liberalisation, which had an adverse effect on agricultural economies of regions growing crops such as cotton and oilseeds; (iv) uneven and slow development of technology; (v) inefficient use of available technology and inputs; (vi) lack of adequate incentives and appropriate institutions; (vii) degradation of natural resource base; (viii) rapid and widespread decline in groundwater table, with particularly adverse impact on small and marginal farmers; (ix) increased non-agricultural demand for land and water as a

4. Farm distress: The definition of the Government of India while announcing the Farm Credit Package on June 18, 2004 (RBI, 2006) was articulated as, "Farmers who have suffered production and income losses on account of successive droughts, floods or other calamities (in districts declared as calamity affected by the State Government concerned) will be treated as farmers in distress and provided debt relief, and their debts will be rescheduled/restructured."

   RBI (2006) working group had considered the following two operational definitions:

   (a) "Distressed farmer is one, who has suffered repeated income/psychological shocks due to failure of investment, weather, crop production or markets, and which has crippled his ability to meet his financial and other family obligations; and feels humiliated by the castigations of the lenders and, in the absence of coping mechanisms, contemplates/takes the extreme step of voluntarily ending his life."

   (b) "Distressed farmer is the one who, in spite of his/her best rightful effort, is not able to earn income enough from his/her farm and non-farm activities to meet his/her genuine family needs. As a consequence, he/she borrows from institutional and/or private sources, but is not able to generate enough additional repaying capacity to retire these loans, thereby landing into intractable indebtedness. In the circumstances, he/she might take desperate step of even suicide."

result of the higher overall GDP growth and urbanisation; (x) aggravation in social distress as a cumulative impact of the above, reflected in an upsurge in farmers' suicides.

During the decade of the 1990s, the situation became aggravated, both due to policy failure and the successive droughts at the end of the 1990s, and because the prices did not pick up even in the event of low production. This was compounded by the economic reforms which took the agricultural sector for granted. The situation was quite alarming in Karnataka, Andhra Pradesh and Maharashtra (Reddy *et al.*, 1998; Vasavi, 1999; Deshpande, 2002; Sainath, 2005a, 2005b, 2005c; Mishra, 2005). A large number of suicides were reported in Karnataka in the first three years of the decade starting 2000-01, while Andhra Pradesh had maximum in 2004-05. In 2006, there was virtually a suicide epidemic in Maharashtra. In fact so alarming was the problem that it attracted nationwide attention and generated frantic debates in the union and state legislatures. The spiraling of suicides in the past decade or so is clearly a symptom of agrarian distress and impoverished condition of farmers. The spate of farmers' suicides that surfaced in these states was naturally associated with the performance of the sector, along with the other factors that were predominant including advent of the World Trade Organisation (WTO), genetically modified (GM) varieties, price collapse and spurious seeds. Agricultural production in these states always has significant fluctuations (Rao *et al.*, 1988) and the prices did not increase despite supply stress. That brought down the gross income flow. On the other side, the cash component in the cost of cultivation has been increasing (Nadkarni, 1988; Reddy, 1994; Deshpande and Prachitha, 2006). As a consequence the net income flow to the farmer households stagnated. The farmer would borrow to meet the increased cost of cultivation or for irrigation well and pumpsets, but the shrinking net income will not allow for payment of debt. These incidents raised serious questions about the state of the agrarian economy and the economic hardships faced by farmers. Distress among the farmers in the country is genuine and the situation is quite depressing in Andhra Pradesh, Karnataka, Maharashtra, Rajasthan, Orissa and Assam (Reddy and Mishra, 2009). Though one cannot draw any 'one to one' correspondence between distress in the farm sector and the present spate of suicides in some of the states, the farm and farm related activities have the largest stake in explaining the unfortunate occurrences (Deshpande and Arora, 2010).

The agrarian crisis has occurred because of multiple reasons, though inadequate income from cultivation is considered to be the prime factor. On the one hand, the decline in public investment in agriculture has increased the transaction cost of the farmers, on the other hand, inadequate institutional credit supply, poor arrangements to supply various inputs required for crop cultivation as well as market for agricultural produce have reduced the cultivation income. It is widely believed now that the agrarian crisis is aggravated since the initiation of economic reforms in India, because the Indian agriculture has been witnessing a few unprecedented shocks and changes over the last one decade. The control on imports of many agricultural products has been gradually removed due to obligations of WTO, which has made significant impact on the domestic prices of certain agricultural commodities (Kalamkar and Narayanamoorthy, 2003). The institutional credit, which is essential for agricultural growth, has slowed down from 6.64 per cent per annum during eighties (1981-1991) to 2.16 per cent during the nineties (1991-1999). The slowdown of institutional credit might have forced the farmers to borrow loan from the non-institutional sources, where the rate of interest is always exorbitant. Studies carried out in those regions where farmers have committed suicides at a large scale have attributed that the inadequate supply of institutional credit is one of the major reasons for the present crisis.[5] In nutshell, the RBI report (2006, Johl Committee) has succinctly summarised the gamut of causative factors of agrarian distress (see, Annexure A-9.1).

Deceleration in Rate of Agricultural Growth and Production

India's agriculture sector has an impressive long-term record of taking the country out of serious food shortages despite rapid population increase. India's performance in agriculture over the past decades has shown considerable progress and the Green, Blue, White and Yellow Revolutions have brought about vast changes in the agrarian scene since independence. The agriculture sector has been successful in keeping with the rising food

5. EPW (2006) has brought out a special issue on suicides by farmers covering regions like Andhra Pradesh, Kerala, Maharashtra and Punjab. Articles included in this volume have clearly highlighted the reasons for agrarian distress and farmers suicides in recent years. Though Karnataka has also experienced farmers' suicides at a large scale, this volume somehow does not include any articles from it.

demand of the growing population. This performance of satisfaction and optimism, however, lost its momentum in the last decade. In the recent past, there has been a sharp deceleration in growth rate of agriculture. The most important manifestations of the crisis are deceleration of agricultural growth combined with increasing inefficiency in input use thereby adversely affecting the profitability of agricultural production. The growth rate in the agricultural sector has always lagged behind the overall growth rate of the economy. Indian economy witnessed the slowing of agriculture at a time when the rest of the economy is growing at unprecedented rates (Table 9.1).

Table 9.1

Annual Growth of Agriculture and Total Economy during Plan Periods (at 2004-05 Prices)

Plan Period	*Average Annual Growth (%)*	
	Agriculture, Forestry and Fishing	*Total Economy*
Fifth Plan (1974-1979)	3.6	4.9
Sixth Plan (1980-1985)	5.8	5.5
Seventh Plan (1985-1990)	3.0	5.7
Eighth Plan (1992-1997)	4.8	6.5
Ninth Plan (1997-2002)	2.5	5.7
Tenth Plan (2002-2007)	2.4	7.7
Eleventh Plan (2007-2012)*	3.5	8.2

Note: *Provisional for the four years period.
Source: GoI (2012a).

The deceleration of agriculture sector growth has sharpened in the late nineties as was evidently reflected by the modest 2.5 per cent growth during Ninth Plan which lowered down to 2.4 per cent in Tenth Plan period. Despite of various efforts, rate of growth of agriculture sector during Eleventh Five Year Plan period was less than targeted growth rate of 4 per cent. In addition to this, weakening agricultural performance during last one decade in a large number of states is a national concern and importantly marked slowdown in growth rates in the traditional green revolution states and breadbaskets of the country, namely Punjab, Haryana and Uttar Pradesh is a major concern (World Bank, 2004). Slow agricultural growth is a concern for policymakers as some two-third of India's people depend on rural employment for a living (World Bank, 2008). The Indian agriculture growth pattern has been highly varied at the state level. It can be seen from Table 4.2 that the growth rate of agriculture national state domestic product (NSDP) slipped from 3.62 per cent during 1984-85 to 1995-96 to less than 2 per cent in the period 1995-96 to 2004-05. This deceleration, although most marked in rainfed areas, occurred in almost all states and covered almost all major sub-sectors, including those such as horticulture, livestock, and fisheries where growth was expected to be high (for details, see Planning Commission, 2008). Further, state-wise trends indicate that the largest slumps occurred in those areas/states that are predominantly rainfed. It can also be seen from Table 9.2 that growth rate in NSDP from agriculture during the period 1995-96 to 2004-05 in every state showed a dismal performance and was negligible or even negative in some states. Also, the growth performance of agriculture of different states during 2000-01 to 2008-09 was not encouraging except in few states like Rajasthan, Gujarat and Bihar (Table 9.3). Consequently, growth of agricultural GDP has been well below the target of 4 per cent set in Ninth, Tenth and Eleventh Plans. Under the circumstances, achieving targeted growth rate of 4 per cent during Twelfth Five Year Plan period will be a herculean task.

Table 9.2

Growth Rate of NSDP from Agriculture (% per annum)

State	*1984-85 to 1995-96*	*1995-96 to 2004-05*	*Rainfed (%)*	*State*	*1984-85 to 1995-96*	*1995-96 to 2004-05*	*Rainfed (%)*
Punjab	4.00	2.16	03	Rajasthan	5.52	0.30	70
Haryana	4.60	1.98	17	Orissa	-1.18	0.11	73
Utter Pradesh	2.82	1.87	32	Madhya Pradesh	3.63	-0.23	74
Tamil Nadu	4.95	-1.36	49	Karnataka	3.92	0.03	75
West Bengal	4.63	2.67	49	Maharashtra	6.66	0.10	83
Bihar	-1.71	3.51	52	Kerala	3.60	-3.54	85
Andhra Pradesh	3.18	2.69	59	Assam	1.65	0.95	86
Gujarat	5.09	0.48	64	All India	3.62	1.85	60

Note: States are ranked by percentage of rainfed area.

Source: Planning Commission (2008).

Indian agriculture is presently in a state of crisis and one of the major challenges is to reverse deceleration in agricultural growth rates so as to successfully achieve a higher broad-based growth. The slacking of agricultural sector growth in the past decade, covering the periods of

Ninth and Tenth Plan, was translated into subdued production response. The lukewarm performance resulted into stagnant production particularly that of food grains during Tenth Five Year Plan period (for details, see Planning Commission, 2006). These results clearly show the signs of acute distress and stagnation in productivity in the sector. A declining growth rate of investment in agriculture, declining efficiency in input-use, no major technological breakthrough and falling prices, have all contributed to the lower agricultural growth in the country. At the grassroots level, the deceleration in the growth rate of agricultural and allied sectors has manifested, inter alia, in widening disparities in the per worker productivity between agriculture and non-agricultural sectors and, in the ultimate analysis, in increasing number of farmers' suicides in different regions of the country. This seems to be a matter of serious concern, especially in view of a growth strategy from globalisation and trade angles.

Table 9.3

Growth Rate of GDP from Agriculture and Allied Sector (% per annum) during 1994-95 to 1999-2000 and 2000-01 to 2008-09

State	*Growth Rate of GDP from Agriculture and Allied (% per annum)*		*State*	*Growth Rate of GDP from Agriculture and Allied (% per annum)*	
	1994-95 to 1999-2000	*2000-01 to 2008-09*		*1994-95 to 1999-2000*	*2000-01 to 2008-09*
Punjab	2.5	2.3	Rajasthan	5.5	8.2
Haryana	2.1	3.4	Orissa	-0.9	3.2
Utter Pradesh	4.1	2.3	Madhya Pradesh	2.7	2.5
Tamil Nadu	1.8	1.4	Karnataka	4.1	-0.7
West Bengal	4.1	2.4	Maharashtra	3.1	2.0
Bihar	8.7	7.1	Kerala	1.9	0.4
Andhra Pradesh	2.8	5.9	Assam	n.a.	n.a.
Gujarat	5.2	7.7	All India	3.0	2.6

Note: n.a.–Not Available.

Source: GoI, 2012a.

The Steering Committee Report on Agriculture for the Eleventh Five Year Plan (Planning Commission, 2008) has indentified the possible reasons for deceleration in agriculture since mid 1990s. According to the report, the major sources of agricultural growth are: public and private investments in agriculture and rural infrastructure, including irrigation, technological change, diversification of agriculture and fertilisers. It looks like that the progress in all these sources slowed down during 1990s, particularly since mid 1990s (Annexure A-9.2). According to the report, the causes of slowdown are increase in subsidies crowding out investment in infrastructure, degradation of natural resources, failure of conservation and improvement of rainfed land, knowledge gap with existing technology, low market infrastructure and too much regulation, institutions not geared to help women farmers, imperfections in land market and plight of small farmers (Dev, 2009).

Decline in Share of Agriculture in Gross State Domestic Product (GSDP) without Matching Decline in Share of Population Dependent on Agriculture

Two prominent features of the Indian economy have hardly changed over the last few decades. These are: i) an overwhelming proportion of the Indian population lives in rural areas and a vast majority of them are dependent on agriculture; and ii) there is no significant diversification of the rural economy (Sidhu, 2010). Thus, farming and related activities constitute basic fabric of rural life and contribute significantly towards income, employment, business opportunities, infrastructure and the quality of life. The share of farming in the rural economy and hence its relative importance as a sector determines its potential economic contribution to rural development. However, the significance of agriculture in terms of the proportion of GDP and its capacity to absorb labour force in adequate measure has been declining. As mentioned earlier, the share of agriculture in the GDP has registered a steady decline over a period of time without matching decline in share of population dependent on agriculture (Table 9.4).

Table 9.4

Share of Agriculture to Total Economy (at 2004-05 Prices)

Sector	*1950-51*	*1960-61*	*1970-71*	*1980-81*	*1990-91*	*2000-01*	*2006-07*	*2010-11*	*2011-12*
Agriculture, Forestry & Fishing	51.9	47.6	41.7	35.7	29.5	22.3	17.4	14.5	13.9
Agriculture incl. livestock	41.8	39.3	34.2	29.8	24.9	18.8	14.7	12.3	n.a.

Note: n.a.–Not Available

Source: GoI (2012a).

One of the major structural constraints witnessed by Indian agriculture has been the continued disproportionate concentration of the labour force in spite of a rapid decline in its share in the national income. While slower growth of GDP in agriculture than non-agriculture is expected, the main failure has been the inability to reduce the dependence of the workforce on agriculture significantly by creating enough non-farm opportunities to absorb the labour surplus in rural areas and equipping those in agriculture to access such opportunities. As its share in workforce has declined marginally over the last four decades but still 73 per cent of rural workforce depends on agriculture, near about half of the agricultural workers being labourers (Table 9.5). It can be also observed that the share of cultivators in the total rural workforce is declining, while that of agricultural labourers has increased marginally at national level. Thus, the crucial dependence of its rural labour force on agriculture is quite evident and it is unlikely to diminish drastically in the future.

Table 9.5

Rural Workforce Distribution in India

Particulars	*Census Year*		
	1981	*1991*	*2001*
Cultivators* (%)	51.10	48.39	40.14
Agricultural Labourers* (%)	29.88	31.64	33.20
Household Industry Workers* (%)	3.08	2.16	3.77
Other Workers* (%)	15.94	17.80	22.90
Rural Main Workers (million)	176.43	222.90	229.67
Rural Main+Marginal workers (million)	197.31	249.03	310.66
Rural Population (million)	523.9	628.9	742.6

Note: *1981 & 1991 figures are percentage to total main workers and data for 2001 are per cent to total workers (main + marginal).

Sources: GoI (1981 and 1991) and *www.censusindia.net*.

Lower Income, Higher Expenditure and Indebtedness of Farmer Households

One of the serious and unrelenting problems faced by the Indian farmers' households has been indebtedness. Studies probing into factors causing recent farmers' suicides have identified indebtedness as one of the major underlying reasons (Assadi, 1998; Revathi, 1998; Deshpande, 2002;

Mishra, 2006). Despite substantial improvement in agricultural output as well as distribution of credit through institutional sources since the introduction of the new agricultural technology, indebtedness among the farmers' households is found to be widespread even today. While studying the Punjab's peasants, Darling (1925) wrote "the Indian peasant is born in debt, lives in debt and dies in debt." Though this was written about eight decades back, the problem of indebtedness not only remains true today but it has aggravated further in recent years. There are many reasons for the persistent indebtedness among the farmers' households in India. First, since the agricultural activities are typically seasonal and heavily dependent upon monsoon, the return from the crop cultivation is not assured most of the time which ultimately affects the repaying capacity of the farmers. Second, though the distribution of institutional credit to agricultural purpose has increased manifold in India since the nationalisation of banks, substantial share of farmers still rely on non-institutional sources (money lenders and others), where the rate of interest is not only exorbitant but the terms and conditions of the loan are often exploitative (see, NABARD, 2001; Athreya *et al.*, 1990). Third, the domination of middlemen in agricultural produce market, which affects the farmers to get remunerative prices for their produce, is also considered to be one of the main reasons for the indebtedness. Fourth, majority of the farmers also take loan for consumption as well as for variety of social obligations, which are unproductive and do not help to generate income. As the surplus income generated through crop cultivation in not assured and often is very less, the farmers are unable to repay the loan in time and thus the burden of debt goes on increasing. Studies based on the decennial surveys show that the proportion of rural as well as cultivator households reporting indebtedness declined over the decades, but the extent of indebtedness (i.e., average debt per household) has increased in current prices (Rao *et al.*, 1997; Rao and Tripathi, 2001). However, a study by Narayanamoorthy and Kalamkar (2005) shows that the incidence as well as extent of indebtedness has considerably increased among the farmers' households over the last one decade across the states and both the incidence as well as the extent of indebtedness is found to be relatively higher among those states that are relatively developed in terms of agriculture.

The intensity of agrarian crisis has been proved beyond doubt when the results of Situation Assessment Survey of farmers carried out by NSSO were released.[6] The results of these surveys candidly show the alarming condition of the farmer households on income, expenditure and indebtedness. It shows that the income that the farmers get from crop cultivation is not even enough to manage the cost of cultivation across different states in India (NSSO, 2005a; Deshpande and Prabhu, 2005; Narayanamoorthy, 2006; 2006a). Another survey initiated by the same Ministry also shows that the indebtedness of the farmers' households has increased substantially between 1990-91 and 2003-04 (NSSO, 2005b; Narayanamoorthy and Kalamkar, 2005). Because of failure in getting higher yield and good prices and resulting inadequate income from crop cultivation results in farmers not able to repay the loan in time that they borrowed mostly from non-institutional sources (money lenders, traders, etc), which in many instances may have forced the farmers to commit suicides (Vaidyanathan, 2006).

According to the NSSO (2005b) survey, at the all India level, the average annual income of the farmer household was ₹ 25,380, which includes cultivation, wage, farming of animals and non-farm business income (Table 9.6). Interestingly, the income from cultivation accounts for only about 45 per cent of the total income and the remaining receipts are from other sources including wage income, which alone accounts for almost 39 per cent of the total income. After excluding the other sources of income, the income from cultivation per annum comes to only about ₹ 11,628 per farmer household. When one looks at whether farmers get any profit from cultivation by comparing the annual income from cultivation

6. The economic status of farmer households was not clearly known to many of us for a long time, because it was not studied using comprehensive survey covering different parts of the country. For the first time, as a part of the millennium study, the Union Ministry of Agriculture initiated a comprehensive socioeconomic survey of the farmers covering educational level, level of living, farming practices, possession of productive assets, awareness and access to modern technology, resource availability, indebtedness and other relevant issues. The NSSO has carried out the Situation Assessment Survey (SAS) of farmers during 2003 (January-December) covering 51,770 households spread over 6,638 villages across the country. The survey reports have already been released in five different volumes namely, (1) Indebtedness of Farmer Households, (2) Access to Modern Technology for Farming, (3) Some Aspects of Farming, (4) Consumption Expenditure of Farmer Households, and (5) Income, Expenditure and Productive Assets of Farmer Households.

Table 9.6

Income, Expenditure and Indebtedness of Farmer Households: July 2002-June 2003

States	Annual Cultivation Income	Annual Total Income	Annual Cultivation Expenditure	Annual Consumption Expenditure	Annual Net Farm Income (2)–(4)	Net Balance after Meeting Consumption Expenditure (3)–(5)	IOI (%)	EOI (₹)
(1)	(2)	(3)	(4)	(5)	(6)	(7)	(8)	(9)
Andhra Pradesh	8916	19608	12085	28632	-3169	-9024	82.0	23965
Assam	21504	37932	3370	32568	18134	5364	18.1	813
Bihar	10152	21720	6809	29508	3343	-7788	33.0	4476
Gujarat	13968	32208	11456	37524	2512	-5316	51.9	15526
Haryana	17928	34584	18270	52968	-342	-18384	53.1	26007
J&K	29112	65856	6342	49308	22770	16548	31.8	1903
Karnataka	15192	31392	10344	31296	4848	96	61.6	18135
Kerala	13440	48048	7793	51000	5647	-2952	64.4	33907
Madhya Pradesh	11952	17160	8886	28068	3066	-10908	50.8	14218
Maharashtra	15156	29556	10793	32268	4363	-2712	54.8	16973
Orissa	4032	12744	3143	20364	889	-7620	47.8	5871
Punjab	33864	59520	25945	58080	7919	1440	65.4	41576
Rajasthan	4308	17976	7668	39456	-3360	-21480	52.4	18372
Tamil Nadu	7908	24864	8597	30072	-689	-5208	74.5	23963
Uttar Pradesh	10032	19596	9151	34788	881	-15192	40.3	7425
West Bengal	8844	24948	8429	32016	415	-7068	50.1	5237
All India	11628	25380	8791	33240	2837	-7860	48.6	12585

Notes: Figures are in ₹ except IOI; EOI: extent of indebtedness; IOI: incidence of indebtedness.

Sources: NSSO (2005a and 2005b).

with the annual expenditure on cultivation, the agrarian crisis comes out vividly. The average annual expenditure on cultivation comes to ₹ 8,791 per household at the all India level. After deducting the average cultivation income (₹ 11,628) of the farmer households from the expenditure on cultivation (₹ 8791), the annual net income (i.e., farm business income) of the farmer household comes to only ₹ 2,837 at the all-India level. It gets further worsened when we look at each individual state. Appallingly, the net income is negative in states like Andhra Pradesh, Haryana, Rajasthan and Tamil Nadu, i.e., the annual expenditure on cultivation is higher than the annual income from cultivation. In fact, it is less than ₹ 1,000 per farmer household in Orissa, Uttar Pradesh and West Bengal. The situation gets further worsened while comparing the annual income with the average annual consumption expenditure of the farmer households (see, Table 9.6). This clearly suggests that the farmers are in severe distress and the income that they get from all sources is not even enough to meet the consumption expenditure of the households. This could be one of the prime reasons for widespread indebtedness among the farmers' households after the introduction of economic reforms in India (see, Narayanamoorthy and Kalamkar, 2005).

The indebtedness of farmer's households in different size class of land possessed presented in Table 9.7 indicates that incidence of indebtedness increased from 46 per cent for marginal and small farmer households to 66 per cent for large farmers, and the share of institutional agencies in the debt increased from 51 to 68 per cent. The average loan amount per farmer also increased with the landholding size. Small and marginal farmer households, which accounted for about 80 per cent of indebted farmer households, absorbed 51.3 per cent of the total outstanding credit from institutional agencies. The sources of financing outstanding debt varied across the landholding size group. The sub-marginal and marginal farmers had taken loan from non-institutional agencies, indicating dependence of small farmers on these sources.

Table 9.7

Indebtedness of Farmer Households (All-India) in Different Size Classes of Land Possessed

Size Class of Land Possessed (ha)	*Total Households (%)*	*Total Indebted Households (%)*	*Incidence of Indebtedness (%)*	*Av. Loan Amount Outstanding (₹/farmer hh)*	*Loan from Institutional Agencies (%)*	*Loan from Non-Institutional Agencies (%)*
<0.01	1.40	1.3	45.3	6121	22.6	77.4
0.01-0.40	32.80	30.0	44.4	6545	43.3	56.7
0.41-1.00	31.70	29.8	45.6	8623	52.8	47.2
1.01-2.00	18.00	18.8	51.0	13762	57.6	42.3
Up to 2.00	83.9	79.9	46.3	8870	51.3	49.7
2.01-4.00	10.50	12.5	58.2	23456	65.1	35.0
4.01-10.00	4.80	6.4	65.1	42532	68.8	31.1
>10.00	0.80	1.2	66.4	76232	67.6	32.4
All Sizes	100.00	100.0	48.6	12585	57.7	42.4

Source: NSSO (2005b).

Underlying Factors of Agrarian Crisis

After having discussed the dimensions and magnitude of agrarian crisis in India, it is important to identify the underlying factors of agrarian crisis. The possible reasons/factors for agrarian crisis in India are discussed below:

Decline in Public and Private Investment

Given the agrarian nature of our economy, agriculture and rural development have always occupied the attention of the planners and policy makers, which is evident from the priorities given to these sectors in terms of resource allocations in earlier five year plans. It can be seen from the Table 9.8 that share of agriculture and allied sectors with irrigation in total plan expenditure has declined from 37.3 per cent in First Five Year Plan period to 10.6 per cent in Tenth Five Year Plan period. During Sixth Five Year Plan period, additional head of rural development was introduced and its share in total plan expenditure has increased from 6.4 per cent in Sixth Plan to 9.5 per cent in Ninth Plan and then marginally declined to nearly 8.0 per cent in Tenth Plan. As mentioned earlier, declining levels of investment in agriculture and allied sectors and irrigation is one of the reasons for deceleration in agricultural growth.

Table 9.8

Plan Expenditure on Agriculture and Allied Sectors, Irrigation, and Rural Development

Plan	*Agriculture & Allied Sectors*		*Irrigation*		*Rural Development*		*Total Plan Exp. ₹ Crore*
	₹ Crore	*% to Total*	*₹ Crore*	*% to Total*	*₹ Crore*	*% to Total*	
I Plan (1951-56)	289.9	14.79	441.8	22.54	-	-	1960.0
II Plan (1956-61)	549.0	11.75	541.6	11.59	-	-	4671.8
III Plan (1961-66)	1088.9	12.70	1019.1	11.88	-	-	8576.5
IV Plan (1969-74)	2120.4	13.44	2415.7	15.31	-	-	15778.8
V Plan (1974-78)	4864.9	12.34	3925.8	9.96	-	-	39426.2
VI Plan (1980-85)	6623.5	6.06	10929.9	10.00	6996.8	6.40	109291.7
VII Plan (1985-90)	12792.6	5.85	16589.9	7.58	15246.5	6.97	218729.6
VIII Plan (1992-97)	24895.7	5.13	31398.9	6.47	40372.1	8.32	485457.2
IX Plan (1997-2002)	37239.0	3.96	67875.0	7.21	88965.0	9.45	941041.0
X Plan (2002-2007)	58933.0	3.86	103315.0	6.77	121928.0	7.99	1525639.0
XI Plan (2007-2012)*	136381.0	3.70	210326.0	5.80	301069.0	8.30	3685973.0

Note: *Estimated (provisional).

Sources: GoI (2007, 2012a).

Investment in agriculture is essential for increasing as well as sustaining the overall growth of agriculture. Public sector investment in agriculture is essential because it stimulates private sector investment, as has been proved by various studies in India (see, Rath, 1989; Mishra and Chand, 1995; Dhawan, 1998). Though many tend to argue that the agrarian crisis started after the introduction of economic reforms in India, one can contest this view. Some symptoms of agrarian crisis could be seen even during the mid-eighties, when the public investment in agriculture started falling down. At that time, some of the researchers cautioned that the reduction in public investment in agriculture will have impact on private investment also, which will ultimately affect the growth of agriculture.[7] But, the policymakers somehow did not take it seriously and therefore, what was cautioned by Rath (1989) earlier turned out to be correct today. The reduction in public investment has affected the private investment in agriculture in many regions in India, as there is a close complementarily between the two (for more details on this issue see, Dhawan, 1998). This reduction in public investment could be the starting point for the agrarian crisis that we have been experiencing currently in different parts of the country. The reduction in public investment may have affected the diffusion of agricultural technology, efficient allocation of resources and have also increased the transactions costs for farmers, all of which one way or the other must have reduced the income of the farmers that they get from crop cultivation (Narayanamoorthy and Kalamkar, 2006).

Capital formation is one of the basic factors for increasing production. This is all the more important in agriculture where we are faced with the task of increasing production to keep pace with the increase in population against the odds of the vagaries of monsoon. As noted earlier, there was a perceptible slowdown of agricultural growth during the post-reform period 1990-91 to 2003-04. There are many reasons for the deceleration in the growth of agriculture both at the national level and state levels during

7. Rath (1989) was one of the economists who cautioned during the late eighties that the reduction in public investment in Indian agriculture would have a serious impact on the overall agricultural growth and rural poverty in the future. When many believed that the reduction in public investment can be compensated by private investment with the support of institutional credit, he contested that "it appears unlikely that without greater public sector investment and other appropriate policy measures, the vast regions characterized by poor agricultural growth can come up and use institutional credit to a greater extent."

this period. The most important reason is that there was a significant deceleration in the public and overall investment in agriculture during this period (Bhalla, 2007). Public investments in the agricultural sector in India declined steadily in absolute terms and as a share of total investments since mid 1980s (Table 9.9). This decline has serious implications for the sector's longer term growth prospects.[8] The growing subsidy requirement of the sector accounts largely for declined public investment in this sector. In 1999-2000, public investments amounted to only about 14 per cent of agricultural subsidies (including foodgrains, fertiliser, canal irrigation and power subsidies). This decline will have a negative impact on private investments over the longer term also, due to the inducement effect of public investments on private investments (World Bank, 2004). However, there has been a significant increase in the Eleventh Five Year Plan period.

Table 9.9

Public and Private Investment in Agriculture and Allied Activities

Year	*Investment in Agriculture & Allied (₹ billion)*			*Share in Agriculture and Allied GCF*		*Share of Agriculture and Allied in Total Gross Investment*		
	Public	*Private*	*Total*	*Public*	*Private*	*Public*	*Private*	*Total*
At 1993-94 prices								
1960-61	2400	2858	5258	45.6	54.4	14.0	12.4	13.1
1970-71	3216	5371	8587	37.5	62.5	13.8	14.6	14.3
1980-81	7301	6932	14233	51.3	48.7	17.7	13.6	15.4
1990-91	4992	11424	16416	30.4	69.6	7.1	11.9	9.9
1993-94	4918	10331	15249	32.3	67.7	32.3	14.6	17.7
2000-01	4417	15161	19578	22.6	77.4	5.3	8.9	7.7
At 2004-05 prices								
2004-05	16182	62666	78848	20.5	79.5	6.7	8.1	7.8
2005-06	19909	73212	93121	21.4	78.6	7.1	8.2	7.9
2006-07	22978	71422	94400	24.3	75.7	7.1	6.9	6.9
2007-08	23040	86966	110006	20.9	79.1	6.1	7.3	7.0
2008-09	24452	114145	138597	17.6	82.4	5.9	10.2	9.1

Source: GoI (2003, 2011). *Economic Survey*.

8. One estimate shows that a 10 per cent decrease in public investments (including irrigation and power) leads to a 2.4 per cent reduction in agricultural GDP growth (World Bank, 2004).

As per the *Economic Survey 2010-11*, the gross capital formation (GCF) in agriculture and allied sectors as a proportion to the GDP in the sector stagnated around 14 per cent during 2004-05 to 2006-07. However, there is a marked improvement in this figure during the current five year plan. It increased to 16.03 per cent in 2007-08, to 19.67 per cent in 2008-09 (provisional) and further to 20.30 per cent in 2009-10. However, the GCF in agriculture and allied sectors relative to overall GDP has remained stagnant at around 2.5 to 3.0 per cent (Table 9.10). As a result, the share of GCF in agriculture and allied sector in total GCF has remained in the range of 6.6 to 8.2 per cent during 2004-05 to 2009-10. There is a need to significantly step up investment in agriculture, both by the private and public sectors to ensure sustained target growth of 4 per cent per annum (GoI, 2011). Ever increasing agricultural subsidies such as fertilisers, power, irrigation, food, etc., have been cited as a main reason behind declining public investment in agriculture (Sharma, 2007). Total agricultural subsidies have increased from ₹ 14069 crore in 1993-94 to ₹ 36514 crore in 2002-03 (1993-91 series) and further increased to ₹ 204668 crore in 2008-09 (base 2004-05). The share of fertiliser subsidies in total agricultural subsidies was about 37.43 per cent in 2008-09.

Table 9.10

Share of Total GCF in Agriculture and Allied Activities to GDP (₹ crore at 2004-05 Prices)

Year	*Gross Domestic Product (GDP)*	*Agriculture & Allied Activities*		*GCF/GDP in Agriculture and Allied Activities*	*GCF in Agriculture as % to Total GDP*	*Share of Agriculture and Allied Sectors in GCF in Total GCF*
		GCF	*GDP*			
2004-05	2971464	76096	565426	13.46	2.56	7.5
2005-06	3254216	86611	594487	14.57	2.66	7.3
2006-07	3566011	90710	619190	14.65	2.54	6.6
2007-08	3898958	105034	655080	16.03	2.69	6.5
2008-09 (P)	4162509	128659	654080	19.67	3.09	8.3
2009-10 (QE)	4493743	133377	656975	20.30	2.97	7.7

Source: GoI (2011a).

Changing Landholding Structure

Land is the prime resource of the vast majority of the poor in India deriving livelihood from agriculture (Planning Commission, 2008). Fragmentation of land is widespread in India and it is believed that fragmented nature of landholdings may play a major role in explaining low levels of agricultural productivity (Jha *et al.*, 2005). With the increasing population, the fragmentation of holdings has increased, resulting in smaller and unviable units of landholdings. Added to this, there has been a decline in the net sown area (Reddy and Mishra, 2010). As per Agricultural Census 2005-06, there were about 71.0 million operational holdings in 1970-71 which increased to 106.6 million in 1990-91 and further increased to 129.2 million in 2005-06 (GoI, 2010). Since alternative sources of employment are not there, the number of operational holdings has been increasing over time. During 2005-06, about 65 per cent of the operational holdings comprised marginal farmers (less than 1 ha). It is generally recognised that a large majority of the cultivators in the country are marginal farmers. However, what is noteworthy is the fact that a progressively increasing share of cultivated land is farmed by this group (Vyas, 2007). Out of total operated area, 20.2 per cent was operated by marginal farmers in 2005-06, which has increased from 9.0 per cent in 1970-71. If we put together marginal and small farmers, they constitute about 83.3 per cent of operational holdings and operate about 41.1 per cent of total area (Table 9.11).

On the other hand, the large farmers constitute hardly 1.0 per cent of operational holdings in 2005-06, but they operate 11.8 per cent of the area. The average size of holdings has also declined over a period of time, from 2.28 ha in 1970-71 to 1.23 ha in 2005-06. In order to address the issue of fragmentation and small holdings, a clear policy with regard to transfer of agricultural land has to be formulated and implemented. The transfer of land has to be made easier to enable the farmers to augment their holdings to viable units (Planning Commission, 2002). Even though small farmers use their resources more efficiently than large farmers, there are still several disadvantages faced by small farmers. There small size often comes in the way of accessing formal credit and insurance, and prevents them from adopting agriculture that enhances yields and returns. Small farmers face high transaction costs and low bargaining power resulting in high input costs and low prices for output which leave them with poor returns. Thus, these factors leave small/marginal farmers often in a state of deprivation.

Table 9.11

Distribution of Operational Holdings in India (1970-71 to 2005-06)

Size Groups	*Numbers of Holdings (in % to total)*							
	1970-71	*1976-77*	*1980-81*	*1985-86*	*1990-91*	*1995-96*	*2000-01*	*2005-06*
Marginal	51.0	54.6	56.4	57.8	59.4	61.6	62.9	64.8
Small	18.9	18.1	18.1	18.4	18.8	18.7	18.9	18.5
Semi-medium	15.0	14.3	14.0	13.6	13.1	12.3	11.7	10.9
Medium	11.2	10.1	9.1	8.1	7.1	6.1	5.5	4.9
Large	3.9	3.0	2.4	2.0	1.6	1.2	1.0	0.8
	Operated Area (in % to total)							
Marginal	9.0	10.7	12.0	13.4	15.0	17.2	18.7	20.2
Small	11.9	12.8	14.1	15.6	17.4	18.8	20.2	20.9
Semi-medium	18.5	19.9	21.2	22.3	23.2	23.8	24.0	23.9
Medium	29.7	30.4	29.6	28.6	27.0	25.3	24.0	23.1
Large	30.9	26.2	23.0	20.1	17.3	14.8	13.2	11.8
	Average Size of Holding (in ha)							
Marginal	0.40	0.39	0.39	0.39	0.39	0.40	0.40	0.38
Small	1.44	1.42	1.44	1.43	1.43	1.40	1.42	1.38
Semi-medium	2.81	2.78	2.78	2.77	2.76	2.73	2.72	2.68
Medium	6.08	6.04	6.02	5.96	5.90	5.84	5.81	5.74
Large	18.10	17.57	17.41	17.21	17.33	17.21	17.12	17.08
All Sizes	2.28	2.00	1.84	1.69	1.55	1.41	1.33	1.23

Source: GoI (2010).

Declining Per Capita Availability of Land

As seen above, because of significant increase in number of operational holdings, the average size of holding has declined over a period of time. Among the workers, number of cultivators has almost doubled (1.82 times) during the period 1951 to 2001, while the number of agricultural labourers has increased by almost four times during the corresponding years. The net sown area (NSA) increased from 118.75 million ha in 1951 to 141.16 million ha in 2001, however, it declined during last decade to 141.16 mha in 2001 (Table 9.12). Due to increase in number of cultivators and agricultural labourers, per cultivator NSA declined drastically. Over time, the gross cropped area (GCA) per agricultural workers has also declined. There are the fundamental issues being faced by Indian agriculture. The over-crowdness in the agriculture sector gave birth to the problem of unemployment, underemployment, low labour and farm productivity, and so on (Sidhu, 2010).

Table 9.12

Per Cultivator NSA and GCA

Sr. No	Particulars	Census Year					
		1951	1961	1971	1981	1991	2001
1	Cultivators (million)	69.90	99.60	78.20	92.50	110.70	127.30
2	Agricultural Labourers (million)	27.30	31.50	47.50	55.50	74.60	106.80
3	Total Workers (million)	97.20	131.10	125.70	148.00	185.30	234.10
4	Net Area Sown NSA (million ha)	118.75	132.20	140.27	140.00	143.00	141.16
5	Gross Cropped Area (million ha)	131.89	152.77	165.79	172.63	185.74	185.71
6	Per Cultivator NSA (ha)	1.70	1.33	1.79	1.51	1.29	1.11
7	Per Agricultural Labours NSA (ha)	4.35	4.20	2.95	2.52	1.92	1.32
8	Per Cultivator +Agril Labours NSA (ha)	1.22	1.01	1.12	0.95	0.77	0.60
9	Per Cultivator GCA (ha)	1.89	1.53	2.12	1.87	1.68	1.46
10	Per Agricultural Labours GCA (ha)	4.83	4.85	3.49	3.11	2.49	1.74
11	Per Cultivator + Agril Labours GCA (ha)	1.36	1.17	1.32	1.17	1.00	0.79

Source: GoI, 2007.

The slowing pace of growth in agricultural output is being further complicated by decreasing per capita availability of arable land. The population growth has also resulted in a downward trend in per capita availability of forest and agricultural land since the 1950s. Per capita availability of forests in India is much lower than the world average (Nagdeve, 2007). Overall, per capita availability of forest land had oscillated around 0.113 hectare during the 1950s, and then has consistently declined (Table 9.13). The per capita availability of forest land declined from 0.124 ha from 1960-61 to 0.074 ha in 2000-01, and further down to 0.064 ha in 2007-08 which is extremely low compared to the world average (0.64 ha). The growth of population is expected to be faster than hoped for improvements in forest cover as well as the quality. Over the last 10 years, despite governmental initiatives of joint forest management, tree grower's cooperative movements and other efforts tangible results are still to be observed, and forest depletion and degradation is still increasing. Similarly, the per capita availability of agricultural land in rural areas has declined consistently from 0.638 ha in 1950-51 to 0.271 ha in 2000-01 and is expected to decline further as population continues to grow.

Table 9.13

Per Capita Availability of Forest and Agricultural Land in India

Year	*Forest Land (ha)*	*Agricultural Land in Rural Areas (ha)*
1950-51	0.113	0.638
1960-61	0.124	0.503
1970-71	0.115	0.410
1980-81	0.099	0.356
1990-91	0.081	0.315
2000-01	0.074	0.271
2007-08*	0.064	n.a.

Notes: *Estimated (Provisional), n.a.: Not Available.

Source: GoI (2006).

Land Degradation

Important changes have taken place in the production base of Indian agriculture, not all of which are conducive to sustainable development. These changes are reflected in the quality of land resource as well as in patterns of land utilisation, and in the availability and use of water for irrigations (Vyas, 2007). Since uncontrolled population growth could accelerate the rate of exhaustion of non-renewable resources and thereby make development unsustainable, issues relating to sustainable agricultural development have been at the forefront of discussions. Especially after the green revolution in India in the early 1970s and the achievement of food security for the economy, controversies started since the mid 1980s, as to whether this agricultural development is compatible with environmental sustainability. The rapid population growth and economic development in country are degrading environment through the uncontrolled growth of urbanisation and industrialisation, expansions and intensification of agriculture, and the destruction of natural habitats. The increasing resource degradation problems such as groundwater depletion, water logging, salinity and land degradation would add to the challenges. Mainly, food security of the poor will be at risk, as they would face severe resource constraints not only in accessing water, but also in investing in land and water management. Already, regions such as Gujarat, which face groundwater depletion problems, land degradation and frequent droughts are highly food insecure. On the other hand, in regions like Bihar and Orissa, low agricultural productivity and output, and high poverty rates leave millions, especially those in rural areas, undernourished (Kumar, 2003). In addition, the climate change risks will have adverse impacts on food production, compounding the challenge of meeting food demand (Kalamkar, 2010).

Land, a non-renewable resource, is central to all primary production systems. Over the years, the country's landmass has suffered from different types of degradations. An ever-increasing population places enormous demands on land resources. This is particularly acute in India, which has only 2.4 per cent of the world's geographical area but supports over 16 per cent of the world's population. It has 0.5 per cent of the world's grazing area but has over 18 per cent of world's cattle population. These pressures have led to drastic changes in the proportion of land utilised for agricultural activities, urbanisation and industrial development. Given the inter-linkages of crop production, livestock economy and environment, land degradation has a major impact on the livelihoods of the people, especially in rural areas (GoI, 2002). Land degradation could indeed be a potentially serious threat to food production and rural livelihoods by the year 2020 (Kalamkar, 2009; 2010). Sehgal and Abrol (1994) estimated that about 187.7 million hectares land is degraded by various sources, which is quite large, even after adjusting the figures cover only agricultural land. The impact of degradation on yield ranges from 4.7 per cent to 10 per cent. Also, the continued spread of degradation could have debilitating effect on food output growth and if the current level of degradation were reduced through investments to improve soils conditions, then presumably yields could rise (see, Kalamkar, 2009). As per Centre Water Commission (CWC) (2004), nearly 173.64 million hectares of land is degraded (see, Table 9.14) and problem of land degradation is most severe in Rajasthan, followed by Madhya Pradesh, Maharashtra and Uttar Pradesh. However, as per GoI (2012a), about 120 million ha land is reported to be degraded in India, and about 5334 million tonnes of soil is lost annually through soil erosion. Out of 120 million ha degraded area, water erosion accounts for 68 per cent, chemical degradation 21 per cent, wind erosion 10 per cent and the rest physical degradation. Another ecological process that is supposed to have eroded the sustainability of agriculture is the soil fatigue caused by this monoculture. The continuous cultivation of wheat, rice, sugarcane over the last four decades, it is asserted, has resulted in a serious deficiency of micronutrients in the soil and has adversely affected its fertility.

Table 9.14

Degraded Land in India (lakh ha)

State	*Water Erosion*	*Wind Erosion*	*Water Logged*	*Salt Affected*	*Ravines*	*Shifting Cultivation*	*Degraded Forests*	*Special Problems*	*Total Degraded Land*
A.P.	103.54	-	3.39	2.4	-	1.5	11.48	-	122.31
Bihar	32.39	-	7.07	0.04	6.0	0.85	10.21	9.0	65.52
Gujarat	94.62	7.04	4.84	10.42	4.0	-	4.84	0.10	125.86
Haryana	15.35	14.0	6.2	5.2	-	-	0.56	0.25	41.62
Karnataka	102.25	-	0.1	4.04	-	-	7.64	-	114.03
M. P.	155.1	-	0.57	2.42	6.83	1.25	41.0	-	207.17
Maharashtra	175.89	-	1.11	5.34	0.2	-	15.92	-	198.46
Orissa	27.71	-	0.6	4.04	1.13	26.48	18.07	2.95	78.03
Punjab	9.14	-	10.9	7.18	1.2	-	0.93	-	32.3
Rajasthan	188.41	156.92	3.48	10.0	4.52	-	10.61	-	373.94
Tamil Nadu	32.42	-	0.18	1.04	0.6	-	3.98	-	38.22
U. P.	58.98	-	19.8	12.95	12.3	-	12.12	15.00	131.15
West Bengal	7.64	-	21.6	9.86	1.04	-	2.69	-	43.03
India	1071.26	177.96	85.26	76.1	39.75	49.16	194.94	27.3	1721.75

Source: CWC (2004).

Depletion of Groundwater and Problems of Water Logging and Salinity

Irrigation would continue to play an unquestionable role in achieving food self sufficiency, creating grain surpluses, stabilising food prices, sustaining agricultural growth, absorbing labour force in rural areas and alleviating rural poverty; all of which are vital for food security (Kumar, 2003) and increasing water shortage would be a major challenge to achieving global food security (Leisinger, 1996). Groundwater accounts for more than 50 per cent of the net irrigated area and its contribution to the nation's food basket is quite significant. Over and above, groundwater is a de-centralised and democratic resource. This is unlike canal irrigation wherein investment is mainly from the State, and access is restricted by topographic constraints. By virtue of this unique characteristic of groundwater, its contribution to local food security is great. In arid and semi-arid areas, the increased demand for water for agriculture and other uses is being met by excessive withdrawal of groundwater, leading to its depletion and quality deterioration. Wheat, rice and sugarcane cultivation cannot be sustained if the groundwater table continues to fall at the current rate, though one of major factor responsible for fast depletion of groundwater resources is cultivation of wheat, rice and sugarcane on large areas. Given India's unique food security policy, there is a growing need to manage water for agriculture. Given the political economy of growth based on urbanisation and industrialisation, there will be a greater pressure to allocate an increasing quantum of water for industrial and municipal uses. This will pose a threat to food security at the aggregate level. The problems would be acute in semi-arid Gujarat, Tamil Nadu, Rajasthan and Maharashtra, which also experience ever-increasing demand for water in all sectors. Out of assessed 5723 numbers of administrative units (blocks/*talukas*/*mandals*/watershed), 15 per cent units are categorised as overexploited, 4 per cent units are critical, 10 per cent are semi-critical, 71 per cent are safe (for details, visit *http://wrmin.nic.in*).

In addition to above, the problems of water logging and salinity pose a serious threat to sustainability of agriculture in command areas. Intensive agricultural practices that rely heavily on water, chemical fertilisers and pesticides have caused water logging and salinity in many parts of the country. Nearly 8.53 million hectares of land is affected by water logging and 7.62 million hectares land is salt affected (Table 9.14). The underlying cause is the incredibly low water rates and the poor control over water delivery. Water logging leads to salinity of groundwater and soil, causing permanent degradation of land and sharp productivity declines. Yield declines are reported from the canal-irrigated areas of Punjab, Uttar Pradesh and Haryana.

II

Food Security[9] in India

India plays a very important role by its contribution in World's as well as South Asian food production. It accounts for 10.6 per cent of world's total cereals production, ranked third next to China and USA in 2008 (GoI, 2011). India's size in terms of food consumers is also many times larger than the average size of the rest of the countries, except China (Acharya, 2007), and accounts for 16.7 per cent of the world's food consumers. Another important dimension of food security in India is that a large number of rural households in India are food grain producers, a fact which has got positive implications for food access. Improving the food security is an issue of considerable importance for the developing countries like India where millions of people suffer from hunger and malnutrition. Since independence, an initial rush to industrialise while ignoring agriculture, two successive droughts in the mid 1960s and dependence on food aid from the US, exposed India's vulnerability to several shocks on the food security front (Nandakumar *et al.*, 2010). That threat reached dangerous proportions in the mid-1960s, leading to the launching of the Green Revolution, achievement of food self-sufficiency, and subsequently, a growing stock of surplus food grains by the mid-1970s. India has made impressive strides on the agricultural front during the last three decades.

9. As defined by World Food Summit (1996), "Food security [is] a situation that exists when all people, at all times, have physical, social and economic access to sufficient, safe and nutritious food that meets their dietary needs and food preferences for an active and healthy life. The concept of food security to include not only food availability, but also economic access to food and biological absorption of food in the body. Thus, food security and food insecurity are the terms used to describe whether or not all people at all time have physical and economic access to sufficient, safe and nutritious food for a healthy and active life. Food security is an essential foundation for meeting various Millennium Development Goals related to hunger, child mortality, mental health, disease, gender equality and primary education (Gill *et al.*, 2003). Availability of sufficient food at the country or local level does not mean that all people within that geographical unit are food secure. Food security implies that food intake must be adequate in both qualitative and quantitative terms. However, the issue of quality of diet has received little attention in the food security debate at all levels (Bose and Dey, 2005). Thus, food security is achieved only when "all people" particularly, the poor have access to sufficient food, which may assured by producing food or by obtaining it through the market."

India is one of the few countries, which have experimented with a broad spectrum of programmes for improving food security (see, Kalamkar, 2011). The food policy and agricultural development strategy adopted by India to improve the food security situation paid rich dividends, and the ensuring improvements in food security can be accessed from several angles. The most significant change was the increase in the domestic output of food grains, particularly cereals (Table 9.15). Food grains production more than doubled from 103.54 mt in TE 1972-73 to about 236.70 mt in TE 2011-12. Virtually all of the increase in the production resulted from yield gains rather than expansion of cultivated area under foodgrains, which remain stagnant at around 125 mha.

Table 9.15

Production of Food Grains in India

(Million Tonnes)

Period	*Cereals*				*Pulses*	*Total Food Grains*
	Rice	*Wheat*	*Coarse*	*Total*		
TE 1952-53	21.59	6.71	17.03	45.33	8.67	54.00
TE 1962-63	34.48	11.28	23.86	69.63	12.00	81.63
TE 1972-73	41.51	24.99	26.10	92.60	10.94	103.54
TE 1982-83	51.33	38.85	29.29	119.47	11.33	130.80
TE 1992-93	73.94	56.01	31.76	161.72	13.03	174.75
TE 2002-03	83.38	69.40	30.18	182.96	11.86	194.81
TE 2010-11	94.53	80.02	38.60	215.61	15.77	231.38

Note: TE-Triennium Endings.
Source: GoI (2011).

Though India was successful in achieving self-sufficiency by increasing its food production and improved its capacity to cope with year-to-year fluctuations in food production, it could not solve the problem of chronic hunger. Though food availability and price stability, which are considered as a good measure of food security till 1970, were achieved through green revolution and PDS, the chronic food insecurity which is primary associated with poverty, still persisted in the country. In fact, the PDS has been criticised for its urban bias and its failure to serve the poorer sections of the population effectively. The targeted PDS (TPDS) is costly and gives rise to a lot of corruption in the process of extricating the poor from those who are not poor. In addition to this, per capita per day availability of food grains in India declined, which was 468.8 gm in 1971 and decreased to 438.6 gms in 2010. Though physical access to food was achieved, economic access at micro-level lagged behind indicating food and nutritional insecurity. One quarter of the world's hungry people live in India. The *State of Food Insecurity in the World 2008* released by the FAO (2012) estimated that India is home to more than 217 million undernourished people in 2010-2012, which was 17.5 per cent of the national population. India is home to the greatest population of severely malnourished children in the world. Accordingly, to the National Family Health Survey of India, 55 per cent children living in rural areas suffer from malnutrition compared to 45 per cent of children in urban areas (*http://www.worldfoodprize.org*). Further, it is estimated that with a population approaching almost 1.2 billion today, and likely to be the most populous country on this planet by 2030 with 1.6 billion people, currently accounting for more than 17 per cent of the global population and 456 million poor or 41.6 per cent living on less than $1.25 a day (Chen and Ravallion, 2008), ensuring food and nutrition security is a challenge for India (Nandakumar *et al.*, 2010).

Declining Per Capita Food Grains Availability

The per capita availability of food grains has fallen substantially during the last decade of reforms, and the maximum decline has taken place in the year 2001 and 2005. Although there is wide variability from one year to the next, broad trends can still be picked up from this figure showing multi-decadal data (Table 9.16). The early years of bounty from the green revolution period were followed by more gradual increases leading up to a peak of 186.2 kg/person/year at national level in 1990-91. Since then, however, food security has steadily declined, throughout what is popularly referred to as the years of liberalisation and 'reforms'. As per 2010 data, per capita annual availability of food grains is typical of availability seen in the late 1970s and early 1980s, which was at lower stage of about 160 kg at national level. The per capita availability of pulses, main sources of protein, has been showing constant decline in the country. The fluctuations in per capita availability are mainly due to fluctuations in production on account of variations in weather conditions, return on competing crops, increasing population. Pulse area has declined relative to other crops, including wheat and oilseeds. Moreover, in contrast with other major crops, pulse production has remained relatively stagnant. According to the data, pulses output has been almost stagnant at 13-15 MT in the past decade baring 2002-03 when it dipped to 11.13 MT. Pulse yields have, on average, failed to show significant gains compared with other crops, a factor that has eroded the profitability and planting of pulses relative to other crops. However, availability of other important articles has increased over the period of time (Table 9.16).

High Incidence of Rural Poverty and Food Insecurity

In order to improve food security, Government of India has adopted a number of strategies, however, problems of food and nutritional security persists in the country. India is home of one-third of the world's undernourished children; almost half of the Indian children are stunted and 40 per cent are underweight; one-third of the Indian women are also underweight (Kumar *et al.*, 2012). IFPRI estimated the hunger index[10] for India as 23.7 per cent in 2011 (rank 67 of 122 countries), which placed it in the category of nations where hunger was 'alarming'. India's score was poorer than that of many sub-Saharan African counties with a lower GDP than India. At the same time, India is the second largest growing economy in the world and therefore, paradox of high economic growth and slow reduction in the number of food insecure persons needs to be understood in proper perspective.

Sufficient food available at the country or local level does not mean that all people are food secure. India has sufficient food production, and yet very high numbers of underweight children, probably due to low incomes, imbalances in household food distribution and weak social networks (UN Task Force on Hunger, 2005). Despite sizeable national food stocks, widespread poverty and hunger remain because agricultural and national economic growth has not adequately benefited disadvantaged regions and the poor (Kerr, 1996). Over the years, the incidence of poverty, both rural and urban, has declined considerably. The percentage of population below

10. Global Hunger Index (GHI) which was designed by the International Food Policy Research Institute (IFPRI) based on three dimensions of hunger: lack of economic access to food, shortfalls in the nutritional status of children, and child mortality, which is to a large extent attributable to malnutrition.

Table 9.16

Net Availability of Food Grains and Other Important Articles in India

Years	*Net Availability of Food (kg/person/year)*							*Other Important Articles**					
	Rice	*Wheat*	*Other Cereals*	*Cereals*	*Gram*	*Pulses*	*Food Grains*	*Edible Oils* (kg)*	*Vanaspati (Kg)*	*Sugar (kg.)*	*Total Cloth (meters)*	*Tea (gm)*	*Coffee (gm)*
1951	58.0	24.0	40.0	122.0	8.2	22.1	144.1	2.5	0.7	5.0	n.a.	362.0	67.0
1961	73.4	28.9	43.6	145.9	11.0	25.2	171.1	3.2	0.8	4.8	15.0	296.0	80.0
1971	70.3	37.8	44.3	152.4	7.3	18.7	171.1	3.5	1.0	7.4	15.6	401.0	65.0
1981	72.2	47.3	32.8	152.3	4.9	13.7	166.0	3.8	1.2	7.3	17.3	511.0	79.0
1991	80.9	60	29.2	171.0	4.9	15.2	186.2	5.5	1.0	12.7	24.1	612.0	59.0
2001	69.5	49.6	20.5	141.0	2.9	10.9	151.9	8.2	1.3	15.8	30.7	631.0	58.0
2005	64.7	56.3	21.7	142.7	3.9	11.5	154.2	10.2	1.2	15.5	33.5	663.0	72.0
2010(P)	67.4	61.3	19.8	148.5	4.9	11.6	160.1	13.3	1.1	17.9	43.1	709.0	86.0

Notes: *Figures for 1951 correspond to 1955-56 and other figures for 1960-61, 1970-71, 1980-81, 1990-91, 2000-01, 2004-05 and 2009-10.

Source: GoI (2012, 2012a).

line declined from 52.3 per cent in 1977-78 to 38.9 per cent in 1987-88, and finally 27.5 per cent in 2004-05. However, the absolute number of poor or food-insecure people continues to be sizeable (Table 9.17). Though estimates provided by Tendulkar Committee are higher than the NSSO figures, the percentage of the population below poverty line has declined by about the same magnitude. Poverty is a major cause of food insecurity and therefore, sustainable progress in poverty eradication is critical to improve access to food. Economic growth is necessary but not sufficient to accelerate reduction of hunger and malnutrition (FAO, 2012). Agricultural growth is particularly effective in reducing hunger and malnutrition. Most of the extreme poor depend on agriculture and related activities for a significant part of their livelihoods. Agricultural growth involving smallholders, especially women, will be most effective in reducing extreme poverty and hunger when it increases returns to labour and generates employment for the poor. Higher agricultural production can improve food security by decreasing food prices for consumers, increasing rural incomes and contributing to economic development.

Table 9.17

High Incidence of Rural Poverty in India

Year	*Poverty Ratio (%)*			*Number of Poor (million)*		
	Rural	*Urban*	*Total*	*Rural*	*Urban*	*Total*
1977-78	53.1	45.2	52.3	264.3	64.6	328.9
1983-84	45.7	40.8	44.5	252.0	70.9	322.9
1987-88	39.1	38.2	38.9	231.9	75.2	307.1
1993-94	37.3	32.4	36.0	244.0	76.3	320.3
1999-00	27.1	23.6	26.1	193.2	67.1	260.3
2004-05	28.3	25.7	27.5	220.9	80.8	301.7
Estimates of Tendulkar Committee						
1993-94	50.1	31.8	45.3	327.7	74.9	403.0
2004-05	41.8	25.7	37.2	298.0	73.1	371.0

Note: Uniform recall Period

Source: GoI (2011a).

Decline in Average Consumption Expenditure and Food Consumption

The economic growth, changes in tastes and preferences and urbanisation has resulted in changing consumption pattern away from traditional food commodities to processed and high value commodities (Murty, 2000; Meenakshi, 1996; Rao, 2000) and importance of food in total expenditures is decreasing over the period. In terms of proportion of expenditure on food items, it can be seen from Table 9.18 that there has been a decline in the proportion of expenditure on food items in last three decades in both urban and rural areas. However, the expenditure on food remained higher in rural areas as compared to urban areas. The more striking feature is that, for all the years, per capita expenditure on cereals has been consistently declining both in rural and urban India, while the one on non-cereals has been rising. However, relative share of cereals in total expenditure on food are highest in the case of both rural and urban bottom class, indicating the dependence of poor people on cereals in fulfilling their nutritional needs. The decline in consumption expenditure on food items has been mainly due to low cost of cereals; in addition there has been a decline in cereal consumption especially among the middle and high-income group. It also observed that dietary diversification increases with increasing family income (for details, see Roy, 2011). The steadily increasing preferences for non-cereals over cereal clearly indicate the directions in which the Indian consumer has been 'directing' and 'propelling' its agriculture to grow (Chadha, 2007).

In physical term as well, there is declining trend in per capita per annum consumption of all cereals from 163.03 kg in 1993-94 to 138.08 kg in 2009-10 in rural area, where as in case of urban area, it declined from 128.97 kg to 114.05 kg during corresponding years (GoI, 2012a). Chadha (2007) noted that reduction in per capita consumption of cereals was extremely sluggish, almost non-noticeable in the case of the bottom 30 per cent of consumer, both in rural and urban India. In contrast, it was highly noticeable in the case of the top 30 per cent of consumers, especially in rural India (see, Chadha, 2007). National Nutrition Monitoring Bureau (NNMB) surveys indicate that the daily intake of foods including cereals and millets (345 gm) in Indian households is lower than the recommended dietary allowance (RDA) (400 gm). The average consumption of pulses and

Table 9.18

Percentage Composition of Consumer Expenditure

Item group	*Rural* Share in Total Consumer Expenditure					*Urban* Share in Total Consumer Expenditure				
	1987-88	*1993-94*	*1999-00**	*2004-05*	*2009-10*	*1987-88*	*1993-94*	*1999-00**	*2004-05*	*2009-10*
Cereals	26.30	24.20	22.20	18.00	15.60	15.00	14.00	12.40	10.10	9.10
Gram	0.20	0.20	0.10	0.10	0.20	0.20	0.20	0.10	0.10	0.10
Cereals substitutes	0.10	0.10	0.10	0.10	0.10	0.10	0.10	0.00	0.00	0.00
Pulses & products	4.00	3.80	3.80	3.10	3.70	30.40	3.00	2.80	2.10	2.70
Milk & products	8.60	9.50	8.80	8.50	8.60	9.50	9.80	8.70	7.90	7.80
Edible oil	5.00	4.40	3.70	4.60	3.70	5.30	4.40	3.10	3.50	2.60
Egg, fish & meat	3.30	3.30	3.30	3.30	3.50	3.60	3.40	3.10	2.70	2.70
Vegetables	5.20	6.00	6.20	6.10	6.20	5.30	5.50	5.10	4.50	4.30
Fruits & nuts	1.60	1.70	1.70	1.90	1.60	2.50	2.70	2.40	2.20	2.10
Sugar	2.90	3.10	2.40	2.40	2.40	2.40	2.40	1.60	1.50	1.50
Salt & spices	2.90	2.70	3.00	2.50	2.40	2.30	2.00	2.20	1.70	1.50
Beverages, etc	3.90	4.20	4.20	4.50	5.60	6.80	7.20	6.40	6.20	6.30
Food total	64.00	63.20	59.40	55.00	53.60	56.40	54.70	48.10	42.50	40.70
Pan, tobacco, intox	3.20	3.20	2.90	2.70	2.20	2.60	2.30	1.90	1.60	1.20
Fuel & light	7.50	7.40	7.50	10.20	9.50	6.80	6.60	7.80	9.90	8.00
Clothing & bending	6.70	5.40	6.90	4.50	4.90	5.90	4.70	6.10	4.00	4.70
Footwear	1.00	0.90	1.10	0.80	1.00	1.10	0.90	1.20	0.70	0.90
Misc. & services	14.50	17.30	19.60	23.40	24.00	23.20	27.50	31.30	37.20	37.80
Durable goods	3.10	2.70	2.60	3.40	4.80	4.10	3.30	3.60	4.10	6.70
Non-food total	36.00	36.80	40.60	45.00	46.40	43.60	45.30	51.90	57.50	59.30
Total expenditure	100.00	100.00	100.00	100.00	100.00	100.00	100.00	100.00	100.00	100.00

Note: *URP estimates shown except for 1999-2000, for which only MRP estimates are available.

Source: NSSO (2011).

legumes was about 31 per cent lower (24 gm) than the RDA of 35 gms/day. However, per capita availability of cereals in 2010 was 407 gms/day and pulses was 31.6 gm/per day. Though the per capita availability of cereals was almost equal to recommended RDA; pulses availability was lower than recommended level. Also distribution of foods, both within the community and the family, may be unfavourable to some vulnerable groups due to low income and purchasing power.

As per recommendations of ICMR[11] Expert Group (1990), on the basis of the present RDA the calorie intake ought to be 2200 kcal per capita per day (*http://www.fao.org*). NSSO data show that in terms of calorie intake, it has declined from 2240 kcal/capita/day in 1983 to 2147 kcal/capita/day in 2009-10 in the rural areas, while it increased in urban area from 2070 kcal/capita/day to 2123 kcal/capita/day during corresponding years. The decline in calorie intake in rural area is attributed, by many scholars, to the declining requirements of energy, because of sedentary lifestyle, increasing mechanisations of agricultural operations, and same domestic activities and increasing use of mechanised mode of transport, particularly in rural areas (Radhakrishna and Reddy, 2004; Rao, 2003). However, level of calorie consumption for poor households is much below the prescribed norms is matter of concern even though the calorie intake of this group has shown an increasing trend in the recent past (see, Kumar *et al.*, 2012). Further NSSO's calorie intake data show that at any given point in time the calorie intake of the poorest quartile continues to be 30 to 50 per cent less than the calorie intake of the top quartile of the population, despite the poor needing more calories because of harder manual work. As the relative price of food items has remained stable over the past 20 years, declining consumption can be attributed to the lack of purchasing power and contraction of effective demand by poor people, who are forced to spend a greater part of their limited incomes on non-food items like transport, fuel

11. Indian Council of Medical Research (ICMR) sets up Nutrition Advisory Committees or Expert Groups and recommend the "Dietary Allowances" in respect of energy (Calories), proteins, fats, minerals, iron, vitamins etc. for various age groups within the population including special groups like infants, pregnant/nursing mothers, children etc., and at the same time recommend dietary allowance by activity groupings also. These Committees/Expert Group also recommend weighted average per capita requirements for the population of the country as a whole, to enable the policymakers plan for production or procurement of the required quantities of various foodstuffs. (*http://www.fao.org*).

and light, health and education, which have become as essential as food. The data also show higher reliance of the poor on cereal-based calories because of a lack of access to fruits, vegetables and meat products. Thus, further reduction in per capita availability of food grains would affect the calorie intake of the poor people.

The institution of PDS[12] in the country is the most important intervention made by the Government of India, towards ensuring food security. However, PDS is facing criticism on several grounds and the major concerns are its failure to serve the poorer sections of the population effectively, urban bias, huge leakages and diversion of PDS grains. Also, the food grains supplied by the ration shops are not enough to meet the consumption needs of the poor or are of inferior quality. Therefore, there is a need to streamline and tightening of food distribution system and also need to understand the reasons for consistent poor performance of PDS in some of the states of India. The lesson learnt from the successful states in PDS management should be replicated in poor-performing states. An alternative system of PDS can also be explored in these states (Kumar *et al.*, 2012).

Demand, Production and Availability of Cereals in India by 2020

India's population is still rapidly expanding. The living standards are rising and are slated to rise faster than in the past. As they rise, both calorific intake and diversification of diet will increase significantly. It is projected that demand for cereals for the year 2020 range from 223.6 million tonnes to 296.2 million tonnes, however, the estimates of availability of cereals from domestic production range from 222.0 to 267.8 million tonnes only (Table 9.19). Although portion of this increase can and will be obtained from abroad, a fulfilling vision of India 2020 depicts this nation with its conducive and varied climate, the largest irrigated area in the world and a vast farming population, as a major food exporter. Both as a challenge and as an opportunity, India can and must do much more to

12. PDS hailed as a means of distribution of essential commodities to a large number of poor people, forms an important constituent of the strategy for poverty eradication and intends to serve the undernourished. However, the average level of consumption of PDS grains in India is only around 1 kg per person/month.

modernise and diversify its agriculture to meet the increased domestic and international demand for a wide variety of food products. Continued growth of the agriculture sector is particularly important because it plays such a vital role in generating purchasing power among the rural population. However, the challenge of achieving food security for all our people remains a real one even today, and will continue to demand our attention in the coming decades.

Table 9.19

Estimated Demand, Production and Availability of Cereals in India for 2020 (million tonnes)

Source of Estimate	*Demand*	*Production*	*Availability*
Rosegrant, Mark W. *et al.* (1995)	237.3	256.2	237.0
Bhalla, G.S. (1999)	296.2	251.0	232.2
Kumar, Parduman	254.5	269.9-309.9	249.7 – 285.9
U.S. Department of Agriculture	267.0	240.0	222.0
Rangi, P.S. *et al.* (for wheat and rice only)	208.92-209.87	188.82- 218.54	174.66 – 202.15
Paroda, R.S.	260-287	-	-
Thamarajakshi	274-287	-	-
Dyson Tim and H. Hanchata	223.6	-	-
Bansal, P.C.	241.35	-	-

Note: Cereals includes wheat, rice and other grains.
Source: Shergill, 2007.

III

Future Strategies

Low production and productivity is the core of problems of Indian agriculture. Low productivity keeps the farmers impecunious and resource starved resulting into a vicious circle of low productivity and poverty in this prime sector of the economy. Despite substantial increase in the use of modern inputs, the agricultural productivity in India has remained low over a long period of time and even toady (Table 9.20). The need to feed teeming millions propelled country to increase the use of fertilisers and adopt better agricultural practices, which have multifarious problems in India. The persistently growing use of external nutrients enormously raised the cost of farming and led the agriculture become an un-remunerative occupation. Secondly, excessive use of chemical fertilisers has increased the amount of alkalinity and salinity in soil causing long-term hazards to soil fertility and sustainable growth of agriculture. Third, growing demand of fertilisers has increased its prices, thus emerges the need of subsidy on domestic and imported budget deficit and adverse balance of payment. As the increase in food production through area expansion is limited, therefore future increases in the food demand will have to be essentially achieved through increase in yield per hectare. India's productivity on major crops ranks far below the world average (Table 9.19). There is enormous scope for accelerating growth in agriculture, through improved soil nutrition and pest management, diversification into higher value-added crops, expansion and more efficient use of irrigation potential, rainwater harvesting, and infrastructure development for agro-processing industries. This low productivity contributes to low farm incomes and wages, lower on-farm employment generation, and relatively high food prices for all Indians. Infact some progressive farmers in India have achieved yields comparable to their foreign counterparts till recently; but these high achievers are a rare exception. Progressive farming must become the rule and the standard. Raising the productivity of our soil and water resources is an effective way to increase the profit margins of farmers, while at the same time reducing the cost of farm produce, so that it is more affordable to the masses. A minimum target should be to double or triple the average yields of major commercial crops. This will reduce pressure on scarce land and water resources as well as enable the country to utilise less fertile areas for other purposes.

Table 9.20

Productivity of Major Crops in India and World (kg/ha) 2007-08

Crop	*India*	*World*	*Highest Productivity (Country)*	*World Crop Yield (% Higher than India)*	*Highest Crop Yield (% Higher than India)*
Paddy*	3370	4309	9731 (Egypt)	27.86	188.75
Wheat*	2802	3086	8281 (UK)	10.14	195.54
Maize*	2324	5109	9658 (USA)	119.84	315.58
Cereals	2619	3380	7916 (New Zealand)	29.06	202.25
Pulses	611	776	37602 (Netherland)	27.00	515.38
Soybean	1235	2445	3525 (Turkey)	97.98	185.67
Groundnut in shell*	1071	1554	3829 (USA)	45.10	257.52
Sugarcane*	68877	71510	121136 (Egypt)	3.82	75.87

Sources: * GoI (2011): refers to the year 2008; other figures from FAI (2009), year 2007.

Tapping the full potential of Indian agriculture to meet the rising domestic demand and to take advantage of the liberalisation of international trade will require, first and foremost, the recognition of the vital role that agriculture can continue to play in national development. The other necessary conditions are: (i) greater public investment in research, (ii) expansion and development of rural infrastructure including roads, storage capacity and organised markets, (iii) improved farmer education, (iv) effective involvement of the private sector to provide technology, investment and organisational expertise for commercialisation, and (v) modification of land regulations to achieve greater production efficiency. To promote sustainable agricultural growth, a regionally differentiated strategy based on agro-climatic conditions and land and water resource availability is needed. Strong measures will also be needed to address the problem of land degradation that affects an estimated 45 per cent of total land area. In order to ensure food security on a sustainable basis, three concerns need to be addressed: (i) the adequate supply of irrigation water to sustain the growth in agricultural production at the national level, (ii) the water security for poor farmers to grow food for subsistence, and (iii) and the adequate economic incentives for farmers to maximise their production from the available land and water with least environmental consequences (Kumar, 2003). Appropriate interventions are required to boost growth in production and reduce the impact of income volatility. Efforts must be made to help them by developing drought resistant seeds, cost-effective dryland farming techniques. In addition, rainwater harvesting techniques, moisture conservation, inter-cropping are imperative to stabilise and improve the production in the dryland areas. India's objective for 2020 must not only be to produce the food its population requires but also to fully exploit the comparative advantages it possesses—agro-climatic variety, irrigation, scientific capabilities and low labour cost to become a low cost, high profit producer for the world market. Agricultural growth involving smallholders, especially women, will be most effective in reducing extreme poverty and hunger when it increases returns to labour and generates employment for the poor. Higher agricultural production can improve food security by decreasing food prices for consumers, increasing rural incomes and contributing to economic development. There is a need to streamline and tightening of food distribution system and also need to understand the reasons for its consistent poor performance. The lesson learnt from the successful states in PDS management should be replicated in poor-performing states.

References

Acharya, Shabd S. (2007). *National Food Polices Impacting on Food Security: The Experience of a Large Populated Country-India*. RP. 2006/70. United Nation University, World Institute for Development Economics Research.

Ahluwalia, M.S. (2000). "Economic Performance of States in Post-Reforms Period", *Economic and Political Weekly* 35(19): 1637-48. May 6.

Assadi, M. (1998). "Farmers' Suicides: Sign of Distress in Rural Economy", *Economic and Political Weekly* 33(14): 747-48.

Athreya, V. B., G. Djurfeldt and S. Lindberg (1990). *Barriers Broken: Production Relations and Agrarian Change in Tamil Nadu*. New Delhi: Sage Publications Private Limited.

Barah, B.C. (2011). "Introduction", in B.C. Barah and Smita Sirohi (eds.), *Agrarian Distress in India: Problems and Remedies*. New Delhi: Concept Publishing Company Pvt. Limited.

Bhalla, G.S. (2007). *Indian Agriculture since Independence*. New Delhi: National Book Trust of India.

Bose, M.L. and M.M. Dey (2005). "Food and Nutritional Security in Bangladesh: Going beyond Carbohydrates Counts", *Agricultural Economics Research Review* 20(2): 203-25.

Chadha, G.K. (2007). "Domestic Demand for Agricultural Commodities: Recent Trends and Expected Future Patterns", Presidential Address delivered at the *20th National Conference on Agricultural Marketing*, at VANAMATI, Nagpur, February 22.

Chen, Shaohau and Martin Ravallion (2008). "The Developing World is Poorer than We Thought, But No Less Successful in the Fight against Poverty, *Policy Research WP* 4703. Washington D.C.: The World Bank Group.

CWC (2004). *Water and Related Statistics, Information Systems Organisation*. New Delhi: Water Resource Information System Directorate, Central Water Commission.

Darling, M.L. (1925). *The Punjab Peasant in Prosperity and Debt*. Delhi: Oxford University Press.

Deshpande R.S. and Nagesh Prabhu (2005). "Farmer Distress: Proof Beyond Doubt", *Economic and Political Weekly* 40(40 and 45), Oct 29-Nov. 4.

Deshpande, R.S. (2002). "Suicide by Farmers in Karnataka: Agrarian Distress and Possible Alleviatory Steps", *Economic and Political Weekly* 37(26): 2601-10.

Deshpande, R.S. and Prachitha (2006). *Agricultural Policy in India: A Policy Matrix in a Federal System*. Research Report. ADRTC, Institute for Social & Economic Change, Bangalore.

Deshpande, R.S. and Saroj Arora (Eds.) (2010). *Agrarian Crisis and Farmer Suicides, Land Reforms in India: Volume 12*. New Delhi: Sage Publication India Pvt. Ltd.

Dev, Mahendra S. (2009). "Challenges for Revival of Indian Agriculture", *Agricultural Economics Research Review* 22(1): 21-46. January-June.

Dhawan, B.D. (1998). *Studies in Agricultural Investments and Rural Savings*. New Delhi: Commonwealth Publications.

EPW (2006). *Economic and Political Weekly* 41(16), April 22-28.

FAI (2009). *Fertilizer Statistics 2008-09*. New Delhi: The Fertilizer Association of India.

FAO (2012). *The State of Food Insecurity in the World 2012*. Rome: Food and Agriculture Organization.

Gill, G.J., J. Farrington. E. Anderson, C. Luttrell, T. Conway, N.C. Saxena, and R. Slater (2003). *Food Security and the Millennium Development Goal on Hunger in Asia*, WP 231. UK: Overseas Development Institute.

GoI (1981, 1991 and 2001). *Primary Census Abstract*. New Delhi: Census of India.

———. (2002). "Development of Wastelands and Degraded Lands", *Tenth Five Year Plan: 2002-2007, Vol. II*, New Delhi: Planning Commission, Government of India.

———. (2003, 2007, 2011). *Agricultural Statistics at a Glance*. New Delhi: Directorate of Economics and Statistics, Ministry of Agriculture, Government of India.

———. (2006). *Selected Socio-Economic Statistics India*. New Delhi: Central Statistical Organization, Ministry of Statistics and Programme Implementations, Government of India.

———. (2007a). *Report of the Expert Group on Agricultural Indebtedness* (Chairman: R. Radhakrishna). Ministry of Finance, Government of India, July.

———. (2010). *Agricultural Census 2005-06*. New Delhi: Agricultural Census Division, Department of Agriculture and Cooperation, Ministry of Agriculture, Government of India.

———. (2011a, 2012). *Economic Survey of India*. New Delhi: Ministry of Finance & Company Affairs, Economic Division.

———. (2012a). *State of Indian Agriculture 2011-12*. New Delhi: Department of Agriculture and Cooperation, Ministry of Agriculture, Government of India.

Jha, Raghbendra; Hari K. Nagarajan and Subbarayan Prasanna (2005). "Land Fragmentation and its Implications for Productivity: Evidence from Southern India", *ASARC Working Paper* 2005/01, Australia South Asia Research Centre, Australian National University, Australia (*http://www.crawford.anu.edu.au/acde/asarc/pdf/papers/2005/WP2005_01.pdf*, accessed on April 18, 2011).

Kalamkar, S.S. and A. Narayanamoorthy (2003). "Impact of Liberalisation on Domestic Agricultural Prices and Farm Income", *Indian Journal of Agricultural Economics* 58(3): 353-64, July-September.

Kalamkar, S.S. (2009). *Globalisation, Food Security and Sustainable Agriculture in India*. The Indian Economic Association, 92nd IEA Conference Volume, Part II, Dec., pp.90-113.

———. (2010). "Climate Change, Environmental Degradation and Sustainable Agriculture in India", *The Indian Economic Journal*, Special Issue: 35-64, December.

———. (2011). "Food Security in South Asia with Special reference to India", *Millennium Asia* 2(1): 93-121. January-June.

Kerr, John (1996). "Sustainable Development of Rainfed Agriculture in India", *EPTD Discussion Paper* No. 20. Washington, D.C., USA: Environment and Production Technology Division, International Food Policy Research Institute.

Kumar, Anjani, M.C.S. Bantilan, Praduman Kumar, Sant Kumar and Shiv Jee (2012). "Food Security in India: Trends, Patterns and Determinants", *Indian Journal of Agricultural Economics* 67(3): 445-63, July September.

Kumar, M. Dinesh (2003). "Food Security and Sustainable Agriculture in India: The Water Management Challenge", *Working Paper* No. 60, Colombo, Sri Lanka: International Water Management Institute.

Leisinger, K.M. (1996). "Food Security for a Growing World Population", Paper presented at the Saguf Symposium *How will the Future World Feed Itself?*. October 9-10.

Meenakshi, J.V. (1996). "How Important are Changes in Taste? A State-Level Analysis of Demand", *Economic and Political Weekly*, December 14.

Mishra, S. and Ramesh Chand (1995). "Public and Private Capital Formation in Indian Agriculture", *Economic and Political Weekly* 30(25): A64-79, June 24.

Mishra, Srijit (2005). *Suicides of Farmers' in Maharashtra* (Interim Report). Mumbai: Indira Gandhi Institute of Development Research.

———. (2006). "Farmers' Suicides in Maharashtra", *Economic and Political Weekly* 41(16): 1538-45. April 22.

Murthy, K.N. (2000). "Changes in Taste and Demand Pattern for Cereals: Implication for Food Security in Semi-Arid Tropical India", *Agricultural Economics Research Review* 13(1): 25-51.

NABARD (2001). *Report of the Expert Committee on Rural Credit*. Mumbai, July.

Nadkarni, M.V. (1988). "Crisis of Increasing Costs in Agriculture: Is there a Way Out", *Economic and Political Weekly* 23(39): A114-19.

Nagdeve, D.A. (2007). "Population and Environmental Degradation in India", Asia Pacific *Journal on Environment and Development* 14(1): 41-63. (paa2007.princeton.edu / download. aspx?submissionId=7192).

Nandakumar, T., Kavery Ganguly, Pravesh Sharma, and Ashok Gulati (2010). "Food and Nutrition Security Status in India: Opportunities for Investment Partnerships", *ADB Sustainable Development Working Paper Series* No 16.

Narayanamoorthy, A. (2001). "Irrigation and Rural Poverty Nexus: A Statewise Analysis", *Indian Journal of Agricultural Economics* 56(1): 40-56.

———. (2006). "State of India's Farmers", *Economic and Political Weekly* 41(14): 471-73. February 11.

———. (2006a). "Relief Package for Farmers: Can it Stop Suicides?", *Economic and Political Weekly* 41(31) 3353-55. August 5.

Narayanamoorthy, A. and R.S. Deshpande (2003). "Irrigation Development and Agricultural Wages: An Analysis across States", *Economic and Political Weekly* 38(35): 3716-22. August 30.

Narayanamoorthy, A. and S.S. Kalamkar (2005). "Indebtedness of Farmer Households Across States: Recent Trends, Status and Determinants", *Indian Journal of Agricultural Economics* 60(3): 290-301. July-September.

———. (2006). "Has Agrarian Crisis made any Impact on Agricultural Wages and Employment in India?: An Exploratory Study", *Indian Journal of Labour Economics* 49(4): 787-798.

NCF (2005). *National Commission on Farmers*. Ministry of Agriculture, Government of India.

NIN (2010). *Dietary Guidelines for Indians- A Manual*. Second Edition. Hyderabad: National Institute of Nutrition.

NSSO (2005). *Situation Assessment Survey of Farmers, Some Aspects of Farming*. Report No. 496, 59th Round (January-December 2003). New Delhi: National Sample Survey Organization, Ministry of Statistics and Programme Implementation, Government of India, December.

NSSO (2005a). *Situation Assessment Survey of Farmers, Income Expenditure and Productive Assets of Farmers Households*, Report No. 497, 59th Round (Jan.-Dec 2003). New Delhi: National Sample Survey Organization, Ministry of Statistics and Programme Implementation, Government of India, December.

———. (2005b). *Situation Assessment Survey of Farmers, Indebtedness of Farmers Households*, Report No. 498, 59th Round (Jan.-Dec. 2003). New Delhi: National Sample Survey Organization, Ministry of Statistics and Programme Implementation, Government of India, May.

———. (2011). *Key Indicators of Household Consumer Expenditure in India, 2009-10*. Report NSS KI (66/1.0), 66th Round (July 2009 to June 2010). New Delhi: National Sample Survey Organization, Ministry of Statistics and Programme Implementation, Government of India, July.

Planning Commission (2002). *Tenth Five Year Plan: 2002-2007*, Vol. II. New Delhi: Planning Commission, Government of India.

———. (2005). Mid-Term Appraisal of the Tenth Five Year Plan (2002-2007). New Delhi: Planning Commission, Government of India.

———. (2006). *Report of the Working Group for the Eleventh Five Year Plan (2007-12) on Crop Husbandry, Agricultural Inputs, Demand and Supply Projections and Agricultural Statistics*. New Delhi: Planning Commission, Government of India, December.

———. (2008). "Agriculture, Rural Development, Industry, Services, and Physical Infrastructure", *Eleventh Five Year Plan (2007–2012)*, Volume III. New Delhi: Planning Commission Government of India.

Radhakrishna, Rao and Ravi Reddy (2004). *Chronic Poverty and Malnutrition in India in the Nineties*. Working Paper. Mumbai: Indira Gandhi Institute of Development Research.

Rangarajan, C. (1982). *Agricultural Growth and Industrial Performance in India*. Research Report No: 33. Washington, U.S.A.: International Food Policy Research Institute.

Rao, C.H.H. (2000). "Declining Demand for Foodgrains in Rural India: Causes and Implications", *Economic and Political Weekly:* 201-206. January 22.

———. (2003). "Declining per Capita Demand for Food grains in India: Why and What Does Employ?", in Dev *et al.* (Eds.) (2003).

Rao, C.H. Hanumatha, S.K. Ray and Subbarao (1988). *Unstable Agriculture and Droughts*. New Delhi: Vikas Publishing House.

Rao, K.S.R. and A.K Tripathi (2001). "Indebtedness of Households: Changing Characteristics", *Economic and Political Weekly:* 1617-1626. May 21.

Rao, Subba, K.G.K; K.S. Ramachandra Rao and A.K. Tripathi (1997). "Indebtedness of Rural Households: A Profile", *Reserve Bank of India Occasional Papers* 18(2 and 3): 525-51. June-Sept.

Rath, Nilakantha (1989). "Agricultural Growth and Investment in India", *Journal of Indian School of Political Economy* 1(1): 64-83. January-June.

RBI (2006). *Report of the Working Group to Suggest Measures to assist Distresses Farmers* (Chairperson: S.S. Johl). Mumbai: Reserve Bank of India.

Reddy, Narasimha D. and Srijit Mishra (2010). "Economic Reforms, Small Farmer Economy and Agrarian Crisis", in R.S. Deshpande and Saroj Arora (Eds.), *Agrarian Crisis and Farmer Suicides, Land Reforms in India*: Vol.12, New Delhi: Sage Pub. India Pvt. Ltd., pp. 43-69.

Reddy, Narsimha and Srijit Mishra (2009). *Agrarian Crisis in India*, New Delhi: Oxford University Press.

Reddy, Sudarshan, S. Vedantha, B.V. Rao, S.M. Reddy and Y.V. Redy (1998). *Gathering Agrarian Crisis-Farmers' Suicides in Warangal District (AP)-Citizens Report*. Warangal.

Reddy, V. Ratna (1994). "New Agricultural Technology: Increasing Costs and Relative Shares of Labour and Other Factors: A Spatio-Temporal Analysis of Andhra Pradesh", *Indian Journal of Labour Economics* 37(2): 215-27.

Revathi, E. (1998). "Farmers' Suicides: Missing issues", *Economic and Political Weekly* 33(16): 1207.

Roy, Satyaki (2011). "Trends and Patterns in Consumption Expenditure: A Review of Class and Rural-Urban Disparities", *Working Paper* No. 2011/04, Institute of Studies in Industrial Development, October.

Sainath, P. (2005a). 'Six Out of Ten"?, *The Hindu*, Banglore Edition, May 31.

———. (2005b). 'Vidarbha: Whose Suicide Is It Anyway"?, *The Hindu*, Banglore Edition, June 25.

———. (2005c). 'As You Sow, So Shall You Weep?', *The Hindu*, Banglore Edition, June 30.

Saleth, M. R., M. Samad, D. Molden and I. Hussain (2003). "Water, Poverty and Gender: A Review of Issues and Policies", *Water Policy* 5(5-6): 385-98.

Sawant, S.D. and C.V. Achutan (1995) 'Agricultural Growth across Crops and Regions: Emerging Trends and Patterns', *Economic and Political Weekly* XXX(12).

Sawant, Shashikala (1983). "Investigation of Hypothesis of Deceleration in Indian Agriculture', *Indian Journal of Agricultural Economics* 38(4): 475.

Sehgal, J. and I.P. Abrol (1994). *Soil Degradation in India: Status and Impact*. New Delhi Oxford and IBH.

Shah, T. and O. P. Singh (2004). "Irrigation Development and Rural Poverty in Gujarat, India: A Disaggregated Analysis", *Water International* 29(2): 167-77.

Sharma, Vijay Paul (2007). *India's Agrarian Crisis and Smallholder Producers' Participation in New Farm Supply Chain Initiatives: A Case Study of Contract Farming*. W.P. No. 2007-08-01, Indian institute of Management, Ahmedabad, August.

Shergill, H.S. (2007). "Sustainability of Wheat and Rice Production in Punjab", *Economic and Political Weekly* 42(52): 81-85.

Sidhu, M.S. (2010). "Globalisation vis-à-vis Agrarian Crisis in India", in R.S. Deshpande and Saroj Arora (Eds.), *Agrarian Crisis and Farmer Suicides, Land Reforms in India*. Volume 12. New Delhi: Sage Publication India Pvt. Ltd., pp.149-174.

UN Task Force on Hunger (2005). *Halving Hunger: It Can be Done*.

Vaidyanathan, A. (2006). "Farmers' Suicides and the Agrarian Crisis", *Economic and Political Weekly* 41(38): 4009-13. September 23-29.

Vasavi, A.R. (1999). "Agrarian Distress in Bihar", *Economic and Political Weekly* 34(32): 2263-68.

Vyas, V.S. (2007). "Marginalised Sections of Indian Agriculture: the Forgotten Millions", *Indian Journal of Labour Economics* 50(1): 1-16.

World Bank (2004). *India Re-Emerging the Agricultural Sector to Sustain Growth and Reduce Poverty*. Agriculture and Rural Development Unit, South Asia Region, The World Bank. New Delhi: Oxford University Press.

———. (2006). *Inclusive Growth and Service Delivery: Building on India's Success*. Washington, U.S.A.: The World Bank and Delhi: Macmillan India Limited.

———. (2008). *India Country Overview 2008* (*http://en.wikipedia.org/wiki/ Agriculture_in_India*, Accessed on April 18, 2011).

Websites referred:

www.censusindia.net;http://wrmin.nic.in;http://www.worldfoodprize.org; http://www.fao.org

Annexure A-9.1

Risk Factors: A Matrix of Issues

Issues	*Supply Side*	*Demand Side*
Credit	• Formal sources: Decline in the number of branches, decline in agricultural credit/direct finance to agriculture as a percentage of NBC, and there is a shift to value addition activities. • Increasing dependence on informal sources—relatively more among smaller farmers.	• Formal sources—not timely. • Repayment difficult during crop loss and price shocks. • Instead of getting them out of credit, the system draws them into it. • Difficulties in meeting consumption requirements and other social obligations. • An increase in market induced consumerism.
Input	• No link between publicly funded research and its extension. This is particularly missing for crops/cultivation in dryland areas. • Technological change is substantial and there is an increasing reliance on the unregulated private suppliers. • Inadequate public investment in agriculture (spread of irrigation in arid regions is a casualty).	• Supplier-induced-demand is on the rise. This is credit-intensive and an important reason for putting the farmer in a quagmire of indebtedness. • There is deskilling. With new technology come new methods of cultivation. Social capital of knowledge in cultivation is rendered redundant. A case in point is the introduction of Bt Cotton seeds. • Greater investments in assets like bore wells in Andhra Pradesh not only increases cost, but has also led to a tragedy when the investments failed.
Output/ Price/ Income	• Increased volatility due to global prices. • Price distortion through subsidies by developed countries. • Low tariff in India. • Minimum support price not functional. A case in point is the Monopoly Cotton Procurement Scheme (MCPS) of Maharashtra. • Forward market: it is a virtual world that will hardly benefit the farmer. In fact, price volatility is the basis through which hedger/speculator can operate.	• Cultivation not profitable. • Income not sufficient for meeting higher education of wards, medical expenditure and other social obligations. • Yield risk because of weather, water and power availability, pests, and spurious quality of inputs among others.
Other Issues	• Interlinked credit, input and output markets. • Non-farm income opportunities limited. • Public health response to occupational health hazards of farming is wanting. • Easy availability of pesticides and other hazardous substances.	• Political dominance of moneylender and/or input dealer and output buyer. • Higher family size: more daughters—greater dowry burden. • Lack of social support.

Source: RBI (2006).

Annexure A-9.2

Growth Rate in Area, Input Use, Credit and Capital Stock in Agriculture: 1980-81 to 2003-04

(Per cent per year)

Sr. No.	*Particulars*	*1980-81 to 1990-91*	*1990-91 to 196-97*	*1996-97 to 2005-06*
1	Technology[a]	3.3	2.8	0.0
2	Public Sector Net Fixed Capital Stock	3.9	1.9	1.4[b]
3	Gross Irrigated Area	2.3	2.6	0.5[b]
4	Electricity Consumption in Agriculture	14.1	9.4	-0.5[c]
5	Area Under Fruits and Vegetables	5.6	5.6	2.7[c]
6	Private Sector Net Fixed Capital Stock	0.6	2.2	1.2[b]
7	Terms of Trade	0.2	1.0	-1.7[b]
8	Total Net Fixed Capital Stock	2.0	2.1	1.3[b]
9	NPK Use	8.2	2.5	2.3
10	Credit Supply	3.7	7.5	14.4[b]
11	Total Cropped Area	0.4	0.4	-0.1
12	New Sown Area	-0.1	0.0	-0.2
13	Cropping Intensity	0.5	0.4	0.1

Note: a-Yield potential of new varieties of paddy, rapeseed/mustard, groundnut, wheat and maize; b-up to 2003-04, c-up to 2004-05.

Source: Planning Commission (2008).

10 Growth Dynamics and Linkages of Livestock and Dairy Development in the SSP[1] Command in Gujarat

Anil Kumar Roy[2]

Introduction

Agriculture development in Gujarat has been remarkably faster as compared to any other states in India in recent decades. Agriculture sector including animal husbandry in Gujarat has registered as high as 9.6 per cent growth during 2000-01 to 2008-09 (Dholakia, 2010). It has registered a three time growth of that of all India figure during the same period. Contribution of gross value of output from agriculture and allied sector to gross state domestic product (GSDP) at current price is 25 per cent in the state (GoG, 2009). Gross value of output from livestock sector has been increasing with compound annual growth rate (CAGR) of 12 per cent during 1999-00 and 2007-08 (GoG, 2009). Livestock and dairy sector contribution to GSDP has currently reached 5 per cent in the state (Sharma and Thaker, 2010). The major drivers of this higher growth in agriculture sector in the state are cotton, oilseeds and livestock. This turnaround in agriculture growth in Gujarat in recent decades has been examined by many researchers (Gulati *et al.*, 2009; Dholakia, 2010; Pathak and Shah,

1. Sardar Sarovar Project of Sardar Sarovar Narmada Nigam Limited, Government of Gujarat.

2. Author gratefully acknowledges helpful comments and suggestions received during the 1st seminar on Evolution and Growth of Sardar Sarovar held at CEPT University during 29-31 October 2010 and subsequently in the 2nd Seminar on the subject held at the same venue during 3-4 January 2011. Author is also grateful to an anonymous referee for the valuable comments in improving this paper substantially.

Research support provided by Ms Anshulika Tewari, Associate Planner at CEPT University is deeply acknowledged.

2010). This miracle in agriculture growth in the state, can be explained by not only favorable policy environments but can also be explained partially in terms of good monsoon in recent years, that might have contributed in the overall performance of agriculture development. However, SSP canal system has also played a role in providing irrigation especially in command areas of the project to boost the higher growth in agriculture. It is expected that upon completion of 3rd phase of SSP in the state, agriculture sector including livestock will have more positive impetus in the long run.

Recent literature reveals that there has been a positive impact of livestock rearing on small, marginal and landless farmers in the state as they (livestock) are considered to be an important asset for farming community and landless farmers especially the rural poor (Shah, 2006). It is also observed that dairy development has been possible on account of the Operation White Flood[3] initiated in the state by the National Dairy Development Board (NDDB) at Anand in Gujarat. The farmers' cooperatives have been working in this direction for a very long time in the state.

Livestock and dairy development in the entire command of SSP in Gujarat has not been studied well yet. Recent literature on agriculture development and allied activities in Gujarat shows a positive high growth leading to higher state domestic product (SDP) (Dholakia and Datta, 2010). The high growth of agriculture is expected to result in a positive influence on the livestock and dairy development in the state both in short and long runs. Therefore, one would like to study: (i) whether the agriculture performance has resulted in a positive impact on livestock and dairy throughout the state and vice versa; (ii) what are the factors that are responsible for livestock and dairy development especially in the catchment area of SSP; (iii) Whether SSP through its irrigational potential has led to positive impact on the sector; and (iv) what could be the approaches and strategies of SSNL[4] for supporting and sustaining the agriculture and livestock development in the state. Various studies conducted in the context of SSP command area development have completely missed the possible livestock development due to the dam. Agriculture development includes

3. Operation White Flood was a cooperative movement started in a small town of Gujarat (India), Anand, for the collective efforts of the dairy farmers there.
4. Sardar Sarova Narmada Nigam Ltd.: A Government of Gujarat initiative.

the development of livestock and related industries; therefore, it would be important to study the development of livestock in the command area of SSP in Gujarat. The present study aims at bringing growth perspectives of livestock farming in the state and in SSP command as well as tries to explore linkages between the livestock and dairy farming with the following objectives:

- To examine livestock and dairy development potential in SSP's command area in Gujarat.
- To find out the growth dynamics and linkages of livestock and dairy development in the SSP command area; and
- To draw policy implication for livestock and dairy development in Gujarat especially in SSP command.

Methodology and Sources of Data

Growth trends of livestock and dairy production both in the state and SPP command during pre and post SSP project periods has been be examined. Livestock data has been obtained from various livestock censuses of 1997, 2003 and 2007 of Gujarat state respectively. The growth of livestock especially the milch animals has been examined taking two time period into considerations—pre and post project period i.e., before 2003 and between 2003 and 2007, both at the districts and *taluka* level using Quinquennial Livestock Census data corresponding to above mentioned period. Growth of milk production in the SSP command districts has been analysed during the entire period of 2000s.

The linkages of livestock and dairy development in the selected SSP command area has been assessed with the help of: (i) composition of livestock in Gujarat as a whole and in five selected districts; (ii) growth of livestock in pre and post SSP scenario in the selected districts and *talukas*; (iii) milk production and per capita availability of milk in the selected SSP districts and the state for year 2000-01 and 2009-10 (*taluka* level milk production data was not available, hence the analysis is limited to district level only); (iv) plan expenditure on livestock and dairy development in the state in recent decades; and (v) livestock and dairy development strategies in the state with reference to the selected districts and *talukas*.

Introduction to Study Area

The present study tries to assess the potential of livestock and dairy development in the culturable command area of SSP. SSP command area consists of 73 *talukas* of 15 districts in Gujarat with a total Culturable Command Area (CCA) of 1.84 million hectares (GoG, 2010). These districts are: Narmada, Vadodara, Bharuch, Panchmahal, Kheda, Ahmedabad, Anand, Gandhinagar, Patan, Mehsana, Surendranagar, Banaskanta, Kutch, Rajkot and Bhavnagar. The largest command area falls in Surendranagar district followed by Ahmedabad and Vadodara.

The SSP command area is one of the largest river valley projects in India. It is expected that once the project is complete, agriculture and livestock development in the state including SSP command will get a major boost to maintain the current high rate of growth. It will also provide both irrigation and drinking water to Gujarat's arid and semi-arid regions including Kutch. Figure 10.1. presents the coverage of SSP command in Gujarat showing its canal networks in 2010.

Selection of SSP Command Districts and *Talukas*

There are 15 districts and 73 *talukas* under the CCA[5] of SSP, however, these districts are partially covered. In order to select the districts which are largely under the CCA, districts having almost 50 per cent or more of their gross cropped area (GCA) under SSP command have been selected. This has been determined by simply working out the percentage of GCA to total CCA in all the 15 districts of SSP command area. The five selected districts are: Ahmedabad, Bharuch, Patan, Surendranagar and Vadodara, which contribute more than 71 per cent of the total CCA of SSP. Out of 1.84 million hectare of SSP's CCA, these five districts altogether constitute 1.32 million hectares of CCA.

The present study tries to understand the growth dynamics and possible linkages of agriculture, livestock and diary development in SSP command in Gujarat. The second section provides a brief introduction to the study area, followed by section three which deals with the growth of livestock and milch animals in the state. Section four brings out growth

5. CCA is the area which can be irrigated from an irrigation scheme and is fit for cultivation.

Figure 10.1

Sardar Sarovar Command Area and Canal Network in Gujarat, 2010

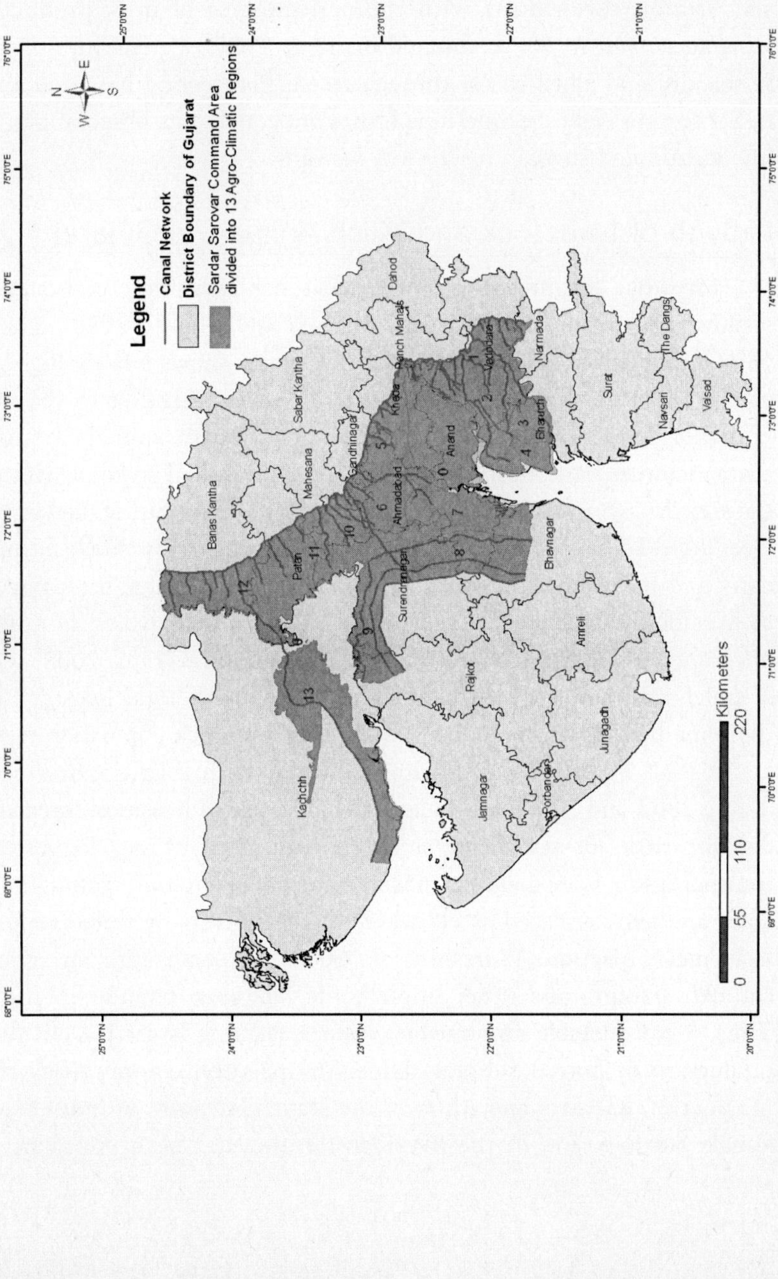

Source: SSNL, Gandhinagar, 2010.

dynamics of livestock and milch animals of SSP command districts and *talukas*. Regional distribution of bovine animals in SSP command districts has been presented in section five. Livestock development and its linkages to agriculture land-use in command districts is presented in the section six. Section seven deals with the performance of milk production of Gujarat as well in SSP command districts. Public expenditure on animal husbandry and allied sector through recent Plan period has been analysed in the section eight. Section nine brings out concluding observations, policy implication and suggestions for way forward.

Growth of Livestock and Milch Animals in Gujarat

Livestock farming in recent decades has been quite consistent with positive growth for over 20 years (Sharma and Thaker, 2010). The latest 18th Quinquennial Livestock Census of Gujarat shows a fairly high growth of 8.5 per cent over the last census. Livestock farming in the state is primarily done for the milk production and draught animals are used for the agriculture activities. The share of milch animal to total livestock in the state has gone up from 21 per cent to 29 per cent during the last 5 years (See Table 10.1.). This indicates the importance of livestock farming for dairy and its related industries in the state. By and large the composition of livestock has changed in favour of milch animal which is a national phenomenon (Kumar and Singh, 2008; Chand and Raju, 2008). Growth of milch animals is much higher than that of the total livestock during the same period in Gujarat. While the total livestock population registered a CAGR of 2 per cent, milch animals grew with a CAGR of 9 per cent during 2003 and 2007 (See Table 10.1). The use of livestock especially, the drought cattle for agriculture activities such as ploughing, harvesting and transportation is evident but not so very prevalent throughout the state. They are being replaced as elsewhere in the country by tractors and other farm mechanisations (Suresh *et al.*, 2008). Mechanisation in agriculture through tractors and other implements has been popular in the state. Thus, a considerable amount of livestock farming is carried out for milk production at household level largely by postural communities such as Maldharies, Rabaries and Ahirs in the state. Generally, at household level female participation in the livestock farming has been prevalent in the state.

Table 10.1

Livestock and Milch Animals: Gujarat, 2003 and 2007

Year	Total Livestock	Total Milch Animals	% of Milch Animals	CAGR of Total Livestock 2003-2007	CAGR of Total Milch 2003-2007
2003	21654621	4450214	20.55	1.66	9.2
2007	23515434	6918507	29.42		

Source: Livestock Census, 2003 and 2007, Gujarat.

The hides and leather industries are potential industries that need to be explored in the long run in the state especially in SSP command areas as livestock are concentrated in this area.

Growth of Livestock and Milch Animal in SSP Command Districts and *Talukas*

It is important to discuss the regional distribution of livestock in the state in order to understand the linkages of SSP command areas and livestock development. The livestock data available for five districts of SSP command areas has been considered in order to make a possible comparison between pre and post SSP implementations in its command areas in the state.

Table 10.2

Growth of Livestock in SSP Command Area, 1997-2007

Sr. No.	Name of the SSP District	Total CCA of District (Ha)	Total Livestock 1997	Total Livestock 2003	Total Livestock 2007	CAGR 1997-2003 (Before SSP)	CAGR 2003-2007 (After SSP)
1	Ahmedabad	340297	584386	671279	714230	2.81	1.25
2	Bharuch	166001	424856	437473	419476	0.59	-0.84
3	Patan	205088	544688	513976	658646	-1.15	5.09
4	Surendranagar	348784	715058	803438	963741	2.36	3.71
5	Vadodara	257779	1091573	1199018	1264358	1.90	1.07
	Total	1317949	3360561	3625184	4020451	1.53	2.09

Source: Livestock Census, 1997, 2003 and 2007, Gujarat State.

Table 10.2 shows growth of livestock in selected five districts of SSP command area for the above mentioned reason. An attempt has been made to make a comparison of before and after the SSP project, as far as the growth of livestock in these districts are concerned. The livestock population in all the five districts of SSP command area has shown a modest CAGR of 1.53 per cent before the project implementation, however, it has registered a higher growth of 2.09 per cent after the SSP project, indicating a positive impact of SSP on livestock. After the SSP implementation i.e., during 2003-2007, Patan district has shown highest growth rate of over 5.0 per cent followed by Surendranagar, Ahmedabad and Vadodara districts with a CAGR of 3.71, 1.25 and 1.07 respectively. Bharuch is the only selected district which has registered a negative growth of livestock during 2003 and 2007.[6] While these five selected districts of SSP command area have shown accelerated growth in the livestock population, the other districts have shown decelerated growth. On the other hand, the selected districts have traditionally been supporting livestock farming which eventually gives impetus to dairy development in the state. The largest milk cooperatives and dairy in the country are located in these districts of Gujarat. This could be partially explained by the perennial cropping system in the districts. The *taluka* level analysis presented in the next section further explains this phenomenon. By and large these districts have shown an accelerated growth rate of livestock population after the implementation of SSP, with on exception of Bharuch district.

In order to further explain the growth dynamic of livestock at *taluka* level, 26 selected *talukas* have altogether shown a modest 2 per cent CAGR in their livestock population during 2003 and 2007 (see Table 10.3). Majority of *talukas* (10) have registered more than 2.5 per cent of CAGR during the same period while 5 of them, 1 from Bharuch and 4 from Vadodara district have shown a negative growth in their livestock population during the same period (see Table 10.3). However, this negative trend in livestock population in these *talukas* has not been reflected in their cattle and buffalo population except for the Dabhoi *taluka*. This means that

6. Bharuch districts and its 4 selected *talukas* seem to be an outlier in the context of livestock and agricultural development in the state. While explaining for the linkages between livestock and agriculture factors, it would be difficult to explain the Bharuch case in particular. The dominance of perennial cropping system has resulted in stagnation in net shown area in Bharuch in all the 4 selected *talukas*. This has probably reduced the scope for further increase in the fodder crop area and the area under permanent pasture and other grazing land. Hence, while exploring the possible association between agriculture intensification and livestock development, all the *talukas* of Bharuch districts have been dropped for the analysis.

by and large composition of livestock is in favour of milch animals which is indicative of a good prospects for dairy farming.

Table 10.3

Growth of Livestock in Selected Taluka of SSP Command Districts, 2003-2007

CAGR	*No. of Taluka*	*Name of the Taluka*					
>2.5	10	Dholka	Barvala	Wagra	Harij	Radhnpur	
		Santalpur	Dasada	Halvad	Chuda	Lakhtar	
1.5-2.5	5	Viramgam	Mandal	Jambusar	Sami	Sankheda	
0-<1.5	6	Detroj-Rampura	Dhandhuka	Shinor	Padra	Waghodia	Limbdi
Negative	5	Bharuch	Dabhoi	Karjan	Savali	Vadodara	

Source: Livestock Census, 2003 and 2007 (Provisional), Gujarat.

It is further observed that the selected districts have also shown much higher growth of cattle and buffalo including the milch animals as compared to total livestock after the SSP implementation. Table 13.4 shows growth of cattle and buffalo in SSP districts during 2003 and 2007. These districts have shown relatively higher CAGR of 3.43 per cent during 2003 and 2007, which clearly indicates a positive impact of SSP in its command areas as far as dairy farming is concerned.

Table 10.4

Growth of Cattle and Buffalo[7] Animals in SSP during 1997 to 2003 and 2007

Sr. No.	*SSP Districts*	*(Cattle+Buffalo), 2003*	*(Cattle+Buffalo), 2007*	*CAGR, 2003-2007*
1	Ahmedabad	502548	560640	2.21
2	Bahruch	275286	275683	0.03
3	Patan	338842	494530	7.85
4	Surendranagar	496697	636974	5.1
5	Vadodara	846489	944230	2.21
	Total SSP Districts	2459862	2912057	3.43

Note: Cattle+Buffalo including the bovine animals.

Source: Livestock Census, 2003 and 2007, Gujarat.

7. Including milch animals (bovine) used for dairy farming.

Patan district has registered highest growth of cattle and buffalo population with a CAGR of 7.85 per cent followed by Surendranagar with 5.1 per cent CAGR during the same period. Bharuch district has shown insignificant CAGR of 0.03 per cent. This could be explained with the help of *taluka* level analysis of cattle and buffalo in the following section.

Table 10.5

Growth of Cattle and Buffalo in Selected Taluka of SSP Command, 2003-2007

CAGR	*No. of Taluka*	*Name of the Taluka*					
>6.5	9	Barvala	Wagra	Sami	Harij	Radhnpur	
		Santalpur	Vadodara	Halvad	Lakhtar		
3.5 to 6.5	6	Mandal	Dholka	Sankheda	Dasada	Chuda	Limbdi
0 to <3.5	10	Viramgam	Detroj-Rampura	Dhandhuka	Bharuch	Jambusar	
		Karjan	Shinor	Padra	Waghodia	Savali	
Negative	1	Dabhoi					

Note: Cattle+Buffalo including the bovine animals.
Source: Livestock Census, 2003 and 2007, Gujarat.

Table 10.5 shows growth of cattle and buffalo in SSP command *taluka* during 2003 and 2007. As mentioned above, despite the moderate growth of livestock, cattle and buffalo population at the *taluka* level has shown much higher growth than that of the livestock. This indicates a positive aspect of dairy farming in the SSP command area. Altogether, the entire 26 *taluka* has registered around 5 per cent CAGR of cattle and buffalo population during 2003 and 2007. This could assure better prospects of dairy development in SSP command area, provided the trend continues like this with the help of irrigation facilities created by the SSP in these *talukas*. Twelve (12) out of 26 *talukas* have shown more that 5 per cent CAGR of bovine/milch animals (cattle and buffalo both dry and in milk) during 2003 and 2007. The notable fact is those *talukas* registering negative growth of livestock population, have shown positive growth of bovine animals in the five years time during the SSP implementations except for Dabhoi as mentioned above. One need to explain the growth trends of milch animals of command area in a regional perspective so as to understand the positive impact and tangible benefit of SSPs in Gujarat.

Regional Distribution of Cattle and Buffalo in Gujarat: District Level Analysis

A regional comparison between the 5 selected SSP command districts, other 10 SSP command districts (partially covered under the CCA of SSP) and non-SSP command districts have also shown a higher growth of livestock especially cattle and buffalo population in favour of the 5 selected SSP command districts. Table 10.6 shows growth of total cattle and buffalo in command and non-command districts.

Table 10.6

Distribution and Growth of Cattle and Buffalo (including Milch Animals) in Gujarat, 2003 and 2007

Sl. No.	*Districts*	*Cattle+Buffalo 2003*	*Cattle+Buffalo 2007*	*CAGR 2003-2007*
1	Total of selected SSP Districts (5)	2459862	2912057	3.43
2	Total of rest of the SSP Districts (10)	6657633	7827854	3.29
3	Total non-SSP Districts (10)	5374412	6009382	2.26
4	Total Gujarat (25)	14491907	16749293	2.94

Note: Figure in parenthesis is the number of districts.
Cattle and buffalo includes bovine animals.

Source: Livestock Census, 2003 and 2007, Gujarat.

It is observed that the group of five selected districts of SSP command area has shown highest CAGR of cattle and buffalo during 2003 and 2007 i.e., after the project implementation period. If one takes into consideration full (5 districts) and partial (10 districts) implementation of SSP project in the CCA districts, these five districts show better prospects for dairy farming (see Table 10.6). The non-SSP command districts registered a CAGR of 2.26 per cent which is lower than the state average for all the 25 districts. This indicates that the in last five years SSP command districts particularly the five selected ones have provided suitable environment for the growth of livestock. It is expected that intensification of agriculture generally lead to intensification of livestock. The expansion of area under irrigation and fodder crop is particularly important for cattle and buffalo rearing in order to improve the dairy development (Kumar and Singh, 2008). Following section presents an overview of agriculture development *vis-á-vis* livestock farming in Gujarat especially in SSP command districts.

SSP is likely to improve the agricultural intensification by increasing the area under irrigation and fodder crops both in short and long run.

Agricultural and Livestock Development in SSP Command: A Linkage

Further aggregate district level analysis shows that livestock pressure on agriculture land use is favourable in SSP command districts than that of the non-SSP command districts (see Table 10.7). This indicates a very good scope for further intensive livestock farming in the command area.

Table 10.7

Livestock and Agricultural Land Use in Gujarat, 2007

Districts	*Total Live-stocks*	*Percentage of Livestock*	*Area under Fodder Crops (Ha)*	*Area under Grazing Land (Ha)*	*Livestock/ ha. of Grazing Land*	*Livestock/ ha. of Fodder Crops*
Total of selected SSP districts (5)	4020451	17.10	37581	151529	26.5	107.0
Total of rest of the SSP districts (10)	11222804	47.73	117016	377022	29.8	95.9
Total non-SSP districts (10)	8272179	35.18	54218	322636	25.64	152.57
Total Gujarat (25)	23515434	100.00	208815	851187	27.6	112.6

Source: For Livestock 2007: *Livestock Census, Gujarat State, 2007*; For Area under Fodder Crops and Grazing Land, 2007: *Season and Crop Report, 2008*. Gandhinagar: Directorate of Agriculture, GoG.

The selected five districts in SSP command area show a favourable scope for further intensification of livestock farming as compared to Gujarat as a whole. These districts constitutes 17 per cent of the total livestock of the state; there is over 26 livestock per hectare of area under the permanent pasture and grazing land, while these five districts have 107 livestock per hectare of area under the fodder crops. The other 10 districts in SSP command constitutes almost 48 per cent of the total livestock in the state, but at the same time its shows a slightly higher pressure of livestock on grazing lands in these districts. Almost 30 livestock are reared over a hectare of grazing land in these 10 districts, while they have 97 livestock per hectare of fodder crops. The rest of the 10 non-SSP command districts show substantially more livestock pressure on fodder crops, while they manage with 26 livestock over a hectare of grazing land. This indicates that area under fodder crops need to increase for sustainable livestock farming as these non-SSP commands contribute 35 per cent of the total livestock in the state. By and large the livestock pressure scenario is favourable in five selected districts of SSP command. Further intensification of livestock farming is possible through overall agricultural intensification, as cropping intensity and livestock intensification go hand in hand with the help of irrigation facilities (Suresh *et al.*, 2008).

Traditionally in Gujarat, livestock farming has been dominated by the agricultural practices and is being generally carried by a particular section of the society. The commercial livestock farming as compared to developed nations is yet to be realised in the state. Hence, agriculture land use pattern play a vital role in the overall livestock farming in the state. Village commons are important factors that help farmers in keeping a sizeable amount of livestock. Besides, fodder banks, milk cooperatives and higher returns from milk production are certain factors that can also play a vital role in promoting a large scale livestock and dairy farming in the state particularly in SSP command districts. Favourable agricultural land use and better managed village milk cooperatives are likely to provide greater emphasis for the sector to grow in the SSP command area in the long run. Diversification in livestock farming in terms of other production such as meat, hides and leather is yet to be realised. As traditional farming system in Gujarat is highly associated with animal husbandry as part of agriculture development, society at large promotes it for the milk production.

Milk Production and its Availability in Gujarat and SSP Command Area

Gujarat has ranked 5th in milk production amongst Indian states after Uttar Pradesh (UP), Rajasthan, Punjab and Andhra Pradesh (AP) in 2008-09 (Sharma and Thaker, 2010; NDDB, 2010). It has improved its ranking from 2001-02 when Gujarat was at 7th place in terms of milk production in the country. Total milk production in the state is 8.84 million tonnes as per the current statistics of 2009-10. The milk production in the state has been constantly increasing during 2000-01 and 2009-10 (see Table 10.8). It has increased from 5.32 million tonnes to 8.84 million tonnes in the last decade (2000-01 to 2009-10) with a CAGR of 5.22 per cent. However, the state has registered a much faster growth of milk production compared to all India figure during the same period. (Sharma and Thaker, 2010).

Table 10.8

Trends in Milk Production and Per Capita Availability in Gujarat, 2001 to 2010

Year	*Per Capita Availability of Milk (in grams/day)*	*Milk Production (in Lakhs Kg.)*
2000-01	291	53173
2001-02	314	58760
2002-03	321	60894
2003-04	333	64207
2004-05	344	67454
2005-06	350	69600
2006-07	373	75331
2007-08	386	79117
2008-09	403	83872
2009-10	421	88428

Source: GoG (2009-10). *27th Survey Report*, Gujarat.

This has led to increase in the per capita availability of milk in the state during the same time (see Table 10.8). The increasing trends in milk production and per capita availability in the state will have positive influence in reducing the acute malnutrition prevailing particularly in dry or drought prone districts in Gujarat (Roy and Hirway, 2006). Out of the total 8.84 million tonnes of milk production in the state, nearly 17 per cent of them i.e., 1.49 million tonnes come from the selected five district of SPP command.

Table 10.9

Milk Production and Per Capita Availability of Milk in SSP and Non-SSP District, Gujarat, 2001-2010

Districts	*Milk Production ('000 Tonnes)*		*Per capita Availability of Milk (Grams/Day)*	
	2001-02	*2010-11*	*2001-02*	*2010-11*
Total SSP Command District (15)	3937.23	6264.78	322.17	434.65
Non-SSP District (10)	1938.78	2874.53	297.84	376.91
Gujarat (25)	5876.01	9139.31	313.71	414.67

Source: 1. *Annual Report, 2010-11*, Dept. of Agriculture and Animal Husbandry (for Milk Production).

2. General Population Table, Census of India, 2001 and 2011.

Table 10.9 shows milk production and per capita availability of milk in SSP command districts and non-SSP district in the state during 2001-02 and 2010-11. It is observed that there has been consistent increase in both milk production and per capita availability of milk in the state during the last decade. The same trends continue in SSP command and non-SSP districts during the same period. However, it is observed that per capita availability of milk is higher in SSP command districts than that of the non-SSP districts and Gujarat as a whole. This reveals that the potential of dairy development in the SSP districts.

Further, share of milch cattle and milk production in SSP districts *vis-á vis* Gujarat reveals that SSP districts are well positioned as far as the dairy and allied activities are concerned.

Table 10.10

Share of Milch Animals (Cows + Buffalos) and Milk Production in SSP District, 2003 to 2007

Districts	*Milch Animals*		*Milk Productions*	
	2003	*2007*	*2003*	*2007*
SSP Districts (5)	16.97	17.39	18.13	16.31
Gujarat (25)	100.00	100.00	100.00	100.00

Note: Figures in parenthesis are the number of districts of the Gujarat.

Source: 1. Livestock Census, 2003 and 2007, Department of Animal Husbandry, Government of Gujarat, Gandhinagar.

2. *Annual Survey, 2011*, Department of Agriculture, Government of Gujarat, Gandhinagar.

The five selected SSP district has shown increase in the share of milch animals during 2003 and 2007 (see Table 10.10), however, their corresponding share of milk production has shown marginal decline. But the important aspect of SSP districts is that the per capita availability of milk has also increased during last decade. The decline in the share of milk production can be explained with the fact that the other SSP districts are the traditional milk producer districts in the state. With the ensured irrigation and animal husbandry services, the SSP districts are also likely to improve their share of milk production in the states. The five selected SSP district have shown relatively higher growth rate in milk production during last decade.

Table 10.11

Growth Trends in Milk Production in Selected SSP Command Districts, 2000-01 to 2009-10

Districts	*Milk Production, 2000-01 ('000 tonnes)*	*Milk Production, 2009-10 ('000 tonnes)*	*CAGR, 2000-01 to 2009-10*
Ahmedabad	220.57	318.11	3.73
Bharuch	103.00	142.01	3.26
Patan	196.24	334.69	5.48
Surendranagar	164.27	313.24	6.67
Vadodara	252.49	379.49	4.16
SSP Districts	936.57	1487.54	4.74
Gujarat State	5317.33	8842.84	5.22

Source: GoG (2009-10) 27th Survey Report, Gujarat.

Amongst the selected five districts of SSP command, Vadodara produced highest amount of milk (i.e., 379.49 thousand tonnes) in 2009-10 followed by Patan and Ahmedabad. The growth trends in the milk production so far in these five districts has been slightly lower than that of Gujarat during the entire period of last decade (2000-01 to 2009-10). However, two of these five SSP command districts namely, Surendranagar and Patan have shown substantially higher growth of milk production during the same period (see Table 10.11). Milk production has registered a CAGR of 6.7 per cent and 5.5 in Surendranagar and Patan districts respectively, during 2000-01 and 2009-10, while Gujarat registered 5.2 per cent CAGR during the same period. Vadodara, Ahmedabad and Bharuch districts have shown 4.1, 3.7 and 3.2 per cent CAGR of milk production during this period. By and large, SSP districts have shown moderate to higher CAGR in milk production in the last decade indicating a positive benefit from the SSP. The faster growth of milk production in the state is also a manifestation of faster growth of livestock sectors particularly the bovine animals. Therefore, the livestock sector has been one of the drivers of agriculture sector growth in the state. The better performance of Gujarat in terms of livestock farming and milk production and per capita availability is attributed to the commitment of public investment in this sector.

Public Expenditure on Animal Husbandry and Dairy Development in Gujarat

The public expenditure on animal husbandry and dairy development in Gujarat through different plan period has shown a fairly good amount of commitment to this sector. Plan period expenditure has so far doubled from Ninth Plan till 2008-09. The total planned expenditure in Ninth plan period was ₹ 8093 lakh which has increased to ₹ 16798 lakh until 2008-09 (see Table 10.12).

Table 10.12

Public Expenditure on Animal Husbandry and Dairy Development, Gujarat

Sl. No	*Plan Period*	*Amount (in ₹ Lakhs)*
1	IX Plan	8093
2	X Plan	13449
3	2007-08 to 2008-09	16798

Source: GoG (2008).

According to the *Bulletin of Animal Husbandry and Dairying Statistics, 2010-11*, planned expenditure on various programmes on animal husbandry and dairy development in the state has gone up to ₹ 12647.95 lakh during 2010-11, of which about 77 per cent were spent on animal husbandry. Out of the total expenditure on animal husbandry, 47 per cent were spent under veterinary services, animal health, and cattle and buffalo development, hence giving emphasis on the milk production in the state. The data on private sector investment on animal husbandry and dairy development was not available with the author while writing this paper. However, it is assumed that Gujarat's private sector investment in agriculture and allied sector would have similar pattern as that of India as major part of the investment in agriculture and animal husbandry at national level comes from private sector and that to from household sector and not from private corporate sector (RBI, 2005).

There are 6404 milk cooperative societies in the state with 1404 thousand members (Directorate of Animal Husbandry, Gujarat, 2010-11). The extension services in animal husbandry are being well taken care at the state level to promote dairy and livestock sector. National dairy development strategies for Gujarat in particular has been pro-farmers and pro-poor. A detailed policy implication of animal husbandry and dairy development has been dealt in the following section.

Concluding Observations and Policy Implication

Gujarat's agriculture development in recent decades have shown remarkable progress which may be replicated by other agrarian states in India. Agriculture sector grew three times faster than that of all India average during 2000s. This turnaround in agriculture growth has been possible primarily because of three drivers i.e., high growth of Bt cotton, oilseeds and livestock including the high growth of livestock in SSP command. The share of livestock to the total value of output in the agriculture and livestock sector is 27.41 per cent as per the quick estimates by the Government of Gujarat, however, the long term annual data for the last decade reveal that the share has been fluctuating between 22 per cent and 30 per cent altogether during 1999-2000 and 2009-10 (Directorate of Animal Husbandry, Gujarat, 2010-11).

Livestock composition has been found in favour of milch animal in the state more so in the selected SSP command districts. Milch animals have shown consistent growth before and after SSP in selected command districts and *talukas*. The data on the share of milk in the total value of output from livestock sector was not available with the author at the time of writing this paper. However, growth has been accelerated after SSP command started receiving water from Sardar Sarovar Dam. This indicates better scope for livestock intensification in the SSP command area.

Milk production in the state has shown relatively higher and consistent growth during 1990s and 2000s (Sharma and Taker, 2010). Per capita availability of milk has been increasing rapidly in the state in recent decades which will have larger positive impacts in rural Gujarat where incidence of malnutrition has been reported. Majority of SSP command districts have shown relatively higher growth rate in milk production during 2000-01 and 2009-10. All the five selected districts together in SSP command have shown around 5 per cent CAGR of milk production during same period.

Higher growth of livestock and dairy development in the state has been sustained due to conducive sociopolitical environment in the state. National policy approach has been encouraging in terms of plan expenditure allotted for animal husbandry and allied sectors for the state. Public expenditure on animal husbandry and dairy development has been consistent and it has doubled after Ninth Five Year Plan expenditure in the state. This shows a positive policy level commitment towards the sector.

Supplying irrigation water to dry (arid and semi-arid) districts will get faster benefits as compared to others. This has been reflected in the growth of livestock, milch animals as well as milk production in the SSP command districts particularly in case of Patan and Surendranagar. Therefore, SSP should aim to provide the irrigation water in the drier districts on a priority basis in order to increase the coverage in terms of better ratio of GCA and command area in the rest of the 10 districts under SSP Command.

The growth of livestock in the state and in particular in the SSP command district has brought and will continue to bring in positive changes in the farming community. It has not only increased the per capita availability of milk in the state, but has also helped sustain agriculture sector's contribution to the state economy. More than a quarter of value out in agriculture sector comes from livestock only. Further an empirical study in the SSP command districts is suggested to explain in details as to how livestock and dairy sector together has improved the livelihood of agrarian community in terms of creating employment opportunities, improving income and crop diversifications etc.

Selected Bibliography

Bagchi, A.K., Panchanan Das and Sadhan Kumar Chattopadhyay (2005). "Growth and Structural Change in the Economy of Gujarat, 1970-2000", *Economic and Political Weekly* XL (28), July 9.

Bardhan, A. (2007). "India's Trade Performance in Livestock and Livestock Products," *Indian Journal of Agriculture Economics* 62(3), July-Sept.

Bardhan, D. and S.K. Tewari (2007). "Risk Attitude and Risk Management Strategies: An Analysis of Dairy Farmers in Tarai Area of Uttranchal State," *Indian Journal of Agriculture Economics* 62(4), Oct-Dec.

Chand, R and S.S. Raju, (2008). "Livestock Sector Composition and Factors Affecting its Growth," *Indian Journal of Agriculture Economics* 63(2), April-June.

Dholakia, R.H. (2010). "Has Agriculture in Gujarat Shifted to High Growth Path," in R.H. Dholakia and S.K. Datta (ed.), *High Growth Trajectory and Structural Change in Gujarat Agriculture*. Ahmedabad: Macmillan Publishers India Ltd.

Dholakia, R.H. and S.K. Datta (ed.) (2010). *High Growth Trajectory and Structural Change in Gujarat Agriculture*. Ahmedabad: MacMillan Publishers India Ltd.

Feroz, S.M. and A.K. Chauhan (2010). "Performance of Dairy Self-Help Groups (SHGs) in India: Principal Component Analysis (PCA) Approach," *Indian Journal of Agriculture Economics* 65(2), April-June.

Government of Gujarat (2008). *Statistical Abstract of Gujarat State-2008*. Gandhinagar: Directorate of Economics and Statistics.

———. (2008a). *Statistical Outline of Gujarat State-2008*. Gandhinagar: Directorate of Economics and Statistics.

———. (2009). *26th Survey Report on Estimates of Major Livestock Products for the Year 2008-09 Gujarat State*. Gandhinagar: Directorate of Animal Husbandry.

———. (2009-10). *27th Survey Report on Estimates of Major Livestock Products for the Year 2009-10 Gujarat State*. Gandhinagar: Directorate of Animal Husbandry.

———. (2010). *Sardar Sarovar Project on River Narmada: Guidelines for District Level Coordination Committee to Expedite Command Area Development for Sardar Sarovar Project*. Gandhinagar: Sardar Sarovar Narmada Nigam Limited. May.

———. (2010-11). *28th Survey Report on Estimates of Major Livestock Products for the Year 2010-11 Gujarat State*. Gandhinagar: Directorate of Animal Husbandry.

———. (2010-11), *Bulletin of Animal Husbandry and Dairy Statistics-2010-11*. Gandhinagar: Directorate of Animal Husbandry.

———. (2010). *Socio-Economic Review 2009-10 Gujarat State*. Gandhinagar: Directorate of Economics and Statistics.

———. (2010). *Meeting the Challenges of Development of Command Area: Sardar Sarovar Project on River Narmada*. (Guidelines for District Level Coordination Committee to Expedite Command Area Development for Sardar Sarovar Project). Gandhinagar: SSNNL.

Gulati, Ashok, Tushaar Shah and Ganga Shreedhar (2009). "Agriculture Performance in Gujarat since 2000: Can it be a Divadandi (Lighthouse) for other States?" *IWMI-IFPRI Working Paper*, May.

Kumar, A. and D.K. Singh (2008). "Livestock Production System in India: An Appraisal Across Agro-Ecological Regions," *Indian Journal of Agriculture Economics* 63(4), Oct-Dec.

Kumar, A., S.J. Stall, N.P. Singh and D.K. Singh (2007). "Livestock Trade of India: Surging Momentum in the New Liberalised Regime," *Indian Journal of Agriculture Economics* 62(3), July-Sept.

Mehta, Lyla (2000). "Drought Diagnosis: Dryland Blindness of Planners," *Economic and Political Weekly*, July 1.

Mian, Rais Uddin Mohammad, J. Fatema and M.H. Rahman (2007). "Impact of Dairy Farming on Livelihood of Participating Women under Grameen Bank in Selected Area of Rangpur District in Bangladesh," *Indian Journal of Agriculture Economics* 62(2), April-June.

NDDB (2010). "Data on Milk Production and Per Capita Availability of Milk by States", Accessed on 31 October 2010 from *http://www.nddb.org/statistics/milkproduction_states.html*, National dairy Development Board, Anand, Gujarat.

Parthasarathy, R. (2010). "Role of Irrigation in the Growth Story of Gujarat", in R.H. Dholakia and S.K. Datta (ed.), *High Growth Trajectory and Structural Change in Gujarat Agriculture*. Ahmedabad: MacMillan Publishers India Ltd.

Pathak, M and V.D. Shah (2010). "Five Decades of Gujarat Agriculture: Some Reflections", in R. Dholakia and S. Dutta (eds.), *High Growth Trajectory and Structural Changes in Gujarat Agriculture*. New Delhi: Macmillan Publishers India Ltd.

Patil, B.R. (2006). *Dynamics of Livestock Development in Gujarat, India: Experiences of an Indian NGO*. PhD thesis. Wageningen University. The Netherland.

Rajarajan, T.R. (2007). "Implication of Trade Liberalisation on Indian Dairy Sector: An Empirical Analysis," *Indian Journal of Agriculture Economics* 62(3), July-Sept.

Ramesh, Chand and S.S. Raju (2008). "Livestsock Sector Composition and Factors Affecting its Growth," *Indian Journal of Agriculture Economics* 63(2), April-June.

Rangaswamy, N. and J.P. Dhaka (2007). "Milk Procurement Cost for Co-operatives and Private Dairy Plants in Tamil Nadu-A Comparison." *Indian Journal of Agriculture Economics* 62(4), Oct-Dec.

Reserve Bank of India (RBI) (2005). *Report of the Expert Group on Investment Credit*. Mumbai.

Roy, A.K. and I. Hirway (2009). "Multiple Impacts of Droughts and Assessment of Drought Policy in Major Drought Prone States in India," *mimeo*. Ahmedabad: CFDA.

Shah, A. (2006). "Changing Interface between Agriculture and Livestock: A Study of Livelihood Options under Dry Land Farming System in Gujarat. *Working Paper* no. 170. Ahmedabad: Gujarat Institute of Development Research.

Sharma, V.P. and H. Thaker (2010). "Livestock Development in Gujarat in 2000s: An Assessment", in R.H. Dholakia and S.K. Datta (ed.), *High Growth Trajectory and Structural Change in Gujarat Agriculture*. Ahmedabad: Macmillan Publishers India Ltd.

Singh, Sukhpal (2007). "Marketing of Liquid Milk: A Case Study of Ahmedabad Milk Market," *Indian Journal of Agriculture Economics* 62(3), July-Sept.

Suresh, A., D.C. Gupta, M.R. Solanki and J.S. Mann (2007). "Reducing the Risk in Livestock Production: Factors Influencing the Adoption of Vaccination Against Bovine Diseases," *Indian Journal of Agriculture Economics* 62(3): 483-91. July-Sept.

Suresh, A., D.C. Gupta, J.S. Mann and U.K. Singh (2008). "Effect of Socio-economic and Agro-ecological Factors on Structure and Ownership of Livestock: Evidence form Rajasthan," *Indian Journal of Agriculture Economics* 63(2), April-June.

11 Futuristic Outlook to Ensure Food Security Through Broad-based Livelihood Activities

Shiv Raj Singh Rathore and K.K. Datta

Introduction

Farmers are engaged in agricultural activities for dual objectives i.e., food self-sufficiency and income generation. Therefore, it is important to know whether farmers meet these two objectives in current agriculture situation or not. Some empirical studies have shown that the farm households derive 72 per cent calories from cereals alone with higher proportion of calories supplemented by own production. Using 1800 kcals/person/day as a threshold level of calorie intake, it is found that around 22 per cent farm households were poor and 25 per cent were food insecure (Singh *et al.*, 2002). However, Suryanarayana (2009) estimated that in rural area around 60 per cent households was food insecure (using 2100 kcals/person/day as a threshold level). The estimates of food insecure farm households in both the studies vary mainly because of calories threshold level to define the food insecurity. On an average a farm household earns ₹ 2115 per month with the average family size of 5.5 (NSSO, 2005). This could mean an average per capita per day income is around ₹ 13, whereas poverty line for rural India during 2004-05 was ₹ 12, implying monthly income to be just enough to sustain the household above poverty line. Overall, it is difficult for the farmers to achieve the objectives of farming i.e., food self-sufficiency and income generation. Though India witnessed a record food production of 245 MT in 2010-11, yet around half of farm households are food insecure. This is a clear paradox. Hence, adequate production does not ensure "access to sufficient and nutritious food at farm household level (Sidhu *et al.*, 2008; Barrett, 2010). Production and income are the two most important determinants of access to food (Sen, 1981) and were defined as 'entitlements' of household, which included endowment (physical, natural, human, social, financial, etc.) and exchange.

The present study seeks to analyse different facets of food security in terms of proportion of food insecure people, their geographic characteristics, how long they remain vulnerable in a year and reasons for such vulnerability. The focus of the paper then turns to assessing the relationship between the food insecurity and self-sufficiency of food production with the special emphasis of seasonality in production. The paper also develops an approach to estimate the marginal effect of an income increase on income distribution, to identify specific income generating activity which ensures the food security and income distribution at household level.

Methodology

Source of Data

The study is based on analysis of 59th NSSO survey round on "Situation Assessment Survey of Farmers" for the year 2002-03. The survey was conducted for rural sector in two rounds, i.e., Visit-I (January to August, 2003) and Visit-II (September to December, 2003).

This survey mainly focused on the rural area by keeping rural farm households as the unit of observation. The NSSO has considered coverage of the farm households on the basis of: (i) if they possessed some land (i.e., land, either owned or leased in or otherwise possessed), and (ii) if they were engaged in some agricultural activities during the last 365 days. The sampling design used in the NSS round was stratified multi-stage random sampling with districts as strata, villages as first stage units and farm households as the second stage units.

Analytical Tools

Food Consumption, Requirements and Gaps

The energy requirements (in calories) as per Indian Council of Medical Research (ICMR) standards for different age and gender groups (see Appendix A-11.1) have been estimated from data available from the Situation Assessment Survey of Farmers in the each state covering both visits. Consumption basket included cereals, millets, pulses, oilseeds, vegetables, milk and milk products, fruits, meat and fish, eggs, and sugar, whereas production basket included all consumption items except vegetables and sugar. By using pre-specified conversion factors (Gopalan *et al.*, 1980) for production and consumption baskets have been separately converted into calories (energy) terms and finally added up to get total calories from food items for both the visits across the households in the each state.

The gaps in calories consumption were computed as difference of requirement and consumption of calories. The method used in computing the gaps in calories is:

Calories Gap = CA – CR

where

CA= Calories available from the item consumed, i.e., sum of the calories of each product, which household consumed.

CR= Calories requirement based on age and sex standards, i.e., normative requirement of the calories (as per ICMR standards).

Based upon calories gap, households were classified into two categories i.e., food secure (calories sufficient) and food insecure (calories deficit). Then food insecure households were further classified into two categories based on calories production from agricultural products as stated below:

Food insecure (-) = Calories deficit households with inadequate production of calories

Food insecure (+) = Calories deficit households with adequate production of calories

The axiom to classify the households into above stated two categories was to verify whether food insecurity at household level was related to the agricultural production system or not.

To examine other equity and distributional aspects of income generation, decomposition method of Gini Index has been applied (Lerman and Yitzhaki, 1985). The analysis determines which income generation activity that is more inclusive and has equalisation effect on income distribution.

Results and Discussion

Region-wise Food Insecurity Status among the Farmers

FAO's *State of Food Insecurity Report* (2002) refers to four elements of food security: food availability, food accessibility, food utilisation and food system stability. Availability focuses on food production whereas accessibility focuses on the ability of people to obtain food, either through production, purchase or transfers. Food utilisation focuses on the nutritional value of food, the interaction with physiological condition and food safety. Food system stability focuses on stability of supply and access, as well as the ability to respond to food emergencies. At farm household level food is generally available from agriculture, but it is not necessary that they consume (accessibility) the sufficient food throughout the year. Therefore, the present research has been focused on all the three elements of food security: food availability, food accessibility and food system stability. Food accessibility in terms of either production or purchase from the market is discussed in detail in this section.

Food accessibility was measured in terms of calories 2108 per capita per day against the requirement of 2033. Almost same estimate (2047 calories per capita per day) was provided for rural area by Government of India (2007) for food accessibility in the 2004-05. By observing these facts it is reflected that overall households are consuming higher calories with respect to requirements, but this is not true.

The maximum number of sample households were located in the Central region (29.37%) followed by Eastern, Southern, Western, Northern and North-Eastern regions (26.37%, 18.08%, 17.54%, 7.31% and 1.33%) respectively. Among them, in the North-Eastern region, the proportion of food insecure households were highest (64.28%) followed by Southern (64.09%), Western (54%) and Eastern (50%) region in Visit-I. As compared to other regions food insecure households were substantially less in Northern region (Table 11.1). The reason of higher food security in the Northern region is steady agricultural growth which has proportional relationship with the food security (Datt and Ravallion, 1998; Ravallion and Datt, 1996; Guhan, 2001).

Table 11.1

Region-wise Food Insecurity among the Farm Households (%)

Regions	*Visit-I*		*Visit-II*	
	Estimated Households	*Food Insecure Households*	*Estimated Households*	*Food Insecure Households*
North	7.31	40.53	7.31	37.33
West	17.54	53.98	17.53	51.73
Central	29.37	48.29	29.41	46.54
East	26.37	50.30	26.35	50.67
South	18.08	64.09	18.08	66.82
North-East	1.33	64.28	1.33	67.02
India	100.00	52.32	100.00	51.81

Source: Authors' estimates based on unit level data of NSSO 59th round on *Situation Assessment Survey of Farmers.*

Seasonal food insecurity was more pronounced almost in all the states of the different regions. For example, in Andhra Pradesh, 5% point more food insecure households were found in Visit-II compared to Visit-I. Same trend was also found in the Central region. If data of Visit-I and Visit-II are compared, it is observed that in the Northern region, seasonal food insecurity was more volatile across the states. It clearly reflected that in the different periods of year, food insecurity among the households varies significantly.

At all India level, 89.35 million farm households were engaged in agriculture activities. The study highlighted that food insecurity among the farm households were 52.32 per cent in Visit-I and 51.81 per cent in Visit-II at the country level. A systematic analysis of study indicates that though there was no significant seasonal change in food insecurity at household level at the country but same may be not true at state level.

The above mentioned facts and figures indicate that at the farm households level, who are supposed to be food secure as they are the real pillar of the agricultural production, are not actually food secure. This has serious implications both at production level as well as consumption level, since same groups of households take both the decision at the household levels. Therefore, an attempt has been made in next section to estimate how many farm households are able to produce enough food from their agricultural activities in both the visits (seasonally).

Food Security/Insecurity with Respect to Agricultural Production System

It is a hypothesis that calories consumption at the farm household level is positively influenced by the agricultural production. However, food self-sufficiency may or may not be a sufficient indicator of food security at the farm household level (Bouis *et al.*, 1998; Maxwell *et al.*, 1998). Therefore, it is important to study the synergy between agriculture and food security among the farm households. In this regard, Table 11.2 gives important inferences and relationship between production of agriculture commodities and food security at the household level.

Table 11.2

Region-wise Food Self-sufficiency among the Farm Households (%)

Regions	*Visit-I*				*Visit-II*			
	Food Insecure		*Food Secure*		*Food Insecure*		*Food Secure*	
	-	+	-	+	-	+	-	+
North	25.34	15.19	24.84	34.63	26.26	11.07	34.47	28.19
West	28.56	25.42	26.06	19.96	45.29	6.44	37.45	10.82
Central	29.17	19.12	27.23	24.48	29.28	17.26	29.78	23.68
East	17.86	32.44	15.80	33.90	40.16	10.51	37.91	11.42
South	37.61	26.48	19.87	16.03	57.78	9.04	28.30	4.88
North-East	17.42	46.86	10.14	25.58	56.39	10.63	28.46	4.52
India	27.20	25.12	22.30	25.38	40.26	11.55	33.31	14.88

Note: (-) = Deficit production of calories w.r.t consumption, (+) = Surplus production of calories w.r.t consumption

Source: Authors' estimates based on unit level data of NSSO 59th round on *Situation Assessment Survey of Farmers.*

In the Southern region, proportional food insecure households (64.09%) were higher as compared to Western (53.98%), Eastern (50.30%), Central (48.29%) and Northern (40.53%) regions. In Southern region, about 37.61 per cent and 57.78 per cent food insecure households could not produce sufficient food to ensure self sufficiency in Visit-I & II, respectively. Similar trend was observed from Western (28.56% and 45.29%), Eastern (17.86% and 40.16%), Central (29.17% and 29.28 %) and Northern (25.34% and 26.26%) regions.

In the Southern region, households were more food insecure as compared to other regions. Since, larger part of the Southern region is arid and semi-arid which is not favourable to grow food crops during the second season (Visit-II) and that may be one of the reasons that food insecurity in Visit-II was much worse than Visit-I in the Southern region.

In Northern region proportional food insecurity among households was low as most of the progressive agriculture states like Punjab and Haryana comprise large number of households, which are specialised in cereal production. The relationship between food security/insecurity with respect to agriculture production did not change in both visits because this region had better irrigation facilities and agriculture prospects.

At all India level, it has been presumed that the food accessibility (consumption) has not strongly related with food availability (agricultural production) at the households level, but our discussion from the evidences shows that at regional level it differs and behave differently across the regions.

Since income and resource endowments plays significant role in the context of accessibility and entitlements to food, in the next section we will try to establish the linkage of these two in order to ensure the food security at the household level.

Income Structure and Food Security

In general, farm households are involved in a range of economic activities for income generation like cultivation, dairying, other livestock, off-farm business and, wages and salaries. Income from some of these activities is more volatile in different seasons/months due to seasonal variations, resource availability and opportunities at household level. Therefore, income from different sources and its magnitude have differential impact in household food security. The activities of households, their degree of specialisation or diversification—and most importantly, incomes that they could derive from their activities—depend on the assets endowments. Some of these sources of income are highly interconnected. For example Lanjouw and Shariff (2002) confirm that land ownership is negatively correlated with non-farm income; and access to land appears to determine whether households remain in agriculture or shift to other activities. It is not surprising that much of the thinking about improving the welfare of the rural poor has focused on access to land and enhancing the productivity of smallholder's agriculture.

Central region of India had the highest food insecure households in absolute term. The average monthly income in this region was lower than national average while in both the cases average land resource per farm is almost identical. The differences in household income (absolute term) was due to lower income from cultivation, off-farm business, wages and salaries, whereas livestock income supplements total family income. Seasonally higher income was observed in Visit-II as compared to Visit-I, despite higher cultivated area in Visit-I. This was because of higher returns from cultivation in Visit-II. Overall it may be inferred that by virtue of lower income in this region the households led to food insecurity. Strategies that increase the income from different activities are the most sustainable means to improving household food security in this region.

In Eastern region, average landholding per household is the lowest. Therefore, absolute income from cultivation and dairying would be lower as compared to other regions. In this region, number of food insecure households is the second largest. Small landholdings is one of most critical factors leading to food insecurity. If income from different sources is analysed among the food insecure households, it is observed that share of income from off-farm activities was higher in this category. Seasonal variation in income from these activities varied drastically; this may be due to the fact that income from off-farm activities was not assured throughout the year.

In Southern region, relatively food insecure households were higher compared to Central and Eastern regions (Table 11.3 and 11.4). However in this region, households earned higher income as compared to Central and Eastern regions. This could suggest that higher income at household level did not contribute to higher food security. Comparatively higher household income was due to higher earnings from cultivation, off-farm business, wages and salaries. However, in food insecure households, income from agricultural activities was lower due to their smallholdings compared to food secure households.

If we look at the income generation activities at household level in North-Eastern region (Table 11.3 and 11.4), it is found that maximum share in household income was contributed by cultivation activity (59-63% in Visit-I). However in subsequent period (Visit-II) this share was reduced, thus resulting in increase in food insecure households. The prime reason of reduced agricultural income was operational landholdings in Visit-I was higher than Visit-II. Peterson (1989) concluded that irrigation is probably the most promising means of reducing food insecurity. In North-Eastern region, the income share from dairying was lower as compared to other regions, but income from other livestock activities was higher. However, in other regions, income from dairying play important role to ensure the food security at household level. Therefore, there is a need to integrate the dairying activities in a progressive way to ensure the long term food security in this region.

Table 11.3

Landholding and Income of the Farm Households in Visit-I

Status		*Income from (%)*					*Total Income (₹/Month)*	*Land (ha.)*
		Cultivation	*Dairying*	*Other Livestock*	*Non-farm Business*	*Wages and Salaries*		
				North				
Food Insecure	-	24.23	11.31	0.56	23.97	39.92	4566	0.36
	+	51.91	18.82	1.32	16.21	11.75	9283	1.40
				West				
Food Insecure	-	31.73	9.53	2.98	20.92	34.85	3235	1.38
	+	47.10	13.13	1.64	23.05	15.07	6059	1.79
				Central				
Food Insecure	-	25.24	10.89	1.20	24.13	38.53	1947	0.89
	+	55.93	13.85	1.29	15.26	13.67	3622	1.30
				East				
Food Insecure	-	14.32	3.20	1.00	43.16	38.31	2586	0.30
	+	52.27	7.09	1.19	20.95	18.50	3296	0.82
				South				
Food Insecure	-	31.26	4.93	1.26	26.86	35.69	3544	0.83
	+	61.15	10.05	0.60	12.41	15.79	5033	1.44
				North-East				
Food Insecure	-	58.82	1.19	2.96	13.53	23.51	5003	1.06
	+	62.79	3.17	3.03	11.39	19.63	4587	1.08
				India				
Food Insecure	-	26.98	7.49	1.50	27.16	36.86	2904	0.83
	+	53.77	11.40	1.24	17.92	15.66	4488	1.25

Source: Authors' estimates based on unit level data of NSSO 59th round on *Situation Assessment Survey of Farmers*.

In the Western region, households were relatively better placed compared to Southern and North-Eastern regions. Household monthly income was higher in comparison to Central and Eastern regions (Table 11.3 and 11.4). Agriculture here is dependent on rainfall and large part of the region is arid/semi-arid. In Visit-I operational landholdings was substantially higher compared to Visit-II, which contributed to higher income in Visit-I compared to Visit-II. Income from cultivation activities varied widely because of variation in irrigation availability. The households reported higher income (₹ 862/month in Visit-I) from livestock activities compared to other regions except the North. Income from livestock activities accounted 13-21 per cent in Visit-I whereas in Visit -II it was 15-33 per cent (Table 11.3 and 11.4). Those households who did not produce sufficient food were more dependent on off-farm business, wages and salaries for income generation.

Table 11.4

Landholding and Income of the Farm Households in Visit-II

Status		*Income from (%)*					*Total Income (₹/Month)*	*Land (ha.)*
		Cultivation	*Dairying*	*Other Livestock*	*Non-farm Business*	*Wages and Salaries*		
				North				
Food Insecure	-	17.11	12.60	0.64	35.37	34.28	5165	0.35
	+	51.10	25.32	0.37	7.02	16.19	8222	1.21
				West				
Food Insecure	-	18.37	12.20	2.30	35.21	31.93	2702	0.62
	+	43.43	27.86	1.59	15.69	11.45	5596	1.26
				Central				
Food Insecure	-	34.41	10.10	0.94	21.95	32.59	2159	0.60
	+	63.72	14.30	1.15	11.68	9.15	3789	1.10
				East				
Food Insecure	-	25.77	4.24	1.49	34.16	34.33	2336	0.31
	+	52.22	14.65	1.54	17.48	14.11	3509	0.63
				South				
Food Insecure	-	22.96	6.13	0.70	33.73	36.48	3125	0.44
	+	42.83	23.40	2.19	16.32	15.25	4712	1.04
				North-East				
Food Insecure	-	43.27	2.27	3.49	21.66	29.30	3512	0.53
	+	38.12	7.46	5.67	21.16	27.59	3768	0.85
				India				
Food Insecure	-	24.50	8.06	1.28	32.00	34.15	2734	0.47
	+	53.74	18.86	1.75	13.32	12.33	4354	1.00

Source: Authors' estimates based on unit level data of NSSO 59th round on *Situation Assessment Survey of Farmers*.

In Northern region, around 37-40 per cent households were food insecure. Food insecurity among the households in this region is important concern, because this region ensures the national food security. Household income here is higher than national average (Table 11.3 and 11.4). The region has lowest proportional food insecure households. Interestingly, the operational landholdings remained unchanged during both visits due to inclusive irrigation of all categories of farmers especially in Haryana and Punjab. This results in unchanged proportional income share from cultivation in both the visits. Income from dairying was highest (₹ 1510/ month in Visit-I) compared to other regions. As discussed earlier, in this region 37-40 per cent households were food insecure—among them around 25-26 per cent households could not produce food self-sufficiently, because of their landholdings were very much small (0.36 ha.). In this region small landholding size was one of the important determinant for household food security. The income from dairying activities was also lower in this category as compared to other categories of households. Small landholding does not allow this category of households to rear more dairy animals because availability of feed and fodder from smallholding is not sustainable in the long run. Therefore, there is a need to buildup new institutions like cooperatives to ensure forward and backward linkages to the dairying enterprises especially to the smallholders.

At all India level, around half of the households were food insecure. Among them, 27.20 per cent (Table 11.2) households were unable to produce food self-sufficiently and their principal source of income was off-farm business, wages and salaries (64% and 66% in Visit-I and Visit-II respectively). The higher dependence of income from these activities was relatively lower as compared to other categories. However, in another food insecure category, 25.12 per cent households were food self-sufficient, but they were not consuming enough calories as per their reported consumption pattern in the survey round. At regional level, there are mixed results that income from agricultural activities in some regions was helpful to ensure the food security at household level. The reason of this pattern may be that there may be 'process at work in the rural economy' which offset the food security at household level by increasing the income from agricultural activities.

Path to Ensure the Food Security and Inclusive Growth

It is important to take note of how economic trends and government policies affect the distribution of income in the society. This section seeks to answers some of these issues. What are the different sources of income inequality? Does a marginal increase in a particular income source increase or reduces the inequality? Does income from dairying, other livestock, wages and salaries work as an income equaliser effect in the rural economy? In order to examine the distribution pattern of the household income from different income sources, and to understand how those are correlated, source-wise decomposition of Gini index was estimated by using Lerman and Yitzhaki (1985) method. Total incomc was divided into five sources of income: cultivation, dairying, other livestock, non-farm business, wages and salaries. The study identifies the contribution of each of the five sources of income to overall income inequality. The share of overall inequality contributed by each income source is also measured.

The Gini inequality of total household income was estimated at 0.7685 (Table 11.6) indicating that income distribution was moderately unequal. The analysis further established that in the total household income, the share of cultivation income was highest (39.40%) followed by non-farm business (25.31%), wages and salaries (19.92%), dairying (14.08%) and other livestock (1.29%) income. It was observed that 1 per cent incremental increase in cultivation and non-farm business income will trigger total income inequality by 0.15 and 4.42 per cent respectively with a caveat that other things are unchanged. On the other hand, the income from dairying, other livestock, wages and salaries has a equalising effect on the distribution of total income for households, which otherwise corroborates the hypothesis of relative income equalising effect through dairying, other livestock, wages and salaries compared to distribution of incomes through cultivation and off-farm business. Though the income share from dairying and other livestock was smaller, but its contribution towards rural livelihood was more secured than other sources. This re-emphasises the importance of dairying and other livestock farming system for its doubly beneficial social impact in improving incomes and reducing income inequality (Mandal *et al.*, 2010). It also confirms that growth through inclusive dairying and other livestock enterprises does not worsen income distribution, but helps in reducing food insecurity. As mentioned earlier, wages and salaries income was important source of income for household food security. Therefore, the Mahatma Gandhi National Rural Employment Guarantee Act (MGNREGA) of the Government of India is a step in right direction to ensure employment guarantee that could be justified from social and distributive standpoints.

Table 11.5

Decomposition of Source-wise Household Level Income Inequality in India

Source	*Income Share*	*Gini of Sources*	*Correlation with Rank of Total Income*	*Percentage Contribution to Total Inequality*	*Marginal Effect*
Cultivation	0.3940	0.8596	0.8974	0.3955	0.0015
Dairying	0.1408	0.8675	0.8147	0.1295	-0.0113
Other livestock	0.0129	0.9612	0.6261	0.0101	-0.0028
Non-farm business	0.2531	0.9771	0.9238	0.2973	0.0442
Wages and salaries	0.1992	0.8734	0.7404	0.1676	-0.0316
Total income		0.7685			

Source: Authors' estimates based on unit level data of NSSO 59th round on *Situation Assessment Survey of Farmers*.

Conclusions and Policy Issues

Using the NSS 59th round data for the year 2002-03, the analysis provides us the evidences that around half of farm households were food insecure in rural India. Spatially, proportional food insecurity was higher in North-Eastern region followed by Southern, Western and Eastern region. However, in absolute terms food insecurity was higher in Central region followed by Eastern, Southern and Western region. Higher food insecurity among the farm households has a serious implication for country's food security because they are the granaries of the India.

The analysis provides the evidence that at farm household level, food availability did not ensure the food accessibility, whereas the accessibility was directly link with the available resources (land) and income. However, higher income always did not ensure the food accessibility. Food accessibility is mainly guided by the nature of returns generated from different income generation activities. For example, daily, monthly, and annual return from different sources of income has different dimensions and magnitudes in the context of household food security. The study points out that seasonal food insecurity was more pronounced in arid, semi-arid and rainfed states like all the North-Eastern states, Southern states and part of the Central, Eastern and Western states of India, where irrigation facilities are not adequate, therefore household consume lower calories. To ensure the food security at household level, there is a need to focus on two broader issues i.e., reorientation of current agriculture production system and providing off-farm income opportunities. Decomposition analysis suggested that income from livestock activities and, wages and salaries is more inclusive and it's ensuring food security at household level because, income from these activities is continuous and regular.

References

Barrett, C.B. (2010). "Measuring Food Insecurity", *Science*, 327, 825-828.

Bouis, H.E., B. Briere, L. Guitierrez, K. Hallman, N. Hassan, O. Hels, W. Quabili, A. Quisumbing, S. Thilsted, Z.H. Zihad and S. Zohir (1998). *Commercial Vegetable and Polyculture Fish Production in Bangladesh: Their Impacts on Income, Household Resource Allocation, and Nutrition*. Washington, D.C.: International Food Policy Research Institute.

Datt, G. and M. Ravallion (1998). "Farm Productivity and Rural Poverty in India", *FCND Discussion Papers* 42. Washington, D.C.: International Food Policy Research Institute.

FAO (2002). *The State of Food Insecurity in the World 2001*. Rome, Italy.

Gopalan, C., Rama Sastri, B.V. and S.C. Balasubramanian (1980). *Nutritive Values of Indian Foods*, 3rd Ed. New Delhi: Indian Council of Medical Research.

Government of India (2007). *Nutritional Intake in India 2004-2005*, NSS 61st Round (July 2004-June 2005), Report No. 513. New Delhi: National Sample Survey Organization, Central Statistical Organization, Ministry of Statistics and Program Implementation.

Guhan, S. (2001). *Rural Poverty Alleviation in India, India's Development Experience* (Edited by Subramanian, S.). Oxford India Paperbacks.

Lanjouw, P. and A. Shariff (2002). "Rural Non-Farm Employment in India: Access, Income and Poverty Impact", *NCAER Working Paper Series* No. 81. New Delhi.

Lerman, R.I. and S. Yitzhaki (1985). "Income Inequality Effects by Income Source: A New Approach and Applications to the United States", *Review of Economics and Statistics* 67: 151-56.

Mandal, S., K.K. Datta and T.D. Lama (2010). "Economic Evolution of Farming System Research in NEH Region: Some Issues", *Indian Journal of Agricultural Economics* 65(1): 118-34.

Maxwell, D., C. Levin and J. Csete (1998). "Does Urban Agriculture Helps to Prevent Malnutrition? Evidence from Kampala", *Food Policy* 23(5).

National Sample Survey Organization (NSSO) (2005). *Report of Income, Expenditure and Productive Assets of Farmer Households*. New Delhi: Ministry of Statistics and Programme Implementation, Government of India.

Pal, Manoranjan and Premananda Bharati (2010). *Development of Methodology Towards Measurement of Poverty*. Kolkata: Indian Statistical Institute.

Peterson, D.E. (1989). *Irrigation and Food Security*. Manila, Philippines: International Rice Research Institute, and Washington D.C.: the American Association for the Advancement of Science, pp.515-32.

Ravallion, M. and G. Datt (1996). "How Important to India's Poor is the Sectoral Composition of Economic Growth?", *World Bank Economic Review* 10(1): 1-25.

Sen, A. (1981). *Poverty and Famines: An Essay on Entitlements and Deprivation*. Oxford: Clarendon Press.

Sidhu, R.S., I. Kaur and K. Vatta (2008). "Food and Nutritional Insecurity and its Determinants in Food Surplus Areas: The Case Study of Punjab State", *Agricultural Economics Research Review* 21: 91-98.

Singh, R.B., P. Kumar and T. Woodhead (2002). "Smallholder Farmers in India: Food Security and Agricultural Policy, RAP Publication 2002/03. Bangkok, Thailand: FAO Regional office for Asia and the Pacific.

Suryanarayana, M.H. (2009). "Food Security: Beyond the Eleventh Plan Fiction", *Indian Development Report 2009.*

Acknowledgements:

We would like to thank the Research Program on Markets, Institutions and Policies of ICRISAT, for research assistance under Village Dynamics in South Asia Competitive Research Fellowship. Many thanks go to Mr T.N. Datta from NDDB for his valuable comments on earlier versions.

Appendix A-11.1

Energy Requirements of ICMR for Different Age Groups Separately for Males and Females

Age Groups (Years)	*Daily Energy Requirements (Kcal/day)*	
	Males	*Females*
0-3	1064	1064
4-6	1690	1690
7-12	2070	1960
13-18	2640	2060
19 or more	2425	1875

Source: Pal and Bharati (2010).

List of States Covered Under Different Regions

Regions	*States*
North	Jammu & Kashmir, Himachal Pradesh, Punjab, Uttaranchal and Haryana
West	Rajasthan, Gujarat and Maharashtra
Central	Uttar Pradesh, Madhya Pradesh and Chhattisgarh
East	Bihar, West Bengal, Jharkhand, Odisha and Assam
South	Andhra Pradesh, Karnataka, Kerala and Tamil Nadu
North-East	Meghalaya, Nagaland, Sikkim, Arunachal Pradesh, Manipur, Mizoram and Tripura

Trends in Costs and Returns in Agriculture

12 Declining Profitability of Paddy in Andhra Pradesh

An Inter-regional Study of Costs and Returns

R.V. Ramana Murthy, T. Satyanarayana and E. Naveen Kumar

Andhra Pradesh (AP) is the second highest producer of rice in India and rice is the major crop grown in the state. Thanks to the Green Revolution strategy, it had not only become a self-sufficient state by early 1990s, but it also emerged as the second highest net contributor to the central pool after Punjab. Paddy is the single major food crop on which majority of farmers depend for their livelihood in the state. An effective state intervention in the past that helped the proliferation of paddy farming in the state now seems to have lost its sheen. In the recent period, there was a growing concern over a steep rise in cost of cultivation of paddy way above the minimum support price (MSP). Surplus production situation was turning nightmarish for the farmers as the flawed procurement system in the state denied even the lowly-fixed MSP to the farmers. This paper deals with three issues: first, it undertakes an independent study of the current cost of production in AP in light of criticism of official estimates; second, it examines the profitability in paddy farming; third, it analyses the emerging regional variations in cost of cultivation in the state. Finally, it suggests certain policy measures regarding pricing of paddy.

Rice Production in Andhra Pradesh: An Overview

Before proceeding to examining costs and returns, we will take a look at the production trends and some issues related to the structure, which have some policy implications. Rice production, according to official data, in the state has grown from a level of 4.8 million tonnes in 1970-71 to 14.45 million tonnes in 2010-11, marking a 300 per cent increase (Ramanamurthy and Mishra, 2012). Such a continuous increase in production is contributed primarily by an increase in area under rice and seconded by an increase in yield. There are three broad phases in which area under rice cultivation has expanded in the state. First, during the Green Revolution period during 1971-1981, when the high yielding varieties (HYVs) were first introduced in the Godavari-Krishna Delta areas. The second phase of this rise is during 1982-1992, which is a period of expansion of canal irrigation. The third phase is period of overwhelming borewell expansion during 1993-2010. The steady growth of area and production is occasionally disturbed by spells of poor monsoon. However, growing demand and overall stability in production and returns made possible by the state intervention encouraged paddy cultivation to proliferate into several non-traditional areas also. Thus, rice production grew from 10.62 million tonnes in 1989-90, to 12.45 million tonnes in 2002-03 and to 14.35 million tonnes in 2010-11 (see Figure 12.1 and Table 12.1). What is more significant is the fact that even the per capita rice production has increased from 132.03 kg/year in 1981-82 to 177.61 kg/year in 2010-11. This growth not only met the public distribution system (PDS) requirements of the state, but led to an increased contribution to the central pool from 1993-94 onwards. The net contribution to central pool increased from 1.75 million tonnes in 1993-94 to 5.63 million tonnes in 2010-11 (Ramanamurthy and Mishra, 2012). Further, the aggregate demand for rice in the state is estimated to be to the tune of 10.45 million tonnes indicating some kind of surplus situation from 2001-02 (Ibid.).

The long term compound growth rate of rice production in the state during 1970-11 was 2.71 per cent (Table 12.2). After the peak growth rate of 3.6 per cent during the initial Green Revolution period during 1973-83, it fell down to 1.35 per cent post-Green Revolution period during 1983-1991. It has once again accelerated to 3.36 and 3.12 per cent respectively during 1991-2001 and 2001-2011 respectively. In the last one decade, the growth is largely helped by a growth in area which grew at 2.03 per cent per annum, while yield grew at 1.07 per cent for the state as whole. In the last one decade during 2001-2011 the *rabi* production has steadily grown due to the growth of areas as well as yield. The productivity of *rabi* has particularly improved more than that of *kharif* crop.

Figure 12.1

Growth of Rice Production in Andhra Pradesh (in million tonnes)

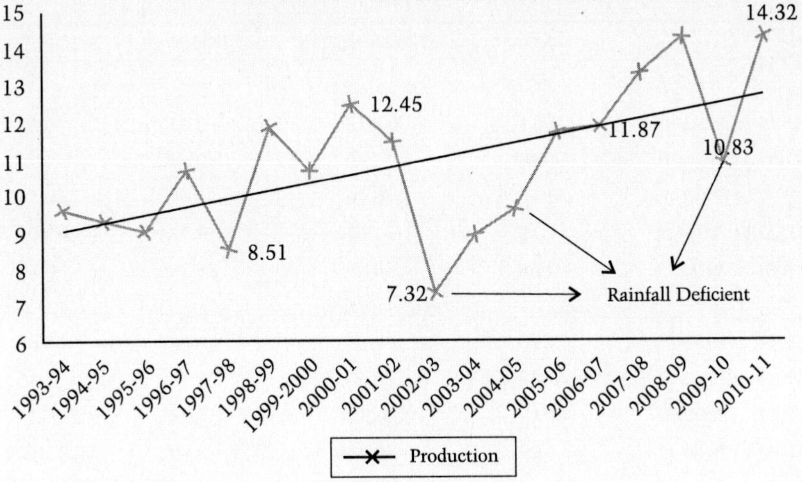

Source: *Statistical Abstract*, DES, GoAP.

Table 12.1

Area, Production and Yield of Rice in Andhra Pradesh

Years	*Area (million ha)*			*Production (million tonnes)*			*Yield (qt per ha)*		*Per cent Rainfall Deviation (mm)*
	Kharif	*Rabi*	*Total*	*Kharif*	*Rabi*	*Total*	*Kharif*	*Rabi*	
2000-01	3.0.	1.29	4.24	8.23	4.24	12.45	27.41	34.01	+0.96
2001-02	2.4	1.39	3.82	6.50	4.89	11.39	26.79	34.96	-6.12
2002-03	2.1	1.77	3.82	5.05	2.23	7.27	23.97	31.89	-38.38
2003-04	2.1	1.87	3.97	5.84	3.11	8.95	27.72	35.94	-22.79
2004-05	2.2	1.87	3.08	6.39	3.28	9.60	28.86	36.83	-33.97
2005-06	2.5	1.45	3.98	6.37	5.37	11.74	25.24	36.59	+8.60
2006-07	2.5	1.37	3.97	6.94	4.31	11.87	26.31	36.81	-7.84
2007-08	2.5	1.41	3.98	8.19	5.13	13.32	31.78	36.50	-1.827
2008-09	2.8	1.58	4.39	8.38	5.86	14.24	29.89	37.00	-12.90
2009-10	2.0	1.38	3.44	5.95	4.88	10.88	28.87	35.43	-30.21
2010-11	2.9	1.76	4.68	7.50	6.79	14.32	25.58	42.67	+22.90

Note: Normal rainfall is about 930 mm.

Source: *Statistical Abstract*, DES, GoAP.

Table 12.2

Compound Growth Rates of Area, Production and Yield of Rice in Andhra Pradesh

Year	*Andhra*	*Rayalaseema*	*Telangana*	*A.P.*
		Area		
1973-74 to 1982-83	1.51	-3.13	0.81	0.82
1983-84 to 1990-91	-0.60	-3.70	0.60	-0.45
1991-92 to 2000-01	0.48	-0.42	1.69	0.83
2001-02 to 2010-11	1.35	-1.62	3.73	2.03
1973-74 to 2010-11	0.66	-0.99	1.34	0.83
		Production		
1973-74 to 1982-83	4.90	-2.29	3.06	3.60
1983-84 to 1990-91	0.33	-0.60	3.73	1.35
1991-92 to 2000-01	3.15	0.95	4.28	3.36
2001-02 to 2010-11	2.09	-0.55	5.99	3.12
1973-74 to 2010-11	2.74	0.38	3.28	2.71
		Yield		
1973-74 to 1982-83	3.34	0.86	2.23	2.76
1983-84 to 1990-91	0.94	3.22	3.11	1.80
1991-92 to 2000-01	2.66	1.37	2.54	2.50
2001-02 to 2010-11	0.73	1.09	2.18	1.07
1973-74 to 2010-11	2.06	1.39	1.91	1.87

Source: Estimated from the data of DES, AP Govt.

It is important to note that paddy grown during *kharifi* season is a six-months crop, belongs to superior quality and has a large local market,[1] because of which open market prices normally tend to be above the MSP for this category. The paddy grown during *rabi* belong to inferior quality,[2] virtually has no local market, and thereby open market price tend to be below the MSP. In either case, millers have been procuring paddy from the farmers, sell the milled *kharif* rice within the state and *rabi* rice to Food Corporation of India (FCI). Good portion *rabi* paddy needs to be parboiled

1. The varieties grown in this season are known as Sona Masuri, Vijay Masuri, (also known as BPT) and Samba Masuri.
2. Varieties such as IR1001, IR1010, Swarna etc., short during 90-day crop.

due low moisture content, such parboiled rice has a market only in Kerala and south Tamil Nadu. The peculiar situation of a steady rise in the area and production of *rabi* rice, which has no local market, has created a surplus situation in the state. FCI after procuring roughly about 8-9 million tonnes of rice, sends back 3.5 million tonnes for public distribution and other schemes. This segmented structure of production is highly relevant for any policy intervention. Though FCI is the default purchaser, it purchases rice through millers in the state, instead of procuring paddy from the farmers, while the latter are expected to pay MSP to the farmers. This arrangement often poses difficulties during bumper crops, when millers depress the same and deny the farmers even MSP. When MSP itself is not covering the full cost of production, the open market prices too seem to betray the farmers in the recent times. We shall examine this in the following section.

Further, the age-old practice in traditional rice growing Coastal districts is that there are virtually no agricultural market yards and millers buy paddy through commission agents, who lift the stock at the farm level. Commission agents extend credit to the farmers and in turn bind the latter to sell the output through them at a fixed commission. Measurement malpractices and arbitrary grading methods earn the former additional revenue. There were no complaints as long as market prices were higher than the MSP. However, with growing surplus rice output in the state, the system is lending itself to the millers in depressing the market prices. In Telangana areas, situation relatively appears to be little better with the existence of agricultural markets yards and official agencies making their presence felt through procurement.

Prices, Costs and Returns to Paddy Farming

Farm Harvest Prices and Minimum Support Prices

In Andhra Pradesh, the farm harvest prices of paddy, which are the market prices received by the farmers in the *mandies*, in the last two decades moved in tandem with the MSP, barring the two years during 2009-2011. The comparison of farm harvest price with MSPs is normally done given the fact it is the latter that singularly raises the former by acting as the floor price and the point to be here is that the market price has mostly remained marginally above MSP for most of the years.[3] Since 1998-99, the farm harvest prices of rice have remained marginally above the MSP for both Grade-A and common variety (see the Figure 12.2a and 12.2b). However during 2010-11, both of them have fallen below the MSP. In fact, a closer look at the data suggests that as far as prices of common variety are concerned, not only have they remained below the MSP since 2008-09 by ₹ 50-60, they have even fallen short by nearly ₹ 200 in 2010-11 (see Table 12.3). The average farm harvest prices is a statistical artefact of the actual prices farmers receive, and can considerable vary since they cover only prices recorded at agricultural market yards.

Table 12.3

Farm Harvest Prices and MSP for Paddy Andhra Pradesh

	Farm Harvest Price			*Minimum Support Price*	
Year	*Kharif Grade-A*	*Rabi Common*	*Average*	*Grade-A*	*Common*
1998-99	510.53	426.17	486.35	470	440
1999-00	559.42	524.44	538.66	520	490
2000-01	507.93	478.58	499.61	540	510
2001-02	565.77	529.40	550.16	560	530
2002-03	623.29	578.08	609.27	560	530
2003-04	569.37	572.27	570.38	560	530
2004-05	605.49	604.07	605.02	580	550
2005-06	648.21	579.59	616.95	600	570
2006-07	670.19	622.22	650.30	610	580
2007-08	777.76	767.17	773.66	675	645
2008-09	1035.75	892.28	963.56	921	950
2009-10	1100.67	955.76	1072.66	1030	1000
2010-11	880.65	850.76	865.70	1080	1030
2011-12	1100.00	1080.00	1090.00	1110	1080

Source: DES, AP Govt.

3. The crucial point that needs to be understood is the fact that the minimum support policy is what basically sustains the market price for the farmers and the absence of it can only imagined.

Figure 12.2a

Market Price and MSP (Grade-A)

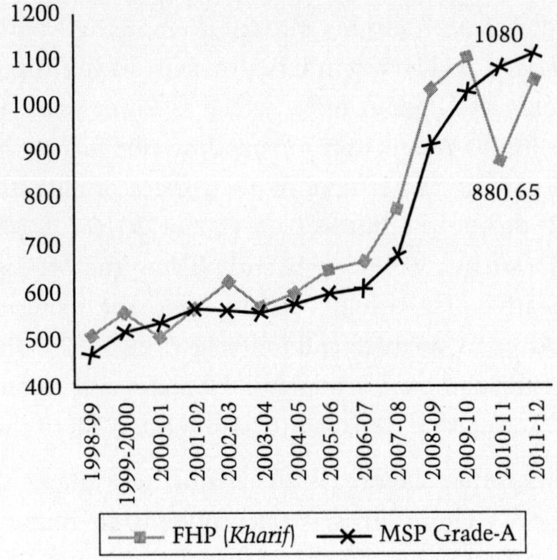

Figure 12.2b

Market Price nd MSP (Common)

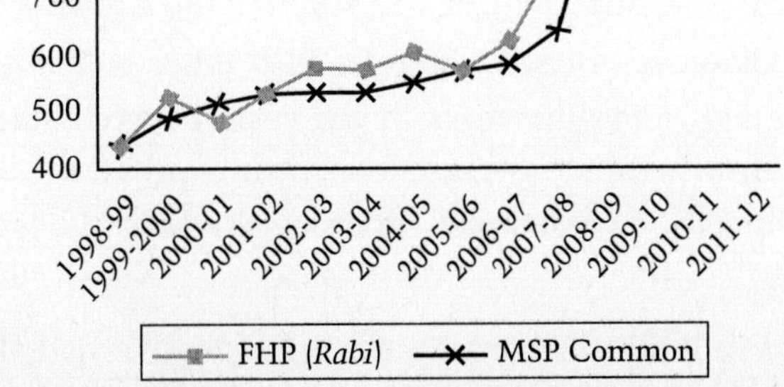

Source: Directorate of Statistics, GoAP.

Curiously, it is not only the prices of common varieties of paddy that have fallen below MSP, but even those of fine and super fine varieties have fallen to the level of MSP (Grade-A) and even below in 2010-11. This is probably indicative of surplus situation emerging even in *kharif* rice. Considerable *kharif* paddy was offered to FCI during this year, and FCI agreed to buy only at Grade-A price, which is lower than superfine price.[4] Thus, current surplus production in superfine rice has pushed their prices below the MSP. At the same time, market prices of all varieties had been below the MSP during the bumper harvest of 2010-11. Market prices of common (*rabi*) varieties were consistently below the MSP since 2008-09. These are indicative of two major issues: first, the rising production has made market prices to stagnate and fall below the MSP owing to a surplus production over the domestic demand of the state; and second procurement mechanisms in the state have failed to assure the MSP to the farmers.

MSP, among other factors, is the major instrument that influences the market prices. There are several problems in fixing the MSP and methodology of Commission for Agricultural Costs and Prices (CACP) in estimating the cost of production. First, the MSP announced by the Central Government, based on all-India average cost of production has often not covered the costs in several states. Second, given a prolonged time-lag of 3-4 years in processing the cost of cultivation data by the CACP, the annual projections are made using wage index and input cost index; both of them are grossly underestimated. Third, the Ministry of Agriculture is fixing MSP lower than what CACP is recommending. The putative concern to contain food grain inflation by pegging MSP conservatively seems to be overriding the other objectives. This has led to a serious situation, particularly in states like AP, where prices fell the costs of production and made the returns negative. We shall marshal the evidence to prove this conjecture.

Besides above mentioned, there are also other problems with CACP methodology. First, the differences in the costs between states are erased in construction of national average, benefitting only few low cost states. Second, the official data is collected from an extremely thin sample, hence,

4. FCI takes only the length of grain as the basic criterion for grading, by this rigid definition even some superfine varieties are graded as common rice, while some superfine are considered as Grade-A.

is unable to represent the true costs. Third, discrepancy between actual and projected costs because of components such as rent, interest, transport costs, fixed costs etc., being underestimated. Lastly, there are differences within agro-climatic regions of a state according to sources of irrigation, which is not analysed and understood well from the official data. For many reasons, the official data renders unhelpful to examine the actual situation.

Cost of Cultivation: The Sample Study for Andhra Pradesh

In order to get a more realistic picture on cost of cultivation in the state, the present study has collected primary data on cost of cultivation, prices, and other variables such as yield, acreage, varieties grown and marketing needed to construct cost of production. The data was collected during 2010-11 in the state through a structured questionnaire,[5] covering 192 households in 13 villages from eight districts, namely East Godavari, Krishna, Karimnagar, Nizamabad, Warangal, Medak, Nalgonda and Mahaboobnagar, based on a stratified random selection method. The districts selected were primarily rice growing ones spanning in three agro-climatic regions namely, Krishna-Godavari Delta, North Telangana and South Telangana. East Godavari and Krishna district in the sample were completely under canal irrigation, while other districts had canal as well as well-irrigation.

The class distribution of 192 sample households constituted 39 marginal, 118 small, and 35 medium operational households (Table 12.4), forming 20.3 per cent, 61 per cent and 18.7 per cent respectively in the total sample. The land held by these three classes was 15 per cent, 49 per cent and 36 per cent respectively. Among these, 69.3 per cent of the farmers were owner-cultivators, 15.6 per cent were owner-tenants,[6] and 15.5 per cent were tenants (which include 1.5% sharecroppers, Table 12.5). Owner cultivation was found to be prominent in Telangana, while tenancy was dominantly observed in Coastal districts. In East Godavari and Krishna district tenancy appeared to be between 60-80 per cent. The pattern observed in the sample was in consonance with the average distribution of operational households in the state as observed earlier. The growing

5. The data is collected during November month and for the rabi season.
6. Owner-tenant is one who owns some land and also leases-in some land.

tenancy in the state was noted by several other researchers (Ramachandran *et al.*, 2009; Vijay, 2006).

Table 12.4

Class-wise Distribution of Sample Households

District	*Marginal Farmers <1 Ha*	*Small Farmers 1-2 Ha*	*Medium Farmers 2-5 Ha*	*Large Farmers > 5 Ha*	Total
East Godavari	9	16	6	0	31
Krishna	3	7	3	0	13
Karimnagar	13	24	5	0	42
Mahabubnagar	3	19	4	0	26
Medak	4	13	3	0	20
Nalgonda	3	25	11	0	39
Nizamabad	1	5	2	0	8
Warangal	3	9	1	0	13
Total	39	118	35	0	192

Source: Field study.

Table 12.5

Types of Cultivators among Sample Households

District	*Own-cultivators*	*Owner-tenant*	*Tenants*	*Share-croppers*	*Total*
East Godavari	4	10	17	0	31
Krishna	5	3	5	0	13
Karimnagar	31	11	0	0	42
Mahabubnagar	21	0	4	1	26
Medak	15	5	0	0	20
Nalgonda	38	0	0	1	39
Nizamabad	7	1	0	0	8
Warangal	12	0	0	1	13
Total	133	30	26	3	192
	(69.3)	(15.6)	(13.5)	(1.5)	100.0)

Source: Field study.

Estimated Cost of Cultivation

The estimated cost of cultivation for the sample in all the six classifications defined by CACP is presented in Table 12.6.[7] The cost categories relevant for owner cultivators are Cost A1, B1 and C1, while those relevant for tenant farmers are Cost A2, A2+FL, B2, C2 and C3. We observed that there are significant differences between that of tenants and owner-cultivators, and between the regions namely, Telangana and Coastal Andhra. The differences in costs between classes of farmers are observed to be negligible.[8] Therefore, our comparisons are presented type-wise and region-wise in our analysis. When paid-out costs (Cost A1) are considered, we observe that they range between ₹ 18,162-24,632 per acre. The average paid-out cost (i.e., Cost A1) is ₹ 20,380 per acre. Barring Mahabubnagar district, differences between the regions are not very large in terms of paid-out costs, though the components of costs differ between the two. The Cost A1 per acre ranged between ₹ 20,196 in Karimanagar district and ₹ 18,162 in Medak in Telangana region. The same in Coastal delta region is ₹ 21,623 in East Godavari and ₹ 24,623 in Krishna.

But there are significant differences in Cost A2 between regions, which includes rent. The Cost A2 on average is 30 per cent higher in Coastal districts compared to Telangana districts. Consequently, these differences remain the same in Cost A2+FL, Cost B2, Cost C2 and the overall cost C3, all being higher in Coastal districts. The reasons are: first, the rental costs are quite high in coastal regions. Second, their labour component is relatively still higher, while farmers in Telangana region are saving labour costs by hiring harvesters. Thereby, even the total cost (Cost C3) is 20-25 per cent higher in Coastal region than in Telangana.

7. The cost of cultivation scheme developed by Commission for Costs and Prices defines five cost concepts, namely, (i) Cost A1 (all fixed and variable paid out costs excluding rent), (ii) Cost A2 (Cost A1+rent), (iii) Cost A2+(FL), (iv) Cost B1 (Cost A1+Interest on value of owned fixed assets excluding land), (v) Cost B2 (Cost A2+imputed rent on own land), (vi) Cost C1 (Cost B1+imputed family labour); (vii) Cost C2 (Cost B2+imputed family labour); (viii) Cost C2* (Cost C2 using Minimum Wages if they are higher than market wages); (ix) Cost C3 (cost C2*+10 per cent managerial input over C2*).

8. The difference in cost of cultivation between the size classes of the sample is negligible due lack of differences in technology between the size classes, and the difference in scale between the marginal, small and semi-medium are not large enough to bring changes in technology and thereby costs. This is well known in the literature (see for instance Rudra, 1982).

Table 12.6

Cost of Cultivation per Acre in Andhra Pradesh in Rabi 2010-11 in Sample Households (₹)

District	East Godavari	Krishna	Karim-nagar	Mahabub Nagar	Medak	Nalgonda	Nizama-bad	Waran-gal	State Average
Cost A1	21643	24623	20196	20909	18162	19725.62	19507	22437	20380
Cost A2	30084	25914	23184	23009	20112	22225	22007	24437	23872
A2+FL	35413	32051	28105	27294	25844	25531	24091	30278	28576
Cost B1	21645	20490	20248	20979	18231	19788	19563	22515	20432
Cost B2	38085	31126	25736	25079	22181	24788	24563	26515	27259
Cost C1	26973	26627	25169	25264	23963	23094	21647	28356	25137
Cost C2	43414	37264	30657	29364	27913	28094	26647	32356	31964
Cost C3	47755	40990	33723	32301	30704	30904	29312	35592	35160
Yield	26	21	27	17	27	25	28	27	

Source: Field study.

The distribution of sample households is found to be satisfactory with relatively few outliers. The coefficients of variation of the various costs across districts in the sample are presented in Table 12.7. The coefficient of variation of the overall sample is about 14.5 per cent. The coefficient of variation of one district, i.e., Nalgonda, is highest among the sample at about 25 per cent for Cost A1 and 31.21 per cent for Cost C1.

Table 12.7

Coefficent of Variation in Cost of Cultivation of the Sample Households

	Cost A1	Cost A2	A2+FL	Cost B1	Cost B2	Cost C1	Cost C2	Cost C3
East Godavari	15.04	11.51	10.54	15.04	9.09	13.15	8.60	8.60
Krishna	14.22	13.40	17.97	14.21	12.33	20.90	15.72	15.72
Karimnagar	12.77	11.91	13.40	12.82	10.78	15.66	12.35	12.35
Mahabubnagar	11.76	11.02	13.05	11.72	10.11	13.80	12.14	12.14
Medak	18.84	17.18	19.20	18.72	15.54	20.77	17.77	17.75
Nalgonda	25.14	22.31	28.20	25.09	20.03	31.21	25.65	25.65
Nizamabad	10.71	9.50	13.36	10.69	8.51	14.88	12.09	12.09
Warangal	6.13	5.63	14.60	6.14	5.21	15.60	13.67	13.67
State Av	14.33	14.38	16.29	14.30	11.45	18.25	14.75	14.75

Table 12.8

Break-up of Cost of Cultivation across the Districts in 2010-11 (Percentages)

	Hired Labour	*Material Cost*	*Machine Labour*	*Rent*	*Interest*	*Fixed Cost*	*Misc.*	*Marketing and Transport*
East Godavari	33.68	20.77	8.89	28.06	5.94	0.08	2.61	0.00
Krishna	33.29	22.61	10.76	21.04	6.52	2.17	3.03	0.58
Karimnagar	20.54	23.43	20.59	12.89	7.19	4.45	4.92	5.88
Mahabubnagar	36.68	17.93	12.21	9.13	7.50	6.10	3.87	6.57
Medak	19.63	29.71	18.93	9.70	7.46	6.87	4.60	3.19
Nalgonda	30.53	18.71	15.93	11.25	7.33	5.68	5.71	4.90
Nizamabad	21.02	19.69	18.80	11.36	7.32	5.08	4.42	8.11
Warangal	23.97	18.96	24.55	8.18	7.58	6.34	5.52	4.89
Average	27.42	21.48	16.33	13.95	7.11	4.60	4.33	4.27

Source: Field study.

On average for the state, the hired labour cost constituted about 27.42 per cent, machine labour cost 16.33 per cent, material inputs cost 21.5 per cent, rental cost 13.95 per cent, interest cost 7.1 per cent and marketing costs are about 4.27 per cent and miscellaneous costs 4.33 per cent (see Table 12.7). According to our field sources, when wage rates had gone up 300 per cent in the last three years, the overall cost had almost doubled, as labour costs constitute one-third of the costs. This view is supported by secondary data on rural wages as well. The fertiliser prices, particularly that of DAP, potash and complex fertiliser have also gone up by 100 per cent in 2010-11.[9]

The important differences between delta regions and the other region is that in the former rental cost is about 28 per cent and labour cost is

9. DAP (Diammonium Phosphate), which is widely used by farmers, has shot up from ₹ 9,724 a tonne in April 2010 to ₹ 19,014 per tonne in 2011 to ₹ 12,300 per tonne in 2012. Similarly, price of MoP (Muriate of Potash) has increased from ₹ 5,362 per tonne in 2011 to ₹ 11,760 in 2012. The price of MoP increased from ₹ 4,633 a tonne to ₹ 11,760-12,000, while the price of NPK went up from ₹ 5,556.8 a tonne to ₹ 14,400-15,540 during the same period. While the price of Diammonium Phosphate (DAP) has increased by 199 per cent from 2010 to 2012, prices of other fertilisers have gone up by up to 250 per cent. In terms of percentage, prices of MoP and NPK increased by 259 and 279 per cent respectively in April 2012, over the prices in April 2010.

about 33.68 per cent (see Table 12.7). The rents are highest in Godavari district followed by Krishna district, to an extent of 30 bags (75 kg) and 20 bags per annum respectively. The *kharif* gives an average yield of 27 bags and the *rabi* a 40 bags, together making 67. After meeting the material, labour, rent and interest on both crops, barely a meagre surplus is left with a tenant. The subsidiary activities like diary, with the help of the grass from the paddy crop and wage labour becomes the major source of livelihood for most tenants. The rents are high due to a larger prevalence of landless labour in Coastal districts who are scrambling to leasing-in land and competing among themselves. When wage rates were low at ₹ 150 and ₹ 40 for male and females respectively, paddy farming used to earn some surplus to tenant farmer, but once the wage rates went up to ₹ 300 and ₹ 125 for male and female respectively, since 2007-08, the cost of cultivation steeply rose overtaking the revenues.

The situation in the Telangana districts is little different. Fixed costs, input costs, marketing and transport costs, machine and miscellaneous costs together constitute 65 per cent. Labour and rent together constitute 35 per cent. In these regions when wages exploded, farmers responded by shifting to harvesters. Though they lose the grass worth ₹ 1500 in using harvester, they save a cost of ₹ 5000 per acre. The decline of bovine in the region also reduced demand for fodder. Harvesters are also amply available in the region which can harvest maize as well as paddy. The rest of the 35 per cent costs constitute labour and material input costs. The interest costs are same across the districts.

Differences between Official and Estimated Costs

Let us examine the differences between our estimates and CACP figures as given in Table 12.9. There is a 100 per cent variation between CACP figures and our estimates. In our view this is because CACP seriously underestimates the costs in the following terms: first, it is making projections for 2010-11 using data of 2006-07, which is not appropriate. Second, the projections made are based on labour cost index and input cost index which considerably lower than the actual costs incurred in the state. Third, CACP underestimates marketing costs, transport costs, interest costs, rental costs, family labour, attached labour and machine labour. What is more disappointing is that the CACP projections on cost of production for 2010-11 are based on 2006-07 data revised up by using labour cost index and input cost index, both turn out to be grossly underestimated.

Table 12.9

Comparative Estimates of Cost of Cultivation (per hectare) of Paddy for Rabi 2010-11

District	*CACP^ Projection*	*Estimated**	*Agri Dept# Estimate*	*NABARD# Estimate*
Cost A1	25782	50340	42844	44201
Cost A2	26546	58964	53128	54621
A2+FL	30557	70583	NA	NA
Cost B1	27740	50469	59982	66302
Cost B2	43279	67331	65982	72302
Cost C1	31413	62088	62482	67802
Cost C2	47485	78951	68482	73802
Cost C3	47746	86846	68482	73802

Source: *Field Study, # Mohan Kanda Committee Report, ^Report of CACP 2009-10.

On the other hand, the cost estimates sent by the Department of Agriculture, Government of Andhra Pradesh (GoAP) to CACP fare slightly better in reflecting the actual costs [See Table 12.9]. The differences between our estimations and those of GoAP are owing to reasons that the latter's figures on cost of fertiliser and pesticide use, labour cost of application of fertilisers, rentals on sprayers, mats, interest on working capital, transport and marketing costs are lower than those observed in our sample. GoAP data did not consider transport costs, marketing costs, labour costs in applying fertiliser and pesticides and the interest costs.[10] For these reasons, there is a 20 per cent variation between our estimates and theirs. As farmers recalled in their narratives, until the year 2006-07 the prices they received were more or less remunerative. However, in their perception the steep escalation of costs occurred with the implementation of MGNREGA.

10. There are no village level market yards so that one can ignore transportation costs, as assumed by the CACP methodology. In the state, farmers hire transport and cover a distance of 10-25 km on average at an expense of ₹ 25-50 per quintal.

Table 12.10

Estimated Cost of Production (per quintal) for Paddy in Andhra Pradesh for 2010-11 (₹)

District	*East Godavari*	*Krishna*	*Karim-nagar*	*Mahabub-nagar*	*Medak*	*Nalgonda*	*Nizama-bad*	*Warangal*	*Average*
Cost A1	832.46	974.39	748.02	1229.98	672.69	789.02	696.71	831.03	846.79
Cost A2	1157.10	1234.01	858.69	1353.51	744.91	889.02	786.00	905.11	991.04
A2+FL	1362.04	1526.25	1040.96	1605.57	957.19	1021.26	860.42	1121.44	1186.89
Cost B1	832.50	975.73	749.93	1234.11	675.25	791.55	698.71	833.90	848.96
Cost B2	1464.84	1482.24	953.19	1475.29	821.55	991.55	877.28	982.05	1131.00
Cost C1	1037.44	1267.98	932.20	1486.17	887.52	923.79	773.13	1050.24	1044.81
Cost C2	1669.78	1774.48	1135.46	1727.35	1033.82	1123.79	951.70	1198.39	1326.85
Cost C3	1836.76	1951.93	1249.01	1900.08	1137.20	1236.16	1046.87	1318.23	1459.53

Source: Field Study.

The cost of cultivation when, expressed as cost per quintal is referred to as cost of production, is useful to compare with the prices and assess the returns. According to the estimates, the costs A1, B1 and C1—all relevant costs for owner cultivators stood at ₹ 846, ₹ 848 and ₹ 1948 per quintal. The MSP for the year at ₹ 1030 covered the paid-out costs of owner-cultivators without considering their family labour. The costs A2, A2+FL, B2 and C2 stood at ₹ 991, ₹ 1121, ₹ 1131 and ₹ 1326 respectively. We can see that except Cost A2, the MSP did not cover any of the costs which are relevant for tenant farmers with the inclusion of cost of family labour (see Table 12.10).

The overall cost C3 is highest in East Godavari and Krishna at ₹ 1836 and ₹ 1951 respectively. These are almost higher by ₹ 600 than costs in other districts, (except Mahabubnagar[11]). In other Telangana districts, the overall Cost C3 range between ₹ 1137-1318. Cost of production in Mahabubnagar is high because of lower yields on hand and higher labour costs on the other.

When we look at the differences in cost of production between the two regions, namely, Coastal Andhra and Telangana (Table 12.11), the difference in terms of Cost A1, it is 10 per cent less in Telangana. This

11. Due to high levels of labour migration in the district, the local wages in the district tend to be high which make labour costs generally higher)

difference widens as we considered Cost A2 because of higher ground rent in Coastal Andhra. Thus even terms of full cost, i.e., Cost C2 and Cost C3, the difference becomes even larger. When looked at the prices, we found that the MSP covered only paid-out cost (Cost A1) of owner-cultivator and not that of the tenant (cost A2 in Andhra). It has also not covered the full cost in both the regions. When it comes to actual price they received, in Telangana, it has covered the paid-out costs of owner-cultivator as well as the tenant. But there too, it has not covered the full cost (Cost C1 and Cost C2).

Table 12.11

Cost of Production in Andhra and Telangana Regions (₹/Quintal)

Regions	*Cost A1*	*Cost A2*	*Cost C1*	*Cost C2*	*Cost C3*	*Received Price*	*MSP*
Coastal Andhra	889	1187	1129	1711	1882	911	1080
Telangana	804	903	976	1168	1285	944	1080
Average	855	1074	1068	1494	1644	902	1080

Source: Estimated.

Table 12.12

Returns to Paddy Production in Andhra Pradesh in 2010-11(₹)

	Returns over				
Regions	*CostA1*	*Cost A2*	*Cost C1*	*Cost C2*	*Cost C3*
		(₹ Per quintal)			
Andhra	21	-276	-218	-800	-971
Telangana	139	40	-32	-224	-341
Average	46	-172	-166	-592	-742
		(In Percentage Terms)			
Andhra	2.44	-31.1	-24.5	-90.0	-109
Telangana	12.0	0.87	-7.36	-29.0	-42.1
Average	6.27	-18.3	-17.7	-65.6	-82.4

Source: Estimated.

Negative Returns

The returns on full cost have become clearly negative in the state for the sample in the year in question. While these losses are huge in Coastal regions compared to Telangana. The losses are far greater in Coastal region compared to Telangana. The average loss in Coastal Andhra per quintal is about ₹ 971, while that in Telangana is about ₹ 341. The average loss for the state is ₹ 742 [see Table 12.12]. In percentage terms, the loss in Coastal Andhra appears to be alarming with a loss of 109 per cent loss over Cost C3, 90 per cent loss over Cost C2 and 24.5 per cent loss over Cost C1. As we know that Cost A2, C2 and C3 include rental component, losses in Coastal Andhra region over these are much higher, because of high rents.

These substantial differences in the returns between the two regions become clear when we look at the composition of cost of cultivation. The rental and labour component in Andhra region together constitute a lion's share of 73.5 per cent of the total cost in the former. In Telangana, the labour-machine-input-fixed costs constitute 72 per cent. The labour use is reduced in the former by the use of harvesters there. The application of fertilisers is higher in Telangana, which possibly explains the marginally higher productivity observed in our sample.

Now coming to analytical aspect of cost conservation, in Coastal Andhra the rental and labour component being exceptionally higher is entirely because of agrarian structure dominated by tenancy relations. The tenants are forced to pay highest rent and avoid use of machinery to save the grass. In Krishna district, they sow green gram even before paddy is harvested. When rents are high, tenants depend on diary as subsidiary occupation, for which the grass is valuable. Use of harvesters would damage the grass. Thus, this structure does not have scope for machines to optimise the cost. What seems to have happened in the Coastal Andhra regions is that while rental costs were already on high, the rise in labour cost since 2007-08, which is the year of implementation of National Rural Employment Guarantee Act, has toppled the apple cart. Along with that we observed the PDS scheme with two-rupee price has substantially increased the bargaining power of labour.[12] In our sample villages, it is reported that

12. The 5-day MNREGA work in a month is observed to provide the critical food supply levels to labour households, thanks to subsidised PDS, this make agriculture labour less desperate to work at prevailing wages. It is now the farmers who are desperate to get labour during the peak days, compete themselves to draw the labour into the work. The peak level labour demand as it is has outweighed the labour supply for quite some time and the latest government programme has led the rise of wages among other reasons.

the male wages have reached ₹ 300 from ₹ 150 and female wages have too doubled from ₹ 75 to ₹ 150. We have further evidence from secondary sources about this rise in nominal wages which supports our field observations. In the light of high rents, the high labour costs have destroyed the prospects of tenant cultivation. At the first place, existence of surplus over the cost of production has led the active tenancy market. Dwindling of this surplus in the absence of commensurate revision of support prices can also lead to downward adjustment of rents, thereby again leaving some initial subsistence wage to tenant is also a possibility. Therefore, there are is no alternative to cost rationalisation by the farmers.

The persistence of farming in spite of losses is because of subsistence nature of farming among small and marginal farmers, who invests considerable amount of unpaid family labour, deriving wage in the residual income. While, family labour is crucial for the enterprise engaged in variety of activities ranging from supervision, day-to-day water management, pest management, transport, to safe-guarding, the self-exploiting nature of farming forces them to depress the wage income for such labour. The opportunity cost of subsistence farming entails complex factors ranging from demographic profiles of farmers, entitlements, alternative livelihoods, expectations about future etc. Thus subsistence farming apparently absorbs the external shocks through starvation and under-consumption and betrays the outward manifestation of latent crisis.

The Increase in Wage Costs

The study so far argued, based on its evidence that the cost of cultivation remained substantially above the official as well as market price. Given the time lag and underestimation of cost of cultivation, we are at loss to show any temporal changes. According our field sources, substantial increase in the nominal wages is the major reason behind cost escalation in the recent period, for which there is some evidence from the state-level data on nominal wages (see Table 12.13), published by Directorate of Economics and Statistics, which supports our conjectures. According to this data, the nominal wages in the state have risen at an annual compound growth rate of 12.6 per cent for men and 13.3 per cent for women during the period 2000-2010. The growth of nominal wages is particularly steep during 2007-2010. In these three years, the cumulative growth of nominal wages for men has been by 91.3 per cent and 102 per cent for women! Even in real terms, the wages have risen for men and women by 52 and 59 per cent respectively during the period.[13] While it is tempting to attribute the sharp rise in rural wages to the implementation of Mahatma Gandhi Rural Employment Guarantee Act since 2007-08 in full-fledged manner in the state, we shall leave this out from our purview.[14] This piece of evidence, however, corroborates our field level observations that wages have substantially gone up since 2007-08. When such price shocks occur in the labour market, which cannot be met from productivity gains which happen in small increments in the long run, there should be a partial adjustment in the price to meet the minimum cost.

Table 12.13

Average Nominal Rural (Daily) Wage in Andhra Pradesh

Years	*Male (₹)*	*% Change*	*Female (₹)*	*% Change*
2000-01	45.90	–	33.2	–
2001-02	48.70	6.18	35.9	8.16
2002-03	49.10	0.75	36.1	0.52
2003-04	52.40	6.60	38.3	6.03
2004-05	55.70	6.30	40.4	5.53
2005-06	59.30	6.40	42.2	4.57
2006-07	66.80	12.50	48.6	15.10
2007-08	66.80	0.00	48.6	0.00
2008-09	99.20	48.50	74.7	52.20
2009-10	119.60	20.60	90.1	21.70
2010-11	150.43	25.70	115.3	27.90

Source: DES, AP Govt.

13. Real wages are computed as nominal wages deflated by consumer price index for agricultural labourers.

14. There could be several explanations for the rise in rural wages such as growing education, increased rural-urban migration, resonance effect of rise in urban wages in construction industry, ageing of traditional rural agricultural labour and so on.

Figure 12.3

Average Nominal Wages in Andhra Pradesh (₹/day)

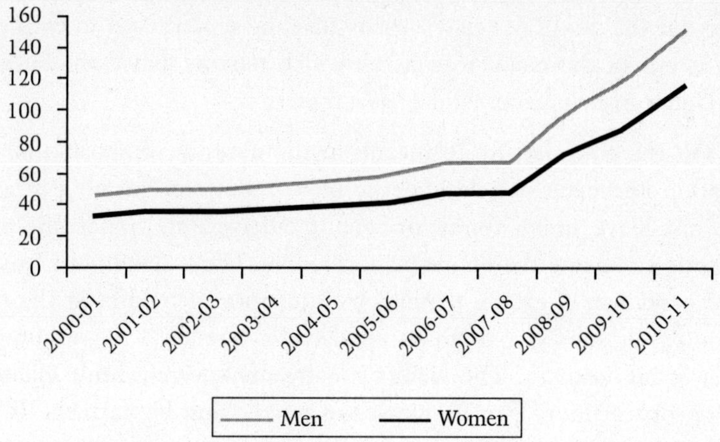

Source: Directorate of Statistics, Andhra Pradesh.

The Regional Differences

As observed earlier, the high cost structure in Coastal districts such as East Godavari and Krishna districts is due high rental and labour cost. The land rents in Coastal Andhra, which constitute about 38 per cent of total cost, are not only in excess over the legal maximum of 16 per cent, it virtually leaves nothing to the tenant (see Figure 12.4). The high rents in region is a result of most landowners, who benefitted from the Green Revolution earlier have diversified[15] into non-agriculture yet hold the land for speculative purposes (Parthasarathy, 2002). The high proportion of landless agricultural labourers and marginal farmers in the region compete among themselves to lease-in the land pushing rents to the hilt.[16] After

15. Rich and middle farmers in Coastal districts, particularly, Godavari districts, Krishna, Guntur and Nellore have diversified into non-agricultural activities such as real estate, trading, finance, rice mills, government jobs etc., and lease out their lands to mostly landless labour. There is great demand in the lease market as lakhs of landless labour and small and marginal farmers compete to lease lands. This has led to bidding up of rents over period.

16. Tenant farmers get consumption and production loans from the commission agents with which they tend to increase their standard of living. Such level of credit is not available for agricultural labour.

the rental cost, the second biggest cost is the labour cost, which is up to 35 per cent and 36 per cent respectively (see Figure 12.4]. These have gone up from earlier 10 per cent to 35 per cent in the last three years and it has wiped out the 20-25 per cent surplus that the tenant used to earn earlier. This is one of the major reasons on which tenants joined the farmers in East Godavari in a `crop holiday' as a protest.

On the contrary, in Telangana in the absence of substantial canal irrigation and extensive dependence on groundwater, farming in general does not leave much room for tenant cultivation. Hence the owner-cultivation remains the dominant formation. The instability of irrigation, the dependence of erratic rainfall, maintenance of motors in the face of erratic power supply, and labour problem have reduced the accumulation levels of big farmers. The labour constraints severely limit viability of supervisory-farming, even in classes of middle and big farmers. It is the small and marginal farmers using family labour who are able to save the paid-out costs, and carry out subsistence-farming through self-exploitation.

Figure 12.4

Break-up of Cost of Production in Coastal Andhra and Telangana

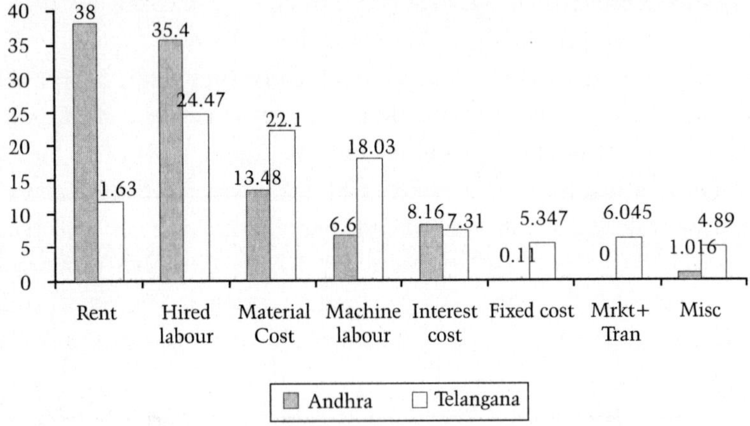

Source: Estimated.

If one seeks to find a solution without altering the present agrarian structure and distribution, revising the MSP will appear as a necessary step. It warrants an increase it to ₹ 2033 per quintal (almost a 100% rise) as demanded by the East Godavari farmers to meet their present cost of cultivation. However this is a short-sighted solution as such a huge revision in MSP will cause a matching rise in retail prices triggering a wage spiral to take the farmers' position back to square one. Therefore, revision of MSP cannot be the sole solution. There must be some cost reducing mechanism that needs to be explored. For instance, mechanisation, as the way Telangana farmers have adopted, can reduce the labour costs by 10 per cent. The institutional credit access to the tenant farmer can reduce the cost by 5 per cent. Rationalisation of rents can save up to 15 per cent. While cost rationalisation could be possible to certain extent, however, rest of the increase needs to be reconciled in fixing the MSP. While there are no objective methods in fixing the price, it is socially optimum nonetheless to cover the costs of owner-cultivator. The full cost in Telangana represents the cost of owner-cultivator at ₹ 1285, thus an increase of ₹ 350 per quintal can be reasonable hike to make paddy farming viable.

Concluding Remarks

The field level evidence clearly suggests that paddy farming in Andhra Pradesh certainly has become unremunerative. The MSP covered only paid-out costs, i.e., namely Cost A1 of owner-cultivators. It does not cover the costs of tenants (reflected by Cost A2) as well as the full costs, namely, Cost C1 and C3. The market price, determined by the MSP, covers only the self-exploiting subsistence farmers, but not that of capitalist farmer. This has primarily occurred, inter alia, due to underestimation of actual cost of production by the official data and the MSP based on this. The CACP not only uses obsolete data, but underestimates the costs of labour, rent, inputs, etc., in the state.

A large amount of cost escalation in the last half-a-decade in the state has occurred due to three factors, namely, increases in labour costs, input costs and rental costs. Labour costs have remarkably gone up since the implementation of MNREGA. The fertiliser costs too have been on the rise due to the changed fertiliser pricing policy in the recent period. There is some scope to rationalise the cost through mechanisation to some extent. The third factor, namely, an exorbitant rental cost in Coastal Andhra is another reason for cost escalation. The agricultural price policy should aim at discouraging costs that arise due to market structure and accommodate the necessary coverage of unavoidable costs.

A steady rise in production of rice in the state has led to situations of excess supply in the state and price stagnation during the *kharif* and price fall in *rabi*. The farm harvest prices are tending to fall below the MSP during bumper harvests during every normal monsoon year. FCI's policy of procurement of rice from millers, instead of paddy from the farmers, is depriving the farmers of its benefit in all bumper crop years. The existing procurement structure in the state allows millers to deny MSP to the farmers and corner the surplus by depressing paddy price.

The findings of the study point at certain immediate policy imperatives. First, there is a need to revise the CACP methodology and its efficacy in assessing the cost of production realistically and revise the MSPs accordingly to preserve the positive incentive structure. Second, there must be some mechanism evolved to address the difference between state level cost and national average cost. Third, in a state like Andhra Pradesh, the strategic role played by middlemen in procurement should be tamed to ensure MSP to farmers. This requires a more dynamic institutional structure, based on an appropriate economic intelligence.

References

Chandrasekhara Rao, N. and Mahendra Dev (2010). "Agricultural Price Policy, Farm Profitability and Food Security", *Economic and Political Weekly* XLV(26-27).

Pathasarathy, B. (2002). "Changing Agrarian Structure and Nature of Transition in Post-Green Revolution Period", in Y.V. Krishna Rao and Subrahmanyam (eds.), *Development of Andhra Pradesh, 1956-2001: A Study of Regional Disparities*. Hyderabad: N.R.R Research Centre.

Parthasarathy, G. *et al.* (1984). "Critical Indices of Transition of Transition to Capitalism within Agriculture and the Changing Agrarian Structure in Godavari Delta of A.P" in Krishna Rao, Y.V. *et al.*, *Peasant Farming and Growth of Capitalism in Indian Agriculture*. Vijayawada: Visalandhra Publication House.

Radhakrishna, R. and Indrakanth (1987). "Effects of Rice Market Interventions in India: The Case of Andhra Pradesh", in *Evaluating Rice Market Intervention Policies, Some Sound Asian Examples*. Manila: Asian Development Bank.

Ramanamurthy, R.V. (2011). "Paddy Glut and Farmers' Crisis in Andhra Pradesh", *Economic and Political Weekly* 46(29).

Ramanamurthy, R.V. and Rekha Mishra (2012). *Pricing Paddy: A Case Study of Andhra Pradesh*. DRG Study. DEPR, Reserve Bank of India [forthcoming].

Ramachandran, V.K., Vikas Rawat and Madhura Swaminathan (eds.). (2009). *Socio-Economic Surveys of Three Villages in Andhra Pradesh*. New Delhi: Tulika Books.

Ravi, C. and Indrakanth (2003). "Food Security and Public Distribution System", in C.H. Hanumantha Rao and Mahendra Dev (edited), *Andhra Pradesh Development:*

Economic Reforms and Challenges Ahead. Hyderabad: Center for Economic and Social Studies.

Rudra, Ashok (1982). *Myths and Realities in Indian Agriculture*. New Delhi: Allied Publishers.

Vijay, R. (2006). "Agrarian Structure and Agrarian Relations: Illustrations from Village Studies", in *Studies in Humanities and Social Sciences* 2(1).

Vamsi, Vakulabharanam, N. Purandra Prasad, K. Laxminarayana and Sudheer Kilaru (2011). "Understanding the Andhra Crop Holiday Movement", *Economic and Political Weekly* XLIV(50): 13-15. Dec 10.

Subba Rao, K. (1978). *Rice Marketing System and Compulsory Levies in Andhra Pradesh: A Study of Public Intervention in Food Grain Markeing*. New Delhi: IEG-Allied Publishers.

Reports:

Mohan Kanda Committee Report on the Problems of Farmers in Crop Holiday Effected Mandals of East Godavari District in A.P. Govt of A.P., (2011).

Kharif Plan for A.P (2011-12), Food Corporation of India, Regional Office, Hyderabad

Reports of Commission for Agricultural Costs and Prices (1983 to 1999)

Socio-Economic Survey (20110-11), Dept of Planning, A.P, Secretariate, Hyderabad.

Statistical Abstracts (2002 to 2009), Directorate of Economics and Statistics, Govt of AP.

Cost of Cultivation Manual (2003). Published by Commission for Agricultural Costs and Prices, Ministry of Agriculture, GoI.

13 Profitability, Crop Productivity and Farm Income in India

Amarnath Tripathi

Introduction

In agriculture development process, India has been centrally focusing on farming and almost ignoring the farmers since Independence. It is replicated from the fact that excepting Situation Assessment of Survey carried out during 2003, the economic and social status of farmer household has never been studied using comprehensive surveys covering different parts of the country (Narayanamoorthy, 2006). We cannot blame the government for focusing on farming deliberately because of two reasons: One, there was famine[1] during the beginning decades of Independence and we depended on other countries for food.[2] There was therefore an urgent need to increase food grain. Two, it was assumed that farmers' situation will be automatically improved with increasing agriculture production. The country has experienced tremendous growth in agriculture production because of increase in agriculture productivity after adaptation of Green Revolution technologies. But this output growth could not much improve farmers' situation except for resource endowed farmers (Dasgupta, 1977).

Nowadays, the situation of both farming and farmer is unsatisfactory in India. It is reflected in slowdown in agriculture output growth

1. India faced a number of threats of severe famines in 1967, 1973, 1979 and 1987 in Bihar, Maharashtra, West Bengal and Gujarat respectively. However these did not materialise in to famines due to government intervention (Dreze and Sen, 1989).

2. In the 1950s nearly 5 per cent of the total food grains available in the country were imported. This dependence worsened during the 1960's when two severe drought years led to a sharp increase in import of food grains. During 1966, India had to import more than 10 million tonnes of food grains as against a domestic production of 72 million tonnes. In the following year again, nearly twelve million tonnes had to be imported (retrieved from *planningcommission.nic.in/reports/sereport/ser/.../agricul.doc* dated 12/01/2013).

(Tripathi and Prasad, 2010), increasing incidence of farmer's suicide (Government of India, 2012), and increasing migration from rural to urban areas (Government of India, 2011). Around 40 per cent of Indian farmers are involved in agriculture without their willingness (NSSO, 2005). Deteriorating status of agriculture has many further consequences, particularly a country like India where agriculture has strong linkages with poverty, employment, food security and economic growth.

In the present paper, profitability (return over total cost) is assessed for three important and widely cultivated crops *viz.*, rice, wheat and sugarcane in India for the period from 1996-97 to 2009-10 to see if Indian agriculture is economically viable or not. These three crops represent Indian agriculture very well. They jointly account for about 53 per cent of net cultivated area in the country. The paper further makes an attempt to build Indian agriculture more viable than present. To fulfill these objectives, farm income is regressed with several covariates at both macro and meso level. At macro level, determinant of farm income is studied at country level involving states, while at meso level it is studied at state level involving districts.

The rest of the paper is organised as: Section II discusses profitability trend in the three crops. Margin to profitability is also discussed for these crops in this section. Section III analyses productivity trend in Indian agriculture to see if reviving trend in agriculture will be sustained in future. Section IV examines factors affecting farm income. Finally, Section V concludes the study and suggests policy implications.

Profitability in Cultivation of Paddy, Wheat, and Sugarcane

Profitability is separately calculated for three chosen crops *viz.*, rice, wheat and sugarcane by subtracting the total cost (Cost C2) from gross value of output (GVO) including both main and by product. Rice and wheat are the main food crops those are widely cultivated in India. Sugarcane is a cash crop which is one of the most profitable crops in India (Sen and Bhatia, 2004). Data on crop yield, value of main product and by product, and cost C2 are collected from annual reports of Commission of Agriculture Cost and Price (CACP), Ministry of Agriculture, Government of India, New

Delhi. The available data are presented in the form of two years average for four periods *viz.*, 1996-1998, 2000-2002, 2004-2006 and 2008-2010 at intervals of two years. The two years annual average estimates of Cost C2, GVO including both main and by product, profitability (return over cost C2), GVO-cost C2 ratio (margin to profitability) and yield for above three crops are shown in following tables (Table 13.1 for paddy, Table 13.2 for wheat and Table 13.3 for sugarcane).

Paddy

In the case of paddy (see Table 13.1), the calculation is carried out in 12 states because consistently data for entire sample period are available only for these states. These 12 selected states are: Andhra Pradesh, Assam, Bihar, Haryana, Kerala, Madhya Pradesh, Karnataka, Orissa, Punjab, Tamil Nadu, Uttar Pradesh and West Bengal. These states jointly account for 80 per cent of total rice cultivated area in the country. All these states except Assam, Bihar, Orissa, Uttar Pradesh and West Bengal show higher GVO-cost ratio in 2008-2010 as compared to 1996-1998. It indicates that growth in value of output is generally higher than increase in total cost of cultivation between the periods 1996-1998 and 2008-2010 in all major paddy growing states except the above mentioned eastern states. In eastern states, the rate of increase in total cost of cultivation is relatively higher as compared to growth in GVO. As reflected from the structure of cost of cultivation, eastern states started adaptation of modern technology at large scale in the above period. It may be one of reason for increasing cost of cultivation in eastern states. Despite widely adaptation of modern technology, crop yield did not increase significantly in eastern states as compared to other states between 1996-1998 and 2008-2010.

Like GVO-cost ratio, the return over total cost at current prices also presents similar trend in eastern states during the study period. It absolutely declined in Assam and West Bengal and marginally improved in Bihar, Orissa and Uttar Pradesh from 1996-1998 to 2008-2010. On the other hand, return over total cost at current prices significantly increased in other than above states. For the period 2008-2010, the highest return over total cost was observed in Punjab followed by Haryana.

Table 13.1

Cost of Cultivation of Paddy in the Selected States

State	Year	Cost C_2 per Hectare (in ₹)	GVO per Hectare (in ₹)	Profitability per Hectare (in ₹)	GVO/Cost	Yield (Kg per hectare)
Andhra Pradesh	1996-98	20988.56	21427.32	438.76	1.02	45.44
	2000-02	26610	26990.45	380.45	1.01	47.84
	2004-06	29156.54	33082.88	3926.34	1.13	51.93
	2008-10	50326.37	60843.29	10516.92	1.21	55.35
Assam	1996-98	9059.675	9971.71	912.04	1.10	21.51
	2000-02	13199.49	12763.59	-435.90	0.97	25.91
	2004-06	14721.74	12632.92	-2088.83	0.86	23.70
	2008-10	21707.57	21145.23	-562.34	0.97	26.29
Bihar	1996-98	9582.915	10616.03	1033.12	1.11	22.02
	2000-02	11937.43	11137.03	-800.40	0.93	23.97
	2004-06	14910.84	13939.82	-971.02	0.93	24.31
	2008-10	18986.93	20780.21	1793.28	1.09	22.81
Haryana	1996-98	18646.55	20286.48	1639.93	1.09	41.13
	2000-02	22755.2	28429.66	5674.46	1.25	38.83
	2004-06	31290.93	36162.07	4871.15	1.16	46.70
	2008-10	47170.21	66190.63	19020.42	1.40	42.81
Kerala	1996-98	10597.15	9839.735	-757.41	0.93	14.45
	2000-02	24306.09	25638	1331.91	1.05	33.61
	2004-06	27255.35	26455	-800.35	0.97	33.40
	2008-10	38037.45	50560.82	12523.37	1.33	42.07
Madhya Pradesh	1996-98	10147.79	10704.97	557.17	1.05	21.42
	2000-02	12279.4	9944.915	-2334.48	0.81	16.44
	2004-06	12331.15	11067.7	-1263.45	0.90	14.45
	2008-10	22112.67	30049.53	7936.87	1.36	24.05
Karnataka	1996-98	0	0	0.00	0.00	0.00
	2000-02	26289.18	28582.29	2293.11	1.09	42.84
	2004-06	29453.35	33166.5	3713.15	1.13	48.00
	2008-10	38776.82	53930.44	15153.62	1.39	47.01
Orissa	1996-98	10978.12	12070.62	1092.50	1.10	26.35
	2000-02	15948.51	15752.28	-196.24	0.99	29.00
	2004-06	18487.4	17279.15	-1208.25	0.93	30.55
	2008-10	27026.47	29672.88	2646.42	1.10	32.02

contd...

...contd...

State	*Year*	*Cost C_2 per Hectare (in ₹)*	*GVO per Hectare (in ₹)*	*Profitability per Hectare (in ₹)*	*GVO/Cost*	*Yield (Kg per hectare)*
Punjab	1996-98	18479.93	22544.59	4064.66	1.22	52.22
	2000-02	22941.59	32035.93	9094.34	1.40	58.49
	2004-06	30888.95	39908.95	9020.00	1.29	65.85
	2008-10	47970.73	69242.48	21271.76	1.44	66.06
Tamil Nadu	1996-98	0	0	0.00	0.00	0.00
	2000-02	28305.09	29499.84	1194.75	1.04	46.94
	2004-06	31448.15	28114.8	-3333.35	0.89	43.25
	2008-10	43601.04	50952.61	7351.57	1.17	45.47
Uttar Pradesh	1996-98	11448.74	13974.55	2525.81	1.22	32.97
	2000-02	15302.48	15378.69	76.21	1.00	32.97
	2004-06	20038.85	19610.05	-428.80	0.98	33.20
	2008-10	30236.14	37245.67	7009.53	1.23	37.10
West Bengal	1996-98	17786.88	19447.83	1660.95	1.09	36.62
	2000-02	20305.42	17378.23	-2927.19	0.86	32.86
	2004-06	24783.25	22945	-1838.25	0.93	36.45
	2008-10	35578.84	36882.4	1303.56	1.04	38.98

Source: Compiled by author using data collected from annual reports of CACP, Ministry of Agriculture, Government of India, New Delhi.

Wheat

Like paddy, profitability in cultivation of wheat (see Table 13.2) is also examined for selected states because of the above reason. These states are: Bihar, Gujarat, Haryana, Himachal Pradesh, Madhya Pradesh, Punjab, Rajasthan and Uttar Pradesh. All chosen states jointly account for around 91 per cent of total wheat cultivated area in the country. In the case of wheat, GVO-cost ratio has gone up significantly from 1996-1998 to 2008-2010 in all states except Himachal Pradesh and Uttar Pradesh. The return over total cost per hectare at current prices in the cultivation of wheat was observed highest in Rajasthan, Haryana and Gujarat during 2008-2010. In 1996-1998, Uttar Pradesh had the highest return but its position slipped down in 2008-2010. It is mainly because of slow improvement in yield in Uttar Pradesh as compared to other states. In 1996-1998, yield of wheat in Uttar Pradesh was 31.53 kg per hectare and it increased marginally to 34.34 kg per hectare in 2008-2010, while growth in yield is observed relatively higher in other high return states like Haryana, Gujarat, Rajasthan and Punjab.

Table 13.2

Cost of Cultivation of Wheat in the Selected States

State	*Year*	*Cost C_2 per Hectare (in ₹)*	*GVO per Hectare (in ₹)*	*Profitability per Hectare (in ₹)*	*GVO/Cost*	*Yield (Kg per hectare)*
Bihar	1996-98	10802.29	12721.02	1918.73	1.18	21.50
	2000-02	14254.98	14827.24	572.26	1.04	24.25
	2004-06	15233.48	16711.59	1478.12	1.10	20.74
	2008-10	20765.13	27384.33	6619.21	1.32	24.40
Gujarat	1996-98	12674.35	18729.83	6055.48	1.48	29.52
	2000-02	17226.88	22913.82	5686.94	1.33	29.41
	2004-06	18246.73	26001.09	7754.36	1.42	32.88
	2008-10	24749.2	39958.44	15209.24	1.61	31.34
Haryana	1996-98	16595.5	22710.44	6114.94	1.37	38.69
	2000-02	21767.26	28671.94	6904.68	1.32	41.78
	2004-06	25757.49	31273.71	5516.22	1.21	39.07
	2008-10	39571.83	55620.26	16048.43	1.41	43.19
Himachal Pradesh	1996-98	7615.4	7240.22	-375.18	0.95	11.23
	2000-02	9540.665	7775.97	-1764.70	0.82	8.93
	2004-06	14983.43	15668.31	684.88	1.05	15.38
	2008-10	15319.13	13221.82	-2097.31	0.86	9.20
Madhya Pradesh	1996-98	9958.245	11837.12	1878.87	1.19	17.40
	2000-02	12367.74	13187.94	820.20	1.07	17.82
	2004-06	15837.23	19125.35	3288.13	1.21	21.05
	2008-10	23353.75	32685.68	9331.93	1.40	24.56
Punjab	1996-98	17662.95	22558.14	4895.19	1.28	39.63
	2000-02	22733.96	31485.36	8751.40	1.38	46.76
	2004-06	25448.54	32319.14	6870.61	1.27	42.50
	2008-10	36853.11	49831.41	12978.30	1.35	40.51
Rajasthan	1996-98	14572.95	20473.95	5901.00	1.40	31.28
	2000-02	18744.63	24153.55	5408.92	1.29	32.18
	2004-06	21012.56	32105.64	11093.08	1.53	33.36
	2008-10	30835.24	50437.14	19601.91	1.64	36.03
Uttar Pradesh	1996-98	13657.35	19082.39	5425.05	1.40	31.53
	2000-02	16442.7	20010.69	3567.99	1.22	31.84
	2004-06	21817.09	23435.33	1618.24	1.07	28.94
	2008-10	32738.72	41348.86	8610.14	1.26	34.34

Source: Compiled by author using data collected from annual reports of CACP, Ministry of Agriculture, Government of India, New Delhi.

For Himachal Pradesh, the situation observed was almost worst in all segment of time except 2004-2006. Return over total cost was observed positive for the period 2004-2006 only, while in other periods it was negative. It means cost of cultivation of wheat is higher than GVO including both main and by product in Himachal Pradesh. This infers that cultivation of wheat in hilly states like Himachal Pradesh is not economically viable.

Sugarcane

In the case of sugarcane, six states namely, Andhra Pradesh, Haryana, Karnataka, Maharashtra, Tamil Nadu and Uttar Pradesh are chosen for the study. These states jointly account for 86 per cent of total sugarcane cultivated area in the country. GVO-cost ratio show increases in output somewhat higher than the increase in total cost in all states during each segment of time except one instance (period 2000-2002) in Maharashtra. In Maharashtra, total cost of cultivation per hectare of sugarcane was observed higher than GVO during 2000-2002 because of increase in input prices. Return over cost per hectare was observed over ₹ 50,000 in Karnataka, Haryana, Maharashtra and Tamil Nadu during 2008-2010, while it was less than ₹ 10,000 in each state except Karnataka during 1996-1998. Uttar Pradesh again recorded unsatisfactory performance in the cultivation of sugarcane like paddy and wheat. In the period 1996-1998, return over total cost in Uttar Pradesh was around ₹ 13,000, next to Karnataka and higher than Haryana, Maharashtra and Tamil Nadu. But, return over total cost was observed lower not only as compared to Karnataka but also to Haryana, Maharashtra and Tamil Nadu. In comparison to other states, growth in total cost of cultivation of sugarcane in Uttar Pradesh is high, but growth in GVO is low. As a result, return over total cost did not increase in Uttar Pradesh as much as in other states. Decline in crop yield in state is the main factor responsible for slow growth in GVO. Crop yield was 490.95 kg per hectare during 1996-1998 and declined to 480.72 kg per hectare during 2008-2010.

Above analysis shows that increase in both total cost of cultivation and GVO was observed over years for each chosen crop in all the study states. In many cases like paddy in Haryana and Punjab, wheat in Rajasthan and sugarcane in Karnataka, the margin of profitability in cultivation per hectare measured by GVO-cost ratio has improved during the overall period under review. But, GVO-cost ratio has deteriorated over years in some cases such as paddy and wheat in Uttar Pradesh. The GVO-cost ratio and return over total cost per hectare has varied crop-to-crop and state-to-state because of variation in technology use and price change.

Table 13.3

Cost of Cultivation of Sugarcane in the Selected States

State	*Year*	*Cost C_2 per Hectare (in ₹)*	*GVO per Hectare (in ₹)*	*Profitability per Hectare (in ₹)*	*GVO/Cost*	*Yield (Kg per hectare)*
Andhra Pradesh	1996-98	0	0	0	0	0
	2000-02	51221.33	56530.82	5309.49	1.10	704.34
	2004-06	59811.67	73110.03	13298.36	1.22	675.225
	2008-10	97300.07	125988.5	28688.39	1.29	773.36
Haryana	1996-98	14916.95	20553.72	5636.775	1.38	258.575
	2000-02	45002.41	64359.43	19357.03	1.43	630.09
	2004-06	50461.75	72485.46	22023.71	1.44	614.7
	2008-10	66176.94	129464.5	63287.6	1.96	592.775
Karnataka	1996-98	31249.73	59043.02	27793.29	1.89	860.99
	2000-02	43295.03	63675.46	20380.44	1.47	755.025
	2004-06	0	0	0	0	0
	2008-10	89472.48	171234.8	81762.34	1.91	1002.085
Maharashtra	1996-98	31641.48	40713	9071.515	1.29	723.93
	2000-02	50481.84	50427.88	-53.955	0.99	768.185
	2004-06	78189.39	96420.42	18231.03	1.23	866.565
	2008-10	106553.3	158475.7	51922.43	1.49	878.91
Tamil Nadu	1996-98	0	0	0	0	0
	2000-02	63143.85	88244.64	25100.79	1.39	1114.39
	2004-06	68296.65	90631.28	22334.63	1.32	913.57
	2008-10	93180.57	144713.2	51532.63	1.55	986.73
Uttar Pradesh	1996-98	22582	35892.31	13310.32	1.59	490.955
	2000-02	29647.57	41789.76	12142.19	1.41	451.775
	2004-06	38756.62	63682.55	24925.94	1.64	549.28
	2008-10	50761.62	98660.43	47898.81	1.94	480.715

Source: Compiled by author using data collected from annual reports of CACP, Ministry of Agriculture, Government of India, New Delhi.

Productivity Trend

In the previous section, we found reviving trend in India agriculture. Both the profitability and margin to profitability of all selected crops has improved in recent years in almost states. The above reviving trend of Indian agriculture has even though enchanted us, yet 'whether this trend will persist in future or not' is still in mind. The present section makes an effort to address this question by examining agriculture productivity. Productivity is strongly associated with profitability. It is clearly reflected from Tables 13.1 to 13.3. In these tables, we can see that both return over cost and margin of profitability move in tandem with crop yield.

After initiation of intensive agriculture approach in late sixties, agriculture productivity improved substantially and it turned to be a dominant factor in determining agriculture growth. Its dominant role in agriculture growth continued till 1980s. After 1980s, the contribution of yield significantly declined in accelerating agriculture output growth (Joshi *et al.*, 2007). Not only contribution, its growth rate was also slashed down in recent decades. Yield was rising at the rate of 2.52 per cent in 1980s and this rate significantly reduced to 1.84 per cent in 1990s and 1.83 per cent in 2000s. This reduction in the growth rate of yield was not only confined to aggregate agriculture, almost all crops have experienced slowdown in the growth of yield in recent decades.

Crop-wise Productivity

Crop-wise productivity trend is presented in Table 13.4. In this table, we compare productivity growth between two periods; 1970-71 to 1989-90 and 1990-91 to 2009-10. Indian agriculture presents different patterns between the above two periods. Agriculture output growth was higher in the earlier period than the later period. Productivity was the main source of output growth in the earlier period, while price has the main source of output growth in the later period (Joshi *et al.*, 2007). Less than 1 per cent productivity growth was observed in only 4 crops in the period 1970-71 to 1989-90, while 11 crops reported less than 1 per cent productivity growth in the later period. Besides, the Table also shows that productivity of all crops except cotton, sunflower, soybean, maize and gram has decelerated in the later period; productivity growth of these five crops has increased in the above period. Despite improvement in productivity growth of many oilseeds and pulses, these are still low yield crops in the country.

Table 13.4

Crop-wise Productivity Growth between Two Periods

Crop	*From 1970-71 to 1989-90*	*From 1990-91 to 2009-10*
Rice	2.29	1.13
Wheat	3.1	0.98
Bajra	0.65	0.24
Jowar	2.19	0.56
Maize	1.59	2.06
Tur	0.49	0.47
Gram	0.49	0.78
Lentils	1.99	-0.1
Soyabean	0.61	0.67
Sunflower	-3.07	0.59
Rapeseed	2.78	1.61
Groundnut	1.03	0.97
Cotton	2.49	3.24
Jute	1.92	1.62
Sugarcane	1.41	-0.1
Tobaco	2.09	0.42

Note: Growth rate is calculated using log-linear model.
Source: *Agriculture Statistics at Glance*, Directorate of Economic and Statistics, Ministry of Agriculture, Government of India, New Delhi.

State-wise Productivity

The spatial pattern of agriculture productivity also presents very surprising behaviour. In the era of wider spread of technology among Indian states,[3] it is expected that dispersion of agriculture productivity across states will decline. But, this hypothesis does not actually hold. Variation in agriculture productivity measured by standard deviation among states, which was 55.0 in 1980, increased to 69.88 in 1990, 88.92 in 2000 and 115.22 in 2005. It shows increasing tendency of variation in agriculture productivity among states over the period of study. Difference in

3. Over years there has been large scale diffusion of technology of high yielding varieties, chemical fertilizers and controlled irrigation water etc (Sen & Bhatia, 2004).

agriculture productivity among states is also reflected from growth pattern of agriculture productivity given in Table 13.5. As productivity growth pattern suggests, Indian states can be divided into four groups: continued acceleration in productivity growth, productivity growth first accelerate then decelerate, productivity first decelerate and then accelerate and continued deceleration in productivity growth (Table 13.6).

Table 13.5

Annual Compound Growth Rate of Agriculture Productivity across Indian States

State	*1980s*	*1990s*	*1990s*	*2000s*
Arunachal Pradesh	3.06	2.39	-	7.49
Assam	1.19	0.39	0.5	3.9
Bihar	2.59	-0.64	1.11	1.93
Manipur	2.94	1.25	-1.31	2.99
Meghalya	0.54	2.71	4.34	5.82
Nagaland	0.39	3.76	2.08	6.43
Orissa	2.22	2.55	-2.23	3.44
Sikkim	3.98	0.11	-	7.64
Tripura	1.4	3.25	2.89	16.65
West Bengal	4.74	4.27	2.03	1.66
Haryana	4.15	1.07	-1.36	0.9
Himachal Pradesh	2.73	0.4	3.43	5.17
Jammu and Kashmir	-0.06	4.45	0.85	4.28
Punjab	4.31	1.37	-1.26	2.26
Uttar Pradesh	2.38	1.83	1.51	2.18
Andhra Pradesh	2.41	2.43	-0.51	3.79
Karnataka	1.91	4.6	2.41	-4.07
Kerala	1.87	2.32	2.59	2.95
Tamil Nadu	2.85	3.96	1.99	0.33
Gujarat	-0.65	11.96	9	12.5
Madhya Pradesh	2.56	2.52	0.88	4.26
Maharashtra	2.79	4.24	0.26	2.32
Rajasthan	4.06	2.75	0.5	3.58

Note: Agriculture productivity was calculated here by dividing value of agriculture output at 1999-2000 prices by gross sown area. Growth rate is calculated using log-linear model.

Source: Data on value of agriculture output and gross sown area are collected from Centre Statistical Organisation and Fertiliser Statistics (Fertiliser Association of India), respectively.

Table 13.6

Classification of Indian States

Groups	*States*
Continued acceleration in productivity growth	Meghalaya, Nagaland, Orissa, Tripura, Jammu and Kashmir, Andhra Pradesh, Kerala, Gujarat, and Madhya Pradesh
Productivity growth first accelerate then decelerate	Arunachal Pradesh, West Bengal, Karnataka, Tamil Nadu and Maharashtra
Productivity first decelerate and then accelerate	Rajasthan, Punjab, Uttar Pradesh, Himachal Pradesh, Sikkim, Assam, Bihar, and Manipur
Continued deceleration in productivity growth	Haryana

Factors Affecting Farm Income

Factors affecting farm income is studied at both macro and meso level. Macro level means country level analysis considering state as observation and meso level means state level analysis considering district as observation. For meso level analysis, Uttar Pradesh (U.P.) is chosen purposely—it is the most populous state in India and agriculture is the main sector that controls most economic activities in the state (Tripathi, 2012). But, its own performance is not very satisfactory. The output growth in agriculture has been decelerating for the past three decades in U.P. During the 1980s, agriculture recorded a 2.43 per cent growth rate, which marginally reduced to 2.40 per cent in the first decade of the 21st century. Agriculture output growth has been always lower in the state than the all-India scenario since the 1980s (Tripathi, 2012). Despite strong agriculture base in the state, the yield per hectare and profitability for not a single crop is the highest in the country. Many small and marginal land holdings, low level of farm mechanisation, inefficient use of irrigation and poor consumption of fertilisers are mainly responsible for the average yield per hectare of almost all crops in the state.

U.P. is divided into four administration regions: Western, Eastern, Central and Bundelkhand. The performance of agriculture varies across these regions. Therefore, factors affecting farm income is also studied at disaggregate region level to see if there is variation among regions. The Western region is agriculturally the most progressive region; the largest chunk of the state's agriculture output comes from this region (around 50%). The Eastern region contributes around 28 per cent, next to Western region, in total value of the state's agriculture output. The Bundelkhand accounts for only 4 per cent of the state's gross value of agriculture output. Agriculture in Bundelkhand region is vastly rain-dependent, diverse, complex, under-invested, risky and vulnerable. The average food grain yield in western region is 2577 kg per hectare, which is much higher than other regions, particularly Eastern (1997 kg per ha) and Bundelkhand region (1067 kg per ha). Irrigation and fertiliser use are the major determinants of the variation in agriculture productivity across regions (Pandey and Reddy, 2012).

In order to assess factors affecting farm income, it is regressed on a set of independent variables using following equations:

$$FI_i = \alpha_i + \sum_{j=1}^{n} \beta_{ij} X_{ij} + \varepsilon_i$$

$$FI_i = \alpha_i + + \sum_{j=1}^{n} \beta_{ij} X_{ij} + \sum_{k=1}^{3} \gamma_{ik} D_{ik} + \sum_{k=1}^{3} \gamma_{ik} X_{ik} D_{ik} \, \varepsilon_i$$

where, FI = Farm income, X_{ij} = Explanatory variables, α_i=Intercept, β_{ij} = Coefficients of explanatory variables, D_{ik} = Dummy, ε_i = Error term, i = observation, j = number of explanatory variables, and k = number of dummy.

The first equation is used for aggregate state level analysis and the second equation is used for disaggregate region level analysis. U.P. is divided into four regions. Therefore, we use three dummies to distinguish the above four regions. A set of independent variables includes: farm size, percentage of small and marginal holdings, crop diversification, technology, quality of soil, rural infrastructure and literacy. The selection of independent variables is based on review of related literatures. Specification and data source of each independent variable is presented in Table 13.7. All data are first collected for three consecutive years 2007-08, 2008-09, and 2009-10 and then converted in the form of three years average. The three years average estimates of all variables are finally used in the above regression analyses.

Table 13.7

Specification and Data Source of Independent Variables

Independent Variables	*Specification*	*Data source*
Technology	Irrigation ratio = Irrigated area/ net cropped area*100	Both area under irrigation and net cropped area collected form Jila Sankhiyaki Patricka
Small and Marginal Holdings	Per cent of small and marginal Farmers	Jila Sankhiyaki Patricka
Diversification	Measured using Simson index	Jila Sankhiyaki Patricka
Literacy	Rural Literacy Rate	Census
Rural Infrastructure	Composite Infrastructure Index was calculated to attach specific weight to five indicators viz access to credit, primary markets, share of government source of irrigation, electricity, and village connected with roads	All required data were collected from Jila Sankhiyaki Patricka
Quality of Soil	Sodic soil	UP Development Report

The estimates obtained from macro level analysis are not presented and discussed in this paper because they are not statistically robust. However, macro level analysis suggests positive and significant impact of non-farm activities on farm income. The estimates obtained from meso level analysis are presented in Tables 13.8 and 13.9. Aggregate state level estimates are presented in Table 13.8 whereas disaggregate region level estimates are presented in Table 13.9.

From Table 13.8, all explanatory variables except quality of soil are found statistically significant. It means these are important determinant of farm income. The Table shows that technology, diversification and rural infrastructure affects farm income positively while literacy and per cent of small and marginal holdings affects adversely. The value of adjusted R2 is 0.51. It means that all these variables together explain 51 per cent variation in farm income. The value of intercept is very high and statistically significant at 1 per cent level. It shows that there are other than above factors that affect farm income. The above relations of explanatory variables with farm income vary region to region, as is evident from Table 13.9. The variation in relation of explanatory variables with farm income is somehow understandable because of distributional performance of agriculture development in the state.

Table 13.8

Factors Affecting Farm Income at Meso Level

Independent Variables	*Coefficients*	*T-stat*
Technology	0.81*	5.66
Small and Marginal Holdings	-2.19*	2.86
Diversification	0.35***	1.77
Literacy	-1.50*	-4.10
Rural Infrastructure	0.08*	2.96
Quality of Soil	-0.04	-0.40
Intercept	19.34*	4.84
Adj R-squared	0.51	
F-stat	7.61*	
Observations	70	
Breusch-Pagan	7.05*	

Note: ***: Significant at 10 per cent level, **: Significant at 5 per cent level, *: Significant at 1 per cent level.

Table 13.9

Variation in Relationship between Farm Income and Other Covariates Observed Above across Regions in Uttar Pradesh

Independent Variables	*Bundelkhand*	*Western*	*Central*	*Eastern*
Technology	1.18	0.68	-0.74	-0.33
Small and Marginal Holdings	1.12	-2.80	3.98	-1.84
Diversification	-0.57	0.40	1.00	-0.36
Literacy	0.01	-1.72	-0.59	0.25
Rural Infrastructure	0.59	0.15	-0.02	0.14
Quality of Soil				
Intercept	-1.04	22.01	-7.50	16.51
Adj R-squared	0.49			
F-stat	3.83*			
Observations	70			
Breusch-Pagan Test	0.17			

Note: ***: Significant at 10 per cent level, **: Significant at 5 per cent level, *: Significant at 1 per cent level.

Surprisingly, estimate of both rural literacy and small and marginal holdings is not in the line of our expectation. The negative impact of rural literacy may be possible because literate people do not like farming. Several village experiences indicate that educated youth does not want to stay in rural areas. The adverse impact of small and marginal holdings indicates that small and marginal farmers are not able to generate farm income. It may be because of subsistence farming. Almost all small and marginal farmers are subsistence farmers. However, farm productivity is higher in small and marginal farm than large and medium farm. All the above speculated issues are matter of in-depth analysis. For this purpose, a comprehensive village level study would be interesting.

Conclusions and Policy Implications

In this paper, trend in profitability and margin to profitability for three crops (rice, wheat, and sugarcane) is examined for the period from 1996-97 to 2009-10. These crops are well representative of Indian agriculture. The above analysis is carried out in the selected states. The chosen states vary crop to crop depending on availability of data for entire sample period. Besides, the paper also examines impact of several factors on farm income at both macro and meso level. For meso level analysis, U.P. is chosen purposely. In order to identify covariates of farm income, it is regressed with a set of independent variables at both macro and meso level.

We observe two distinct trends in both profitability and margin to profitability in almost states during the period from 1996-97 to 2009-10. In late nineties and beginning years of the first decade of 21st century, both profitability and margin to profitability have declined, whereas they have improved in recent periods from 2004-2006 to 2008-2010. It shows revival of farming in India. But, the above trend may not sustain in future because the productivity which is closely and positively associated with profitability has decelerated in recent years.

For factors affecting farm income, the study does not find robust estimates at macro level. However, the macro level analysis reveals that non-farm income has positive and significant effect on farm income. At meso level, technology, crop diversification and rural infrastructure are found significant variables encouraging farm income. However, the results show negative relationship of farm income with literacy and percentage of small and marginal farmers. The study also observed variation in above relationships across regions in U.P. The study, on the basis of above observations, finally suggests following policy implications.

- There should be more disaggregate policy for agriculture development than homogenous policy. It is because of reason that strong variations exist in the performance of Indian agriculture.
- Productivity trend observed in this paper emphasises need for a third breakthrough of technology with the condition that it should be more resistant to climate change. Climate change is likely to affect crop productivity adversely.
- Promote non-farm activities at village level because it helps to reduce overburden on agriculture and support farming through providing more investment. There is a strong linkage observed between farm and non-farm activities.
- Make available all basic facilities like access to credit, input and output market, irrigation etc., to each and every farmer. Collapse of above systems is mainly responsible for deteriorating conditions of the farmers.
- Induce agriculture diversification because it not only increases farm income, also reduces farm risks.

References

Dasgupta, B. (1977). "India's Green Revolution", *Economic and Political Weekly*: 241-60.

Dreze, J. and A.K. Sen (1989). *Hunger and Public Action*. New York: Clarendon Press.

Government of India (2012). *Crime in India: 2011*. New Delhi: National Crime Record Bureau, Ministry of Home Affairs.

Government of India (2011). Drop-in-arcticles: Office of The Registrar General & Census Commissioner, India. Retrieved January 15, 2013. New Delhi: Office of The Registrar General & Census Commissioner, India. *http://censusindia.gov.in/Ad_Campaign/drop_in_articles/08-Migration.pdf*

Joshi, P.K., A. Gulati and P.S. Birthal (2007). "Agriculture Diversification in India: Status, Nature and Pattern", P.K. Joshi, A. Gulati and J. Cummings, *Agriculture Diversifcation and Smallholders in South Asia*. New Delhi: Academic Foundation.

Narayanamoorthy,. A. (2006). "State of India's Farmers", *Economic and Political Weekly*: 471-73.

NSSO (2005). *Situation Assesment of Farmers: Some Aspect of Farming*. New Delhi: National Sample Survey Organization, Ministry of Statistics and Programme Implementation, Government of India.

Pandey, L.A. and A.A. Reddy (2012). "Farm Productivty and Rural Poverty in Uttar Pradesh: A Regional Perspective", *Agriculture Economics Research Review*: 25-35.

Sen, A. and M. Bhatia (2004). *State of the Indian Farmer: Cost of Cultivation and Farm Income*. New Delhi: Academic Foundation.

Tripathi, A. (2012). "Economic Growth in Uttar Pradesh during High Growth Regime of India", *Indian Economic Journal*.

Tripathi, A. and A.R. Prasad (2010). "An Overview of Agrarian Economy of India: The Performacne and Determinants", *The ICFAI Journal of Agriculture Economics*: 62-82.

Acknowledgment:

This work was completed under the Think Tank Initiative of IDRC, Canada. Author is thankful to the IDRC, Canada and the Institute of Economic Growth for Institutional and Financial support. Author would also like to thanks referee for comments on the earlier version of the paper.

14 Imbalances in Costs and Returns in the Cultivation of Major Crops in India

A State Level Study

Chirala Sankar Rao

Everything else can wait, but not agriculture.

— Jawaharlal Nehru

Introduction

Agriculture, as the largest private enterprise in India, is the lifeline of the economy. Agriculture provides the underpinning for our food and livelihood security, and support for the economic growth and social transformation of the country. During 2008-09, the agricultural sector contributed to approximately 15.7 per cent of India's GDP (at 2004-05 prices) and 10.23 per cent (provisional) of total exports besides providing employment to around 58.2 per cent of the work force. The continued high growth of agriculture is essential to meet the food and nutritional security requirements of the nation and to provide livelihood and income to the people in rural areas. Agriculture will continue to have the pride of place in national policies and plans since the nation's food security depends on the performance of the agricultural sector (Ministry of Agriculture, 2010).

The agrarian crisis that has ravaged India's rural countryside during the post-reform era has grown on a three pronged set of symptoms: rising input costs, dwindling produce price realisation and the inability of farmers to abandon cultivation without alternative livelihood sources. It is a well known fact that the crisis started when the government decided to demolish, without giving adjustment time, the elaborate mechanism that was built up, in stages, in the post-independence period to the beginning of the 1990s to protect the peasantry from the vicissitudes of the market (Patnaik, 2003). These protectionist arrangements, consisting essentially of a system of input price subsidies and output price support, despite weaknesses in implementation, had enabled farmers to take up cultivation in a predictably stable price environment. During the post-reform period, the government not only slashed the subsidies on major inputs but also absolved itself of the responsibility to produce or procure and distribute these inputs at farm gates. When prices of farm inputs thus went up, private operators seized the opportunity and pushed up prices further (Bagchi, 2004).

The sustainability of agriculture depends on its profitability which is the measure of difference between the cost and return of the cultivation of crops. The present crisis of Indian agricultural sector is the result of the growing gap between the costs and returns of major crops in the direction of rapid increase in the cost of cultivation but with less proportionate increase in the returns. The root cause of this imbalance lies in the irrational agricultural policies in the recent past i.e., liberalisation and deregulation of agricultural input market and trade but control and regulation of output market and trade, decreasing ratio of public investment in agriculture for infrastructure, R&D, extension services etc. This imbalance in costs and returns of the cultivations varies across the crops and states in India.

Hitherto, the work done by Sen and Bhatia (2004), in their millennium study on the state of the Indian farmer uses the data on costs of cultivation and makes an intensive analysis of the performance of Indian agriculture during the second half of the 20th century. The authors examine the structure of costs and returns to farming and farm incomes in relation to 20 major crops in 14 states for the period from 1970-71 to 1998-99. Significantly, the study traces the history of the cost of cultivation surveys in India even before independence and the farm management studies (FMS) of 1954-55 to 1972-73, before taking up the comprehensive scheme for analysis. Their review of the survey methodology, concepts and the structure of costs of cultivation provides for the first time a firm basis for further research in this area. The focus of the study being the trends in farm business income across crops, states and over time, the authors analyse the cost of cultivation and its breakups from the vantage of crop technology and its diffusion.

Raghavan (2008) in his paper estimated and compared how the paid-out cost of cultivation of wheat in the states of Haryana, Madhya Pradesh,

Punjab, Rajasthan and Uttar Pradesh underwent changes before and during the era of economic reforms by using the data based on the costs of cultivation of wheat from 1970-71 to 2004-05. But there has not been any major research work to study the trends and patterns of cost, revenue, profit and yields of major crops in India across the time and space in the recent period. The present study makes an attempt to fill up this research gap.

This paper consists of following specific objectives and hypothesis:

Objectives

- To study the trends and patterns of the cost of production of major crops in India
- To study the trends and patterns of yields and returns/revenues of major crops in India
- To study the trends profitability of the cultivation of major crops in India

Hypothesis

- There will be relatively higher imbalances in costs and returns of cultivation of commercial /non-food crops than non commercial/ food crops
- The imbalances in costs and returns of cultivation of crops are relatively high during post-reform period
- Fast increase in input cost over output price coupled with stagnation in yields caused losses in cultivation of crops

Methodology and Data Source

Basic statistics such as mean average, coefficient of variation and exponential growth rates are used to study the trends and patterns of costs, revenues, profitability and yield of the crops. The exponential growth rates of cost, value of output, yield of crops is worked out by fitting log linear equation of $\ln Y = a + bt$, where, Y is cost/value of output/yield of a crop, a is constant, t is time, b and c are coefficients of Y. The coefficient b gives the exponential growth rate of Y.

The study is based on the both primary and secondary level data. The secondary level data was obtained from various reports of Cost of Cultivation of Principle Crops in India, Ministry of Agriculture, Government of India. The missing year data of cost of cultivation of crops was replaced with the data generated by the interpolation method of linear trend at point. For the purpose of comparison over time, the value term data of cost, revenue and profit of crops which were at current prices have been converted in to constant prices at 2004-05 by using the respective WPI-Agricultural Commodity Index

The cost concepts used in the study are A2 which is considered as paid-out cost and C2 which is total cost. The details of these concepts are given in Appendix A-14.1. The cost of cultivation of crops data is available in two break-up form such as: (i) cost of cultivation per hectare, and (ii) cost of production per quintal. The study uses the former for the analysis.

The main focus of the study is the crop-wise analysis. The study has taken seven major crops such as paddy, wheat, maize, *urad* (Black gram), groundnut, cotton and sugarcane for the purpose of analysis. The selection of these crops was done based on the following reasons such as: (i) Representational crops from each crop group like rice, wheat and maize represent the cereals, *urad* (black gram) represents the pulses, groundnut represents oilseed and cotton and sugarcane represent non-food and pure commercial crop group, (ii) Availability of cost of cultivation of data at state levels in a continuous time series from 1981 to 2009 in more number of states. The time period of the entire study is from 1981 to 2009, again this has been divided in to three decennial sub-periods such as 1981-1990, 1991-2000 and 2001-2009 for the analysis of trends of costs, returns, profits and yields of the crops. The analysis is made at state and all-India levels. The primary data was obtained from the field survey done by the author in three villages from three districts i.e., Krishna, Warangal and Kurnool (a village from each district) of Andhra Pradesh in November 2011 as the part of PhD work. The description about the field work is given in the Appendix A-14.2.

The scheme of analysis goes in the following way. Section I contains the introduction, objectives, hypothesis and methodology, section II deals with analysis of cost, revenue, profit and yield of major crops in India, section III deals with analysis of cost, revenue, profit and yield of major crops in Andhra Pradesh based on field work and section IV contains the summary and policy suggestions.

Analysis of Cost, Return, Profit and Yield of Major Crops in India

Paddy

Tables A-14.3.1 to A-14.3.4 in Appendix A-14.3 give the details about the paddy crop cultivation in India. The total cost (C2) of paddy per hectare in India has gradually increased over the time during the study period. The annual exponential growth rates and cost variations (CVs) of C2 have declined over the decades which are positive trends in paddy cultivation in India. The cost varies across the states from ₹ 10,769 in Madhya Pradesh to ₹ 30,297 in Tamil Nadu during the study period. Andhra Pradesh, Punjab, Karnataka and Haryana are highest cost states. The striking point is that across the states the CV is much higher than the yield variation which shows there is no strong relation between the cost of cultivation and yield of the crop across the states. So there must be factors other than yield to explain the CVs. The data shows, during 1981-2009, the instability is relatively high in total cost of paddy than paid out cost. This instability gradually decreases over the three decades where the sub-period of 1981-1990 shows relatively high volatility of the paddy cost among the states. The annual exponential growth rates of paddy cost shows that the growth is relatively high during first sub-period i.e., 1981-1990 but declined during later decades.

The returns of paddy produce per hectare in India gradually increased during all the sub-periods; it stands highest during last sub-period i.e., 2001-2009. It also varies across the states from lowest ₹ 11,510 in Madhya Pradesh to highest ₹ 32,011 in Punjab during 1981-2009. This has gradually increased across all the states during all the three decades. The growth rate of paddy produce is highest during third sub-period followed by first sub-period. The second sub-period i.e., 1991-2000 has witnessed the negative growth rate in majority states with few exceptions. The instability of value of paddy produce in India and among states has gradually declined over the time during 1981 to 2009.

C2-cost based profit represents the true economic logic of profit. All the states have register positive profits in the three decades, but vary across the states ranging from ₹ 7,237 in Karnataka and ₹ 741 in Madhya Pradesh. During the entire study period, this profit has declined in majority of states except in few states like Punjab, Haryana and Andhra Pradesh where it has increased. It can be understood from the CV values, when compared to costs and returns, the profits of paddy crop are more instable but they have increased over the time in majority of the state with few exceptions. The data shows the last sub-period i.e., 2001-2009 has witnessed even the negative profits of paddy in few states. But the A2–profits (Farm Business Income (FBI)) of paddy are positive among all the states also during all the sub-periods. The A2 and C2 profits are shown in Figure 14.1 and Figure 14.2 below.

The data shows that, over the last three decades, the revenue growth of paddy has slightly fallen behind the yield and cost growth which is the clear indication of falling profitability of paddy crop in India but with some exceptions where third sub-period has witnessed an opposite trend in which the profitability of paddy has started increasing.

Figure 14.1

Paddy: A2-Profit per Hectare (in ₹)

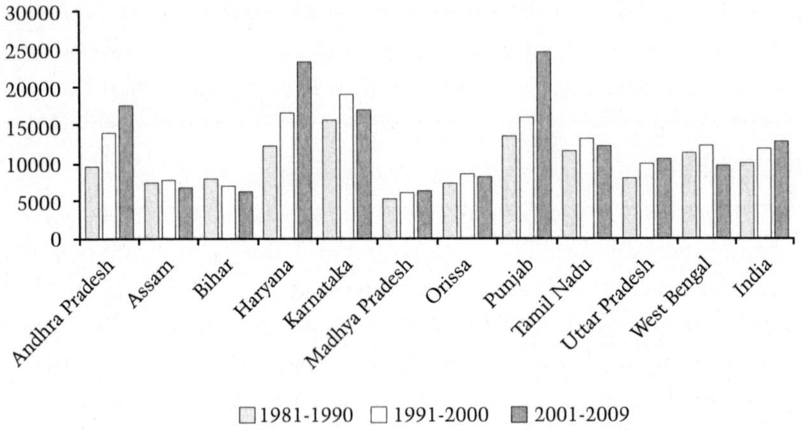

Source: Computed from various reports of *Cost of Cultivation of Principle Crops in India*, DES, MoA, GoI.

Figure 14.2

Paddy: C2-Profit per Hectare (in ₹)

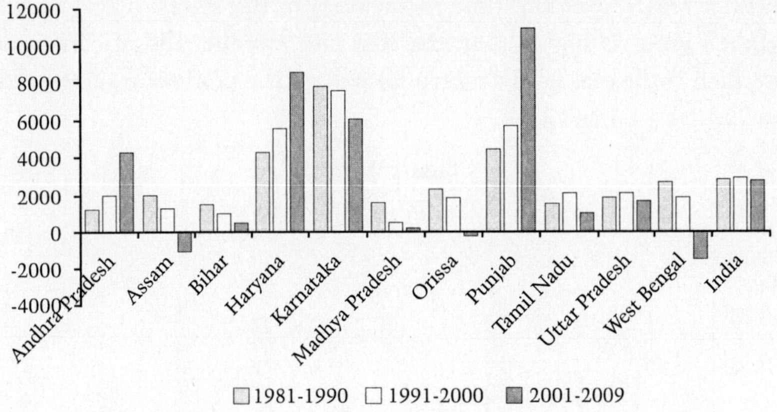

Source: Computed from various reports of *Cost of Cultivation of Principle Crops in India*, DES, MoA, GoI.

Wheat

Tables A-14.3.5 to A-14.3.8 in Appendix A-14.3 gives the details about the wheat crop cultivation in India. Wheat is the second major crop in India. The average C2 of wheat in India for the period of 1981-2009 stands at ₹ 17,048 with the annual growth rate of 1.31 per cent. It gradually increased over the three decades, reported highest growth in first decade (1981-1990) but negative growth rates in later two decades. The results vary across the states where Himachal Pradesh reports lowest cost of ₹ 11,733 while Punjab reports highest. The CV results show that the instability in wheat's cost of cultivation has decreased over the time.

The revenue of wheat varies across the states during the study period. Haryana with ₹ 28,877 stands at first place followed by Punjab, Rajasthan, Uttar Pradesh, Gujarat and Bihar, while Himachal Pradesh has lowest revenue per hectare. The exponential growth rates of wheat's revenue per hectare reveal that the revenue has increased slightly greater than the cost in majority of the states during last 29 years which shows the profitable nature of wheat crop in India. The yield growth rates are less than the revenue which shows the falling yield levels is the matter of concern in the cultivation of wheat.

All the states report the profit in wheat crop but it varies widely across the states as shown in the tables. The amount of profit per hectare has slightly increased over the time in majority of the states. The instability of wheat's profit is higher than the cost and revenue, and also rises over time which is the matter of concern for policy. The profits are shown in the below Figures 14.3 and 14.4.

Figure 14.3

Wheat: A2-Profit per Hectare (in ₹)

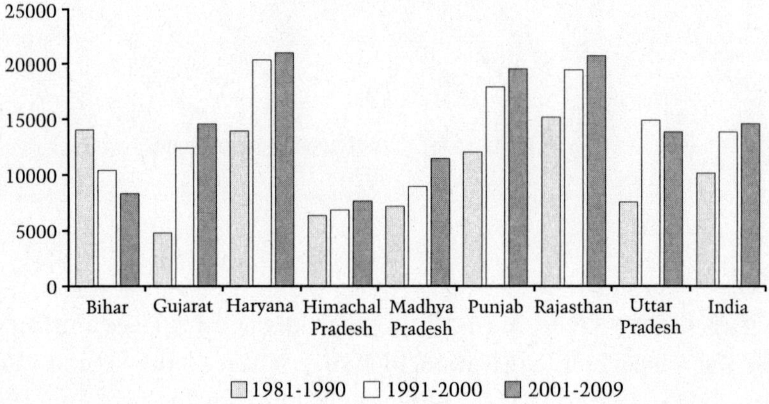

Source: Computed from various reports of *Cost of Cultivation of Principle Crops in India*, DES, MoA, GoI.

Figure 14.4

Wheat: C2-Profit per Hectare (in ₹)

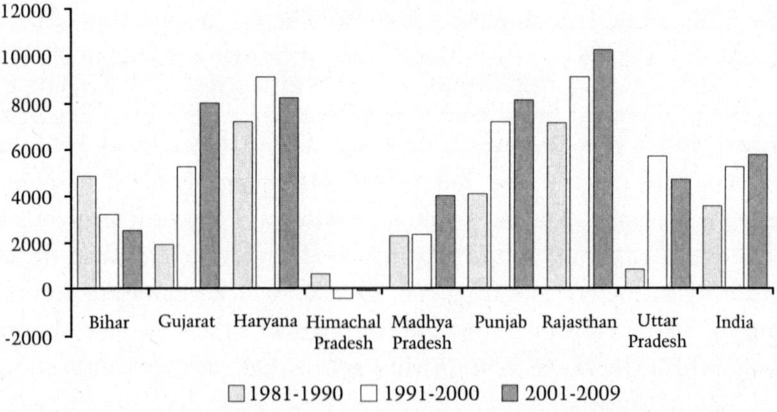

Source: Computed from various reports of *Cost of Cultivation of Principle Crops in India*, DES, MoA, GoI.

Maize

Tables A-14.3.9 to A-14.3.12 in Appendix A-14.3 give the details about the maize crop cultivation in India. The C2 of maize crop widely varies across the states during study period i.e., 1981 to 2009 which ranges from ₹ 6,975 in Madhya Pradesh to ₹ 17,633 in Andhra Pradesh. The variation across the states in maize's cost is larger than the variation in yield. The cost has gradually increased over the time in majority of the states with the exception in Andhra Pradesh where the cost has declined. The CV results show that the instability of maize's cost across the states has decreased over the time in the last 29 years. It is high during first sub-period but declined in latter sub-periods. The annual growth rates reveal that cost of maize has increased relatively high during first sub-period but declined in latter sub-periods.

The average revenue of maize output in India during the study period is ₹ 10,964; this varies across the states from ₹ 6,223 in Madhya Pradesh to ₹ 16,278 in Bihar. The data shows that the value of maize output has gradually increased over the time from 1981 to 2009 and has increased relatively more during third sub-period. The instability has slightly increased over the time. The yield growth rates of maize are on par with the revenue growth rates with increasing trend over the time and substantial increase in last sub-period.

Profit based on C2-cost show that maize has been always a loss making crop at India and at the majority of states with few exceptions. Bihar turns out to be largest profit making state. But the amount of loss in maize cultivation has gradually downed over the time period. But the A2–profit (FBI) data show that maize is profitable crop among all the states also during all the sub-periods. These profits are shown in Figures 14.5 and 14.6.

Figure 14.5

Maize: A2-Profit per Hectare (in ₹)

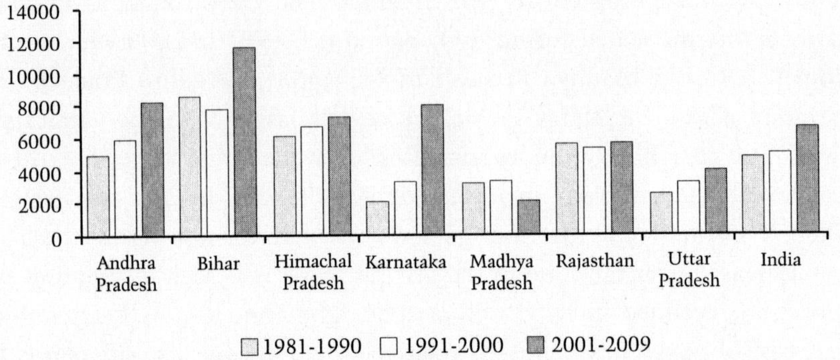

Source: Computed from various reports of *Cost of Cultivation of Principle Crops in India*, DES, MoA, GoI.

Figure 14.6

Maize: C2-Profit per Hectare (in ₹)

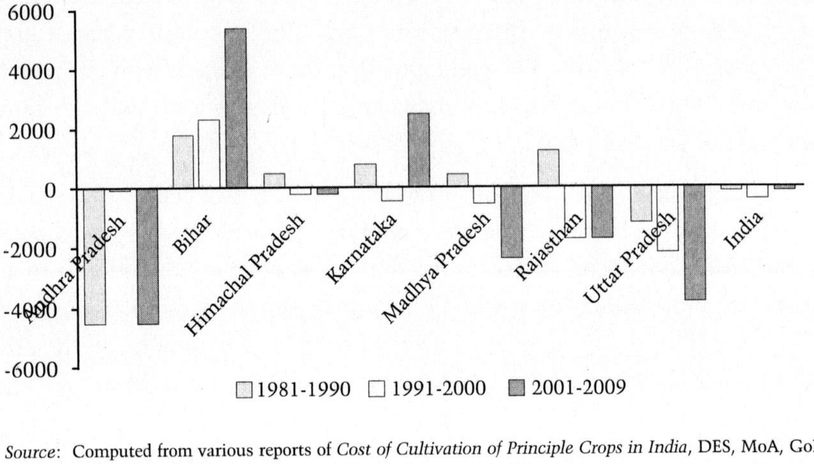

Source: Computed from various reports of *Cost of Cultivation of Principle Crops in India*, DES, MoA, GoI.

Urad (Black Gram)

Tables 14.3.13 to 14.3.16 in Appendix A-14.3 gives the details about the *urad* crop cultivation in India. The C2 of *urad* stands at ₹ 6,745.7 in India during 1981-2009. It has relatively less variation across the states when compare to other crops. It varies from ₹ 5,363 in Orissa to ₹ 8,645 in Tamil Nadu. The cost has declined during second sub-period but rose during last sub-period in majority of states. The CV data shows that the cost was relatively instable in most of the states during 1991-2000. The exponential annual growth rate of this cost is very minimal during the entire study period i.e., 1981-2009, it also reported negative growth during last two decades which explains that the total cost of *urad* has declined during 1981 to 2009 in majority of states. Unlike in other crops, *urad* has less variation in cost and revenue compared to yield variation among states which are relatively high.

The revenue of *urad* stands at ₹ 8,505.6 per hectare at all-India level for the period of 1981-2009. It has decreased during 1991-2000 but slightly increased during 2001-2009 which is also reflected in zero exponential annual growth rate. The growth rate was positive during the first decade but reported negative rate in later decades which explains the falling value of produce of *urad* crop in most of the states in India. The CV results show that the instability was relatively high for the value of output of *urad* in most of the states in India during the study period and it is more instable during recent decade i.e., 2001-2009. The very low growth rates of yield, also negative in first two sub-periods, shows the stagnancy in the *urad* yields in India in the last three decades period.

Urad crop in India is less profitable crop during the study period. Based on total cost concept (C2) it varies from ₹ 16 in Maharashtra to ₹ 4,379 in Andhra Pradesh. The profit has gradually declined over the time period and even reported loss in Madhya Pradesh and Maharashtra in the recent decade. But in case of paid out cost concept it is positively profited crop among all the states in all the periods. The profit of *urad* crop exhibits very high instability in all the states also in all the time periods. The falling trends of cost, revenue and profit with stagnancy in the yields show the declining nature of *urad* crop in India. The profits are shown in Figures 14.7 and 14.8.

Figure 14.7

Urad: A2-Profit per Hectare (in ₹)

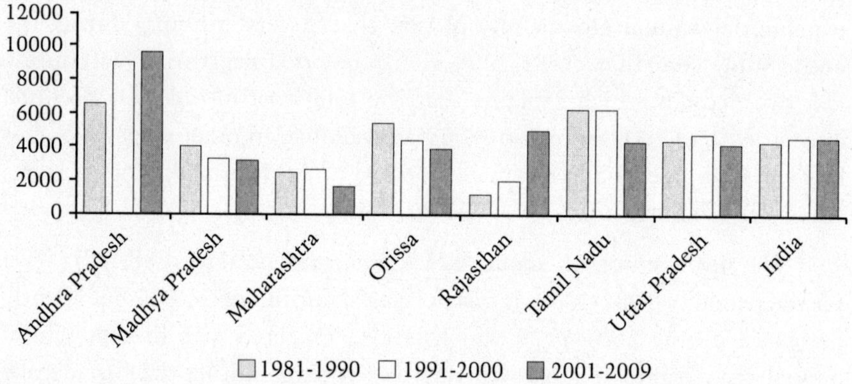

Source: Computed from various reports of *Cost of Cultivation of Principle Crops in India*, DES, MoA, GoI.

Figure 14.8

Urad: C2-Profit per Hectare (in ₹)

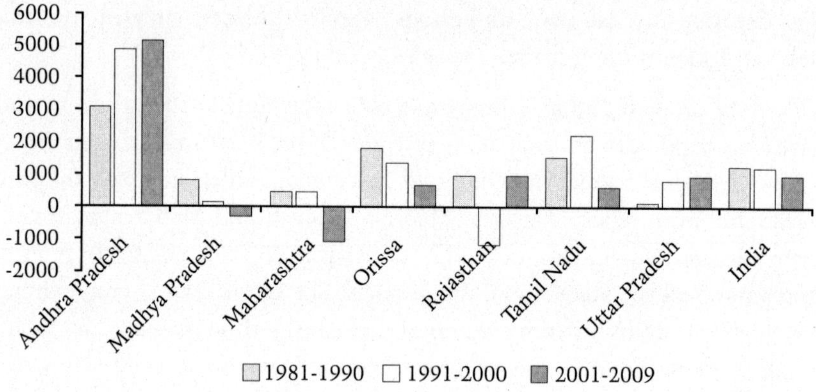

Source: Computed from various reports of *Cost of Cultivation of Principle Crops in India*, DES, MoA, GoI.

Groundnut

Tables A-14.3.17 to A-14.3.20 in Appendix A-14.3 gives the details about the groundnut crop cultivation in India. Groundnut is the major oilseed crop in India. It is high cost and low profitable crop in most of the states in India during 1981 to 2009. At all-India level, the C2 stands at ₹ 16,655 and the paid cost (A2) stands at ₹ 9,732.5 for the period of 1981-2009. The total cost gradually increased in all the sub-periods in most of the states in India. The cost grows more during first sub-period but declined during rest of the sub-periods in the majority of the states in India. But the instability has declined over the time period. The same trends report in case of paid-out cost too.

The revenue of groundnut stands at ₹ 17,885 per hectare at all-India level and it has increased over the time during the study period. Tamil Nadu stands first in income of groundnut crop followed by Orissa and Gujarat, while the lowest one is Karnataka. The instability in value of produce of this crop has declined over the time across majority of state during 1981 to 2009. But the annual exponential growth rate has declined over the time and it was very high during the first decade i.e., 1981-1990. The very low growth rates of yield indicate the stagnancy in the yield of groundnut in India during last three decades time.

The total cost based profit results show that groundnut is less profit making crop in most of the states in India. It is also loss making crop in few states like Andhra Pradesh for all the sub-periods, Karnataka, Maharashtra and Tamil Nadu for second and third sub-period. The instability is very high for the profit of this crop among all the states in all the sub-periods. The profits of groundnut are shown in Figures 14.9 and 14.10.

In case of groundnut, the variations in cost, revenue and profit among the states and across the time are consistent with variations in the yields. The data shows that fast growth of cost over revenue coupled with the stagnancy in the yields resulted in falling profits in groundnut cultivation in India during the last three decades.

Figure 14.9

Groundnut: A2-Profit per Hectare (in ₹)

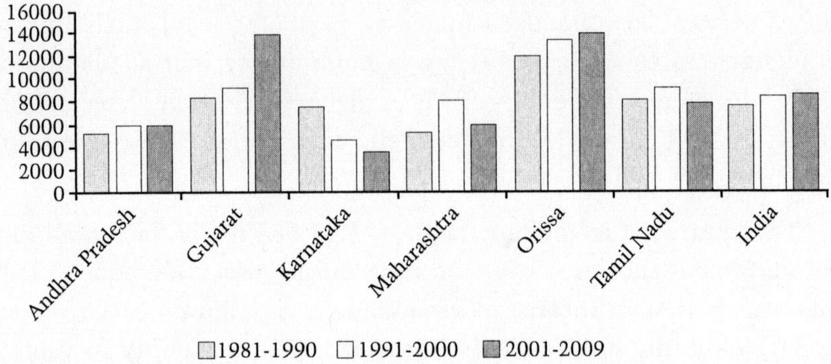

Source: Computed from various reports of *Cost of Cultivation of Principle Crops in India"*, DES, MoA, GoI.

Figure 14.10

Groundnut: C2-Profit per Hectare (in ₹)

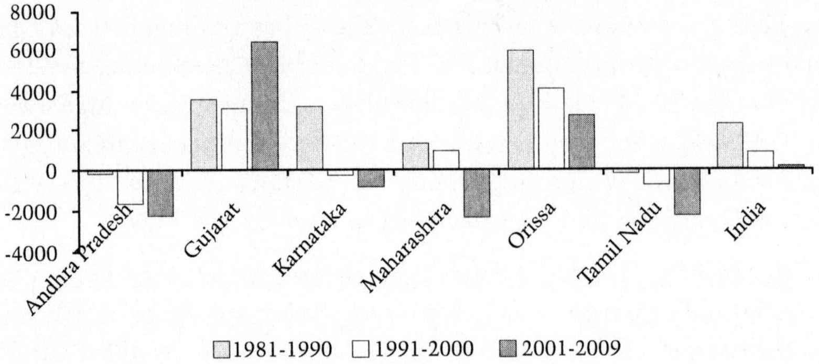

Source: Computed from various reports of *Cost of Cultivation of Principle Crops in India*, DES, MoA, GoI.

Cotton

Tables A-14.3.21 to A-14.3.24 in Appendix A-14.3 gives the details about the cotton crop cultivation in India. Cotton is major emerging commercial crop in India. At all-India level, the average C2 stands at ₹ 19,689 per hectare while paid-out cost stands at ₹ 10,383 during 1981-

2009. They have increased over all the sub-periods among majority states in India. The instability of cost is relatively high in cotton when compared to other crops but it has decreased over the time. The exponential growth rate of cost of cotton is relatively high during first and third sub-periods, but registered negative growth during second sub-period in most of the states in India which shows the fall of growth of cost in cotton during 1991-2000.

The revenue of cotton produce stands at ₹ 22,768 in India during the study period and it has gradually risen over the time across all the states. Among states, Punjab stands first followed by Andhra Pradesh, Gujarat and Tamil Nadu, while Karnataka is lowest cotton revenue state. The instability of revenue per hectare in cotton crop is also high among all the states in India but it has reduced over the time in majority of the states. The annual exponential growth rates of cotton revenue show that at all-India level it stands positive of 2.3 per cent during the entire study period and it is very high during first and third sub-periods but reports slightly negative growth rate in second sub-period which explains the fall of cotton revenue per hectare during 1991-2000 but increased during 1981-1990 and 2001-2009. The same trend follows in majority states in India.

At all-India level, the total cost based (C2) profit of cotton crop stands at ₹ 3,079 during the study period i.e., 1981-2009. It has decreased during 1991-2000 but increased during 2001-2009. Among states, Punjab reports highest level of profit while Tamil Nadu reports even loss in all the sub-periods. The CV results show that there is high level of fluctuations in cotton's profit during the study period. The paid-out cost based profit of cotton cultivation always stands positive and high. The profits of cotton are shown in Figures 14.11 and 14.12.

The variation in cost, income and profit per hectare in cotton cultivation in most of the states in India during 1981 to 2009 are in consistent with variations in yield levels. The second sub-period i.e., 1991-2000 has witnessed crisis for cotton cultivation where all the cost, revenue, profit and yield levels have shown negative growths.

Figure 14.11

Cotton: A2-Profit per Hectare (in ₹)

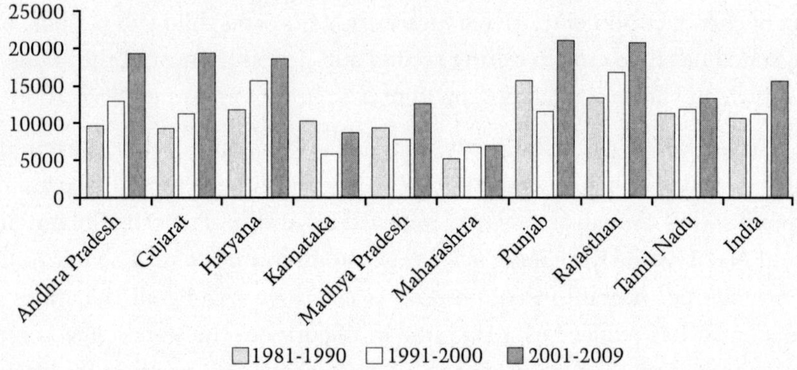

Source: Computed from various reports of *Cost of Cultivation of Principle Crops in India*, DES, MoA, GoI.

Figure 14.12

Cotton: C2-Profit per Hectare (in ₹)

Source: Computed from various reports of *Cost of Cultivation of Principle Crops in India*, DES, MoA, GoI.

Sugarcane

Tables A-14.3.25 to A-14.3.28 in Appendix A-14.3 gives the details about the sugarcane crop cultivation in India. Sugarcane is the highest cost, revenue and profit making crop than others in all its growing states in India. At all-India level the C2 stands at ₹ 59,291 per hectare during the study period i.e., 1981-2009 and it has gradually increased over the period. Among the states, Tamil Nadu reports highest cost with ₹ 75,105 per hectare followed by Andhra Pradesh, Maharashtra, and Karnataka, while Uttar Pradesh reports lowest cost with ₹ 35,382 per hectare. The cost has increased in all the states during all the sub-periods. The CV results show that the instability in cost has declined in majority of the states during 1991-2000 but slightly increased during 2001-2009. The annual exponential growth rates reveal that the cost has increased 1.8 per cent during the entire study period. The second sub-period has recorded negative growth of cost of sugarcane in majority of the states which shows the slowdown in the cost rise. But the last sub-period has recorded relatively high growth rates than first sub-period which shows the fast growth of its cost in most of the sugarcane grown states in India.

Sugarcane reports highest revenue per hectare than any other crops in the study. At all-India level it stands at ₹ 91,054 per hectare during 1981-2009. It has increased during second sub-period but decreased during third sub-period. Among states, Tamil Nadu records highest revenue per hectare i.e., ₹ 1,17,427 followed by Karnataka while Haryana records lowest value i.e., ₹ 73,139 during the entire study period. The CV results of the value of produce of sugarcane crop report that the instability has gradually increased over the time in the study period in majority of the states. The exponential growth rate of it is reported relatively less than other crops and it has reported even negative growth rate during second sub-period but slightly higher growth rates during the rest of the sub-periods among all the states.

The profit levels (both A2 and C2) are relatively very high in case of sugarcane crop ranging from ₹ 14,990 in Andhra Pradesh to ₹ 47,255 in Karnataka which also shows the high level variations of profits among states in India during last three decades period. Over the time the profit has increased during second sub-period but slightly declined during third sub-period in majority states in India. The CV results show the increasing instability of profits of sugarcane over the time in India. The profits are shown in Figures 14.13 and 14.14.

Figure 14.13

Sugarcane: A2-Profit per Hectare (in ₹)

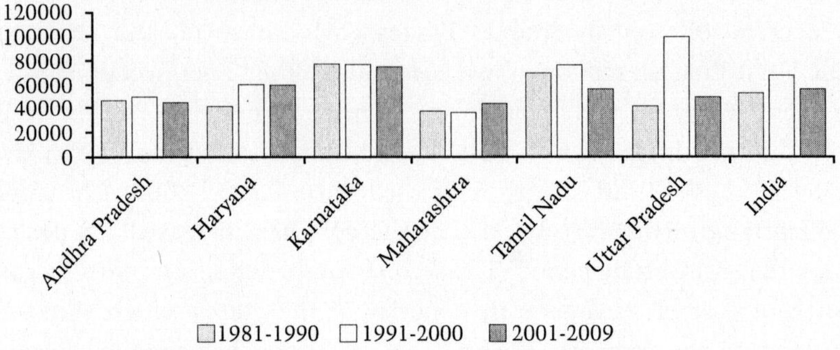

Source: Computed from various reports of *Cost of Cultivation of Principle Crops in India*, DES, MoA, GoI.

Figure 14.14

Sugarcane: C2-Profit per Hectare (in ₹)

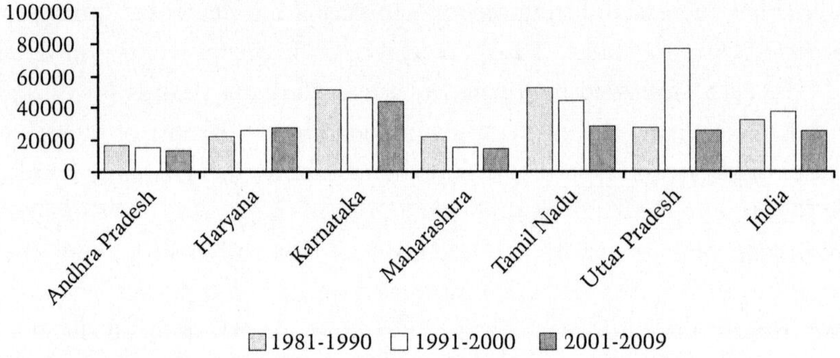

Source: Computed from various reports of *Cost of Cultivation of Principle Crops in India*, DES, MoA, GoI.

Cost of Cultivation, Revenue and Profit of Major Crops in Andhra Pradesh: A Field-work based Study

Paddy

Paddy is a staple crop in Andhra Pradesh. The cost details of paddy are given in Table 14.1. Based on total cost concept (C2), paddy is loss making crop in Andhra Pradesh in 2011. But it varies across the regions where it reports loss of ₹ 3,951.7 per acre in Coastal village, a small profit of ₹ 614 per acre in Telangana village and relatively high profit of ₹ 7,234.5 per acre in Rayalaseema village. This profit in Rayalaseema village is due to its cultivation in relatively virgin lands which have high fertility and high yield. The cost is relatively high in Rayalaseema village followed by Coastal and Telangana.

But in case of paid-out cost (A2) concept, it is profitable in all the regional villages and more profitable in Rayalaseema village followed by Telangana and Costal Andhra villages. These differences in the profit are mainly due to differences in the use of own labour from which we can state that, in paid cost concept, higher the share of the own labour higher is the profit in paddy cultivation. These profits are shown in Figure 14.15.

Table 14.1

Paddy: Cost, Revenue, Profit and Yield Levels in Andhra Pradesh

Paddy	*A2–Cost per Acre*				*C2–Cost per Acre*			
	Coastal	*Telangana*	*Rayalaseema*	*A.P*	*Coastal*	*Telangana*	*Rayalaseema*	*A.P*
Cost (₹)	17192.3	13502.2	14799.5	15171.7	25097.7	22504.3	26824.5	23975.5
Revenue (₹)	21088.9	23118.4	34059.0	22949.1	21146.0	23118.4	34059.0	22972.3
Profit (₹)	3896.6	9616.2	19259.5	7777.4	-3951.7	614.1	7234.5	-1003.2
Yield (Kg)	1809	2240	2499	2140				

Source: Field-work based.

Figure 14.15

Paddy: Profit per Acre (₹) in Andhra Pradesh

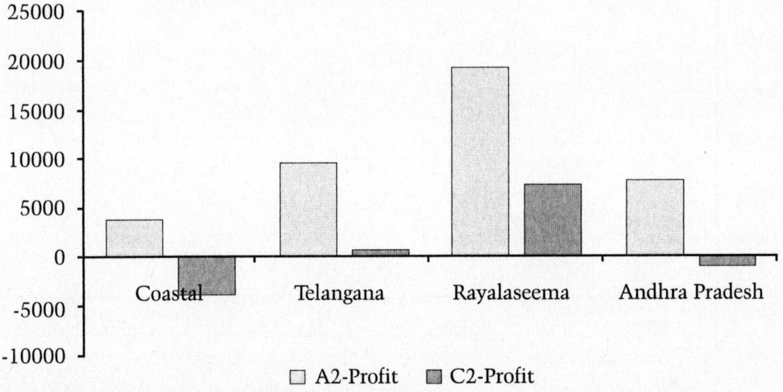

Source: Field-work based.

Maize

Maize is an emerging crop in all the regions in the state due to its high yield and easy farm management. Table 14.2 provides the details of cost of cultivation of maize in Andhra Pradesh. In case of total cost concept (C2), it is slightly loss making of ₹ -416 per acre but in case of paid cost it is a good profit making crop of ₹ 7,436 per acre at the state level. The similar conditions of total and paid-out costs of maize prevail in all the regional villages too. But it is more profitable crop in Telangana village in terms of paid cost due to higher share of own human labour and manure. These details are given in Figure 14.16.

Table 14.2

Maize: Cost, Revenue, Profit and Yield Levels in Andhra Pradesh

	A2–Cost per Acre				*C2–Cost per Acre*			
	Coastal	*Telangana*	*Rayalaseema*	*A.P*	*Coastal*	*Telangana*	*Rayalaseema*	*A.P*
Cost (₹)	15727.1	10536.0	10725.0	11732.6	18295.7	21014.0	14692.5	19584.5
Revenue (₹)	17590.7	20694.0	14302.5	19168.5	17590.7	20694.0	14302.5	19168.5
Profit (₹)	1863.6	10158.0	3577.5	7436.0	-705.0	-320.0	-390.0	-416.0
Yield (Kg)	1621	2055	1313					

Source: Field-work based.

Figure 14.16

Maize: Profit per Acre (₹) in Andhra Pradesh

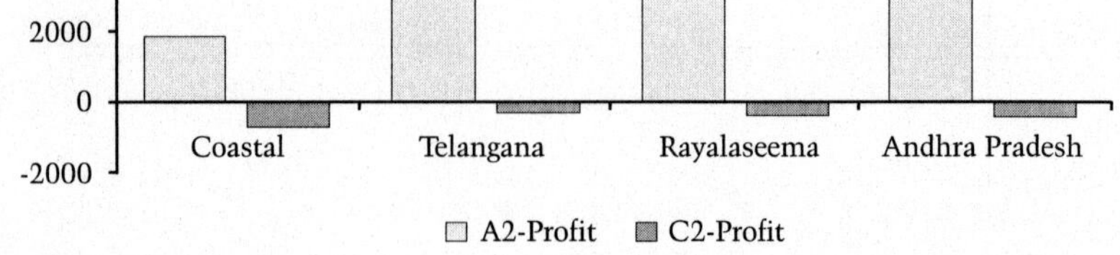

Source: Field-work based.

Seed-Paddy

Table 14.3 gives the detail about the cost of cultivation of seed-paddy. It is grown in Telangana village where farmers are contracted out by the seed companies to grow the paddy for seed purpose. An agreement would be signed between the company and the farmers in seed-paddy cultivation regarding the quality of output and insurance. Only few farmers who are experts in farming will be selected for this contract. This crop is high labour intensive which forms more than 50 per cent of total cost and profitable too. The profits are shown in Figure 14.17.

Table 14.3

Seed-Paddy: Cost, Revenue, Profit and Yield Levels in Andhra Pradesh

	Seed-Paddy Per Acre	
	A2	*C2*
Cost (₹)	28732.5	38870.0
Revenue (₹)	62250.0	62250.0
Profit (₹)	33517.5	23380.0
Yield (Kg)	1638	

Source: Field-work based.

Figure 14.17

Seed-Rice: Profit per Acre (₹) in Andhra Pradesh

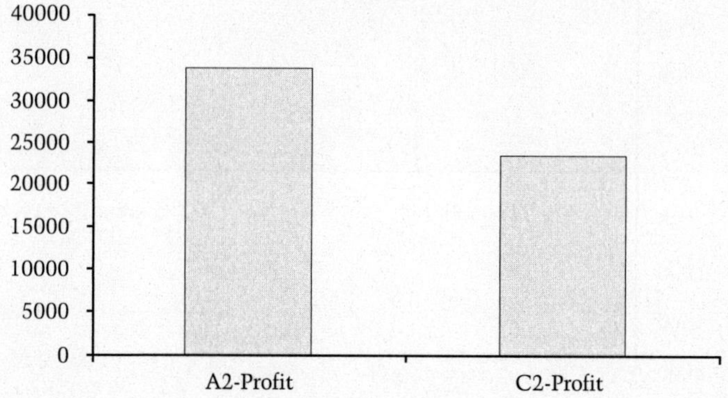

Source: Field-work based.

Coarse Cereals *(Jowar+Korra)*

Jowar and *Korra* are the fast declining crops in the state, grown mostly in Rayalaseema village. These are less expensive food crops hence the cost of cultivation and the revenue and profit are very low. These crops are loss making in terms of total cost concept (C2 but are less profitable in case of paid cost concept (A2). The details are shown in Table 14.4 and Figure 14.18.

Table 14.4

Coarse Cereals (Jowar+Korra): Cost, Revenue, Profit and Yield Levels in Andhra Pradesh

Per Acre	*Coarse Cereals per Acre*	
	A2	*C2*
Cost (₹)	2592.5	5731.3
Revenue (₹)	4275.0	4275.0
Profit (₹)	1682.5	-1456.3
Yield (Kg)	225.0	

Source: Field-work Based.

Figure 14.18

Coarse Cereals: Profit per Acre (₹) in Andhra Pradesh

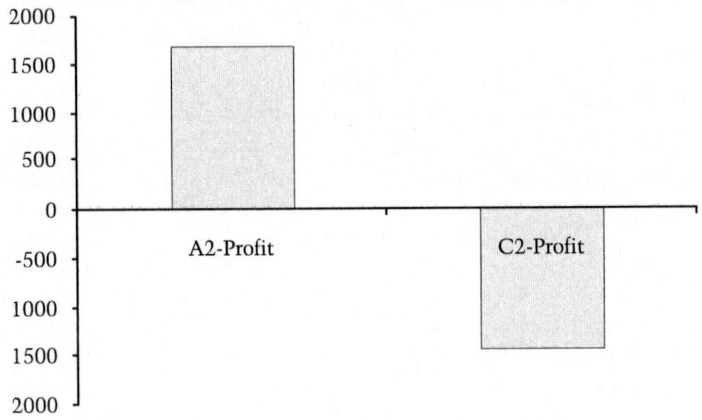

Source: Field-work based.

Pulses

Red Gram

Red gram is the main pulses crop in Andhra Pradesh, mainly grown in Rayalaseema region. This crop is low cost crop which is cultivated in the dry lands. This is also a risky crop since it depends on weather and is also more prone to pest attack. The spurious seeds also hampered the productivity of red gram in the field work village of Rayalaseema during 2011. Hence, there was a loss of ₹ 3,479 per acre in red gram cultivation. This is not a profit making crop even in the concept of paid cost. The data regarding these cost details are shown in Table 14.5 and Figure 14.19.

Table 14.5

Pulses: Cost, Revenue, Profit and Yield Levels in Andhra Pradesh

Per Acre	*Red Gram per Acre*		*Green + Black Gram per Acre*	
	A2	*C2*	*A2*	*C2*
Cost (₹)	4856.4	7030.2	13532.7	17225.9
Revenue (₹)	3551.0	3551.0	19106.3	19106.3
Profit (₹)	-1305.4	-3479.1	5573.6	1880.4
Yield (Kg)	140		421.0	

Source: Field-work Based.

Figure 14.19

Red Gram: Profit per Acre (₹) in Andhra Pradesh

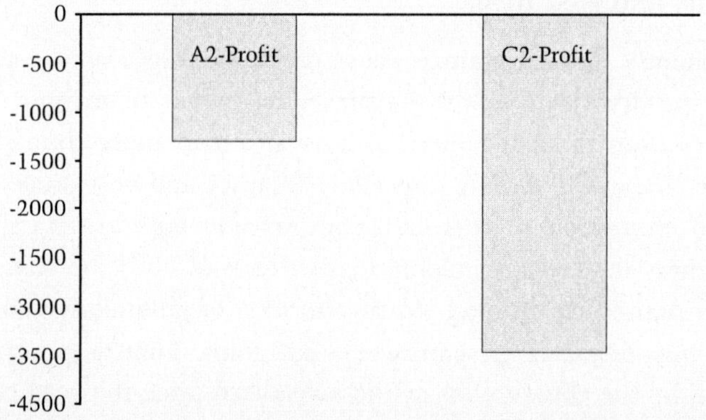

Source: Field-work based.

Green Gram and Black Gram

Green gram and black gram are other important pulses crops in the state, grown in Coastal Andhra village. These crops are preferred due to their easy farm management and less labour intensity. They are marginally profitable crops in case of total cost concept but profitable crop in case of paid cost. But the increasing pest attack is adversely affecting the productivity and the profitability of these crops in the state. The cost details are given in Table 14.5 and Figure 14.20.

Figure 14.20

Green and Black Gram: Profit per Acre (₹) in Andhra Pradesh

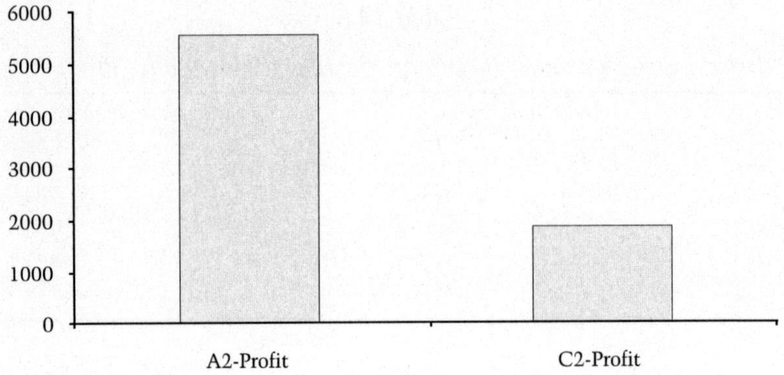

Source: Field-work based.

Groundnut and Red gram

Groundnut and red gram is most prevalent mixed crop combination grown in the dry lands of the Rayalaseema region mainly to reap high productivity from the same piece of land and to manage the risk of crop failure due to various reasons especially weather and pest attack. Despite this mixed crop strategy, this crop combination fails to reap the profits to the farmers instead it rendered loss in case of both the cost concepts which was mainly due to bad weather in case of groundnut and spurious seeds and uncontrollable pest in case of red gram. Though groundnut seed is provided by the government at the subsidised price the seed cost forms the significant share (10.9%) in the cost of cultivation. The cost details are given in Table 14.6 and Figure 14.21.

Table 14.6

Groundnut and Red Gram: Cost, Revenue, Profit and Yield Levels in Andhra Pradesh

	Groundnut+Red Gram per Acre	
	A2	*C2*
Cost (₹)	6241.9	9685.5
Revenue (₹)	6430.6	6430.6
Profit (₹)	188.7	-3254.8
Yield (Kg)	216	

Source: Field-work based

Figure 14.21

Groundnut and Red Gram: Profit per Acre (₹) in Andhra Pradesh

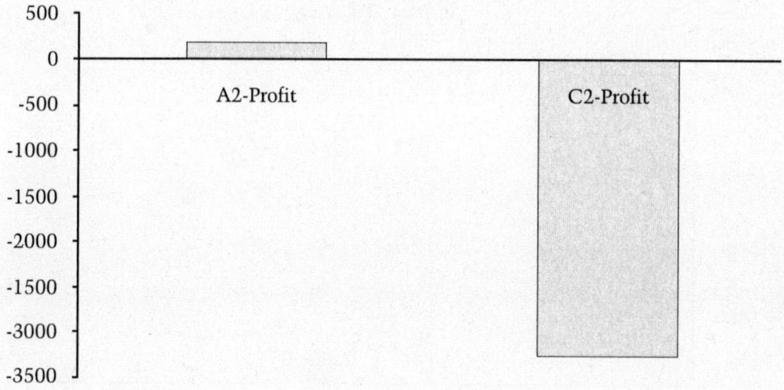

Source: Field-work based.

Cotton

Cotton is fast emerging commercial crop in the state especially in Telangana and Coastal Andhra region due to its high profitability. The cost details are given in Table 14.7. In case of total cost it is marginally profitable crop in the state but a high profitable crop in case of paid cost due to use of own labour. It is a high labour intensive crop hence the promotion of this will create more agricultural employment. Data also reveal that, in case of total cost, cotton yield profit in Telangana village but yield marginal loss in Costal Andhra region mainly due to unseasonal excessive rains during 2011. But in case of paid cost, it is relatively high profitable crops in both the regions especially in Telangana region. These profits are shown in Figure 14.22.

Table 14.7

Cotton: Cost, Revenue, Profit and Yield Levels in Andhra Pradesh

	A2-Cost per Acre			*C2-Cost per Acre*		
	Coastal	*Telangana*	*A.P*	*Coastal*	*Telangana*	*A.P*
Cost (₹)	28162.9	18427.7	22675.8	34524.2	31831.0	33006.2
Revenue (₹)	33101.7	34193.5	33717.1	33101.7	34193.5	33717.1
Profit (₹)	4938.8	15765.8	11041.3	-1422.5	2362.6	710.9
Yield (Kg)	789.0	850.0	819.5			

Source: Field-work based.

Figure 14.22

Cotton: Profit per Acre (₹) in Andhra Pradesh

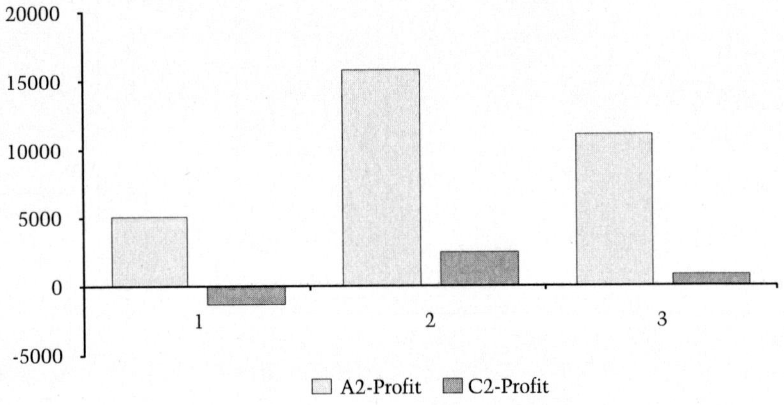

Source: Field-work based.

Red Chilly

Red chilly is the important and highly profitable commercial crop in the state, cultivated especially in Coastal Andhra region where irrigation is available. The cost of cultivation of this crop is very high than any other crops so the revenue too. In 2010, it is highly profitable crop (₹ 63,509 per acre) due to high yield and the output price. But unlike in other crops, the output price of red chilly is highly volatile because of its dependency on the international market for export. This is highly labour intensive crop where more than 52 per cent of cost goes to human labour. These details are given in Table 14.8 and Figure 14.23.

Table 14.8

Red Chilly: Cost, Revenue, Profit and Yield Levels in Andhra Pradesh

	Red chilly per Acre	
	A2	*C2*
Cost (₹)	72601.3	83680.4
Revenue (₹)	147189.2	147189.2
Profit (₹)	74587.9	63508.8
Yield (Kg)	2009	

Source: Field-work based.

Figure 14.23

Red Chilly: Profit per Acre (₹) in Andhra Pradesh

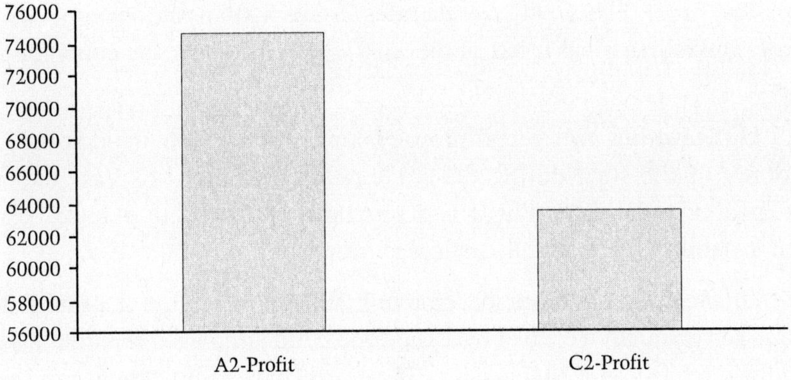

Source: Field-work based.

Summary and Policy Suggestions

Summary

In paddy crop's cultivation, the cost, revenue, yield and profit widely vary across the states. Across the states, the cost variation is much higher than the yield variation which shows there is no strong relation between the cost of cultivation and yield of the crop across the states. So there must be other factors than yield to explain the cost variations. The rate of profit in paddy cost cultivation has gradually come down during the last three decades in majority of the states with few exceptional states like Punjab, Haryana and Andhra Pradesh where it has increased. The annual exponential growth rate of paddy's cost of cultivation is relatively higher than revenue in majority of states during the study period of 1981 to 2009. This mismatch between the cost and revenue in paddy crop has resulted in fall in profits in majority of the states. Over the time the instability in profit of paddy has increased which shows the growing risk in paddy cultivation in recent decade.

Wheat is the profitable crop in India and in majority of the states too. Though there is rise in both cost and revenue along with yield in wheat during 1981 to 2009 in all the states in India, the growth rates have gradually declined over the time. Unlike in case of paddy, the cost variation of wheat among the states is consistent with yield variations. The CV results show that the instability in wheat's cost of cultivation has decreased over the time. The yield growth rates are less than the revenue which shows the falling yield levels is the matter of concern in the cultivation of wheat.

Unlike other two cereal crops, maize is not profit making crop in majority of the state during the study period. But the loss has gradually declined over the time. But it is a profitable according to paid-cost (A2). The instability has gradually increased over the time.

Urad crop is less profitable crop in India also in most of the states. The profit has gradually declined over the time period and even reported loss in Madhya Pradesh and Maharashtra in the recent decade. The instability is relatively high for cost, revenue and profit of *urad* crop in India. The growth rates all are minimal in most of the states in India. Unlike in other crops, *urad* crop has less variation in cost and revenue in comparision to yield variation among states which are relatively high. The very low growth rates of yield also negative in first two sub-periods shows the stagnancy in the *urad* yields in India in the last three decades period. The falling trends of cost, revenue and profit with stagnancy in the yields show the declining nature of *urad* crop in India

Groundnut is high cost and low profitable crop in most of the states in India during 1981 to 2001. The total cost gradually increased in all the sub-periods in most of the states in India. In case of groundnut, the variations in cost, revenue and profit among the states and across the time are in consistent with variations in the yields. The fast growth of cost over revenue coupled with the stagnancy in the yields resulted in falling profits in groundnut cultivation in India during last three decades time

Cotton is the profit making crop in most of the states in India during 1981 to 2009 with exceptions where negative profit in Tamil Nadu during all the periods and in Maharashtra during recent decade. Most of the states report fall in cost, revenue and profit during the second sub-period i.e., 1991-2000, while in the rest of the period they reported high growth. The variation in cost, income and profit per hectare in cotton cultivation in most of the states in India during 1981 to 2009 are in consistent with variations in yield levels. The instability levels are relatively high in cost, revenue and profit of cotton crop in the state during 1981 to 2009 which shows the high volatile nature of cotton crop cultivation in India. The second sub-period i.e., 1991-2000 has witnessed crisis for cotton cultivation where all the cost, revenue, profit and yield levels have shown negative growths.

Sugarcane is the highest cost, revenue and profit making crop than others in all its growing states in India during entire study period also during all the sub-periods with few exceptions. Karnataka reports highest profit of ₹ 47,255 per hectare followed by Uttar Pradesh and Tamil Nadu while Andhra Pradesh reports lowest profit of ₹ 14,990 per hectare during the study period. The profits of sugarcane are also associated with highest instability which is the matter of concern. In the sub-period of 1991-2000, cost, revenue and profits report lowest growth rates in most of the states. The variations in cost, revenue and profits are nearly in consistent with the variations in the yield in most of the sugarcane states in India during 1981 to 2009.

It can be observed from the CV data of all crops that the instability is greater among the revenues and profits than the costs and yields of all the crops in the study which explains that increasing risk associated with output market than with input market in Indian agriculture in the recent three decades.

The crop-wise cost, revenue and profit analysis from the field work in Andhra Pradesh shows that cereals, pulses and oilseed crops are still loss making crops in terms of total cost. In terms of paid cost, they turn out to be profitable when the farmer posses more of own resources like human and animal labour, land, manure etc., hence the cultivation crops pay the economic rent to these owned resources. Otherwise in total cost concept, cultivation of these crops is loss oriented.

The commercial crops i.e., cotton and red chilly are profitable crops. But the fast growing area of these commercial crops at the cost of food crops will have adverse consequences on the economy in terms of shortage of food grain availability and the resultant high food inflation.

Cultivation becomes profitless because of the fast rise in input cost over output price. In other words, farming is profitless because of high production cost and the low output price. The stagnancy in the yield levels in last two decades also fuelled the problem.

Policy Suggestions

Indian agriculture is operating in the framework of controlled/regulated system in output market but decontrolled system (with few exceptions) in input market. The available output price is insufficient to meet the ever increasing input cost. Hence, there should be well-balanced policy either controlled or decontrolled system in both the input and output markets. We would prefer the decontrolled system in both the input and output market sides but provided that the government is taking care of the livelihood needs such as food, health, education and social security of the all agriculturally dependent people in the country. Agricultural extension services should be strengthened to economise the input cost and realise the yield gaps. The group approach can be an effective method for this. To address the growing risk in the output price and profitability of crops, there should be proper risk management mechanism in the form of crop insurance and cooperative farming between the farmers and companies. To address the problem of regional variations in the yield and profitability of crops, the agro-climatic centric regional planning should be strengthened.

References

Bagchi, Amitya Kumar (2004). "The Common Minimum Programme", *Social Scientist*, July-August, pp.3-16.

Kapre, B.N. (1974). *Comprehensive Scheme for Studying the Cost of Cultivation of Principal Crops*. New Delhi: Directorate of Economics and Statistics, Government of India.

Ministry of Agriculture (2010). *Annual Report 2009-10*. New Delhi: Department of Agriculture and Co-operation, Govt of India.

———. (various years). *Cost of Cultivation of Principle Crops in India (1991, 1996, 2001, 2007)*. New Delhi: Directorate of Economics and Statistics, Department of Agriculture and Co-operation, Govt of India.

Patnaik, Prabhat (2003). "Agricultural Production and Prices under Globalisation" in *The Retreat to Unfreedom*. New Delhi: Tulika Books. pp.198-210.

Raghavan, M. (2008). "Changing Pattern of Input Use and Cost of Cultivation", *Economic and Political Weekly*, June 28.

Sen, Abhijit and M.S. Bhatia (2004). "Cost of Cultivation and Farm Income", *State of the Indian Farmer: A Millennium Study*. Vol 14. New Delhi: Government of India and Academic Foundation.

Appendix 14.1

Cost Concepts

The cost of production of agricultural crops is the most important factor that Commission for Agricultural Costs and Prices (CACP) considers while making its recommendations on minimum support prices (MSPs) to the Government. The Directorate of Economics & Statistics (DES), Ministry of Agriculture, has been implementing a Comprehensive Scheme for Studying the Cost of Cultivation of Principal Crops in India since 1970-71 to generate estimates of cost of cultivation/ production of the principal crops in order to meet cost of cultivation data requirement.

Costs are generated following certain cost concepts. These cost concepts and the items of costs included under each concept are given below:

Cost A1 :

(i) Value of hired human labour.

(ii) Value of hired bullock labour.

(iii) Value of owned bullock labour.

(iv) Value of owned machinery labour.

(v) Hired machinery charges.

(i) Value of seed (both farm produced and purchases).

(ii) Value of insecticides and pesticides.

(iii) Value of manure (owned and purchased).

(iv) Value of fertiliser.

(v) Depreciation on implements and farm buildings.

(vi) Irrigation charges.

(vii) Land revenue, cesses and other taxes.

(viii) Interest on working capital.

(ix) Miscellaneous expenses.

Cost A2 : Cost A1+ rent paid for leased-in-land.

Cost B1 : Cost A1 + interest on value of owned fixed capital assets (excluding land).

Cost B2 : Cost B1 + rental value of owned land (net of land revenue) and rent paid for leased-in land.

Cost C1 : Cost B1+ total value of family labour.

Cost C2 : Cost B2 + total value of family labour.

*Cost C2** : Cost C2 + Additional value of human labour based on use of higher rate i.e., statutory wage rate or the actual market rate. This is an intermediate concept.

Cost C3 : Cost C2*+ 10 per cent of cost C2* to account for managerial input of the farmer.

Appendix A-14.2

Brief Description of the Fieldwork Methodology

Politically Andhra Pradesh can be divided into three regions such as Coastal Andhra, Telangana and Rayalaseema. While the Coastal Andhra region represents the agro-climatic zone of East Coast Plains and Hills Region, other two regions represent the agro-climatic zone of Southern Plateau and Hills Region.

Selection of the District:

The district selection was done in a way that district should have similar type of the cropping pattern of the region. For this the study used the area-GCA ratio analysis of major crops at region and district wise. Based on these criteria, Krishna from Coastal Andhra, Warangal from Telangana, and Kurnool from Rayalseema were selected.

Mandal (Block) Selection:

One *mandal* from each district was selected by using the same criteria which was used for district selection such as the *mandal* should have similar type of cropping pattern of the district. Area-GCA ratio analysis of major crops was used to identify the similar cropping pattern *mandal* in the particular district. Base on this, G. Konduru from Krishna district, Hasanaparti from Warangal district and Dhone from Kurnool district were selected.

Village Selection:

One village from the each *mandal* was selected by on the criteria of that village shall cultivate maxim number of crops. Mandal Agricultural Officer's suggestions were taken to identify this type of village. Based on this method, Kavuluru from G. Kondurumandal, Sitanagaram from Hsanapartimandal and Kanapakunta from Dhonemandal were selected.

Selection of Farm Households in the Village:

Once the village is selected, the stratified random sampling method was used for dividing farm households into sizes. All the farm operational holdings in the village was enumerated and classified into three farm size classes such as: Small (0–1.0 ha), medium (1–2 ha.), and Large (2 ha & Above). In each size class, 15 house holdings were selected by using simple random sampling method and total 45 samples were surveyed in a village. The same method was followed in all three villages. Total 135 farm households from three villages were surveyed.

The interview scheduled method was used for the collection of data from farm households. The main content of farm household data collection was private resources, educational levels, and cost of cultivation of crops and factors of cropping decision. The data regarding the common property resources, irrigation, roads, agricultural markets and other infrastructural facilities of the village was collected from the village officials.

Appendix A-14.3

Table A-14.3.1: Paddy: A2-Cost Details

State	A2-Cost Details												A2-Profit Details							
	A2-Cost per Hectare (₹) at 2004-05 Prices				CV (%)				Growth Rate (%)				A2-Profit per Hectare(₹) at 2004-05 Prices				CV (%)			
	1981-2009	1981-1990	1991-2000	2001-2009	1981-2009	1981-1990	1991-2000	2001-2009	1981-2009	1981-1990	1991-2000	2001-2009	1981-2009	1981-1990	1991-2000	2001-2009	1981-2009	1981-1990	1991-2000	2001-2009
Andhra Pradesh	14795	13267	14953	16315	13.0	15.0	6.2	7.7	1.2	4.8	-1.9	2.2	13685	9682	14005	17776	29.8	24.5	10.6	17.9
Assam	4960	4355	4597	6036	20.1	21.9	13.1	4.5	2.0	6.8	2.6	0.8	7433	7422	7943	6878	18.1	21.2	12.7	16.8
Bihar	6334	5178	6534	7396	19.2	21.7	7.2	9.4	1.9	6.3	-1.8	0.6	7213	8076	7164	6308	22.2	12.3	14.7	32.8
Haryana	12276	10516	11738	14830	17.6	9.2	6.5	11.8	1.5	1.7	-1.5	-0.7	17246	12376	16617	23354	40.3	23.3	18.7	36.3
Karnataka	14068	8882	16703	16901	33.3	19.7	23.1	13.7	2.9	3.4	-3.8	-4.3	17314	15681	19185	17048	30.3	24.7	34.6	23.5
Madhya Pradesh	5518	4333	6051	6243	19.8	20.1	9.1	8.2	2.1	6.1	1.0	0.3	5992	5510	6182	6317	45.2	23.7	26.1	68.0
Orissa	7614	6462	7389	9145	17.9	18.2	9.1	3.2	1.9	5.3	1.3	0.0	8146	7534	8711	8198	20.7	19.8	12.2	26.4
Punjab	14085	14534	12181	15703	13.7	6.1	4.7	12.7	0.2	-0.9	-0.6	-0.4	17926	13680	16201	24561	31.3	8.9	14.4	20.9
Tamil Nadu	19343	19737	19359	18889	12.8	19.9	6.8	3.0	0.1	6.9	-2.2	-0.6	12483	11705	13312	12425	21.1	26.7	6.0	24.7
Uttar Pradesh	7848	6650	7580	9477	18.4	17.0	10.0	6.0	1.8	4.4	-2.1	1.1	9552	8126	9900	10749	24.7	20.7	12.3	28.3
West Bengal	10777	9445	10660	12388	16.7	20.0	11.2	4.3	1.5	5.7	-0.9	0.1	11174	11364	12298	9713	22.8	21.6	18.5	22.6
India	10693	9396	10704	12120	18.4	17.2	9.7	7.7	1.5	4.6	-0.9	-0.1	11651	10105	11956	13030	27.9	20.6	16.4	28.9

Source: Computed from various reports of *Cost of Cultivation of Principle Crops in India*, DES, MoA, GoI.

Table A-14.3.2: Paddy: C2-Cost Details

State	C2-Cost Details												C2-Profit Details							
	C2-Cost per Hectare (₹) at 2004-05 Prices				CV (%)				Growth Rate (%)				C2-Profit per Hectare (₹) at 2004-05 Prices				CV (%)			
	1981-2009	1981-1990	1991-2000	2001-2009	1981-2009	1981-1990	1991-2000	2001-2009	1981-2009	1981-1990	1991-2000	2001-2009	1981-2009	1981-1990	1991-2000	2001-2009	1981-2009	1981-1990	1991-2000	2001-2009
Andhra Pradesh	26066	21792	27013	29763	16.4	17.7	5.0	7.8	0.0	5.5	-1.4	2.2	2413	1158	1945	4328	82.7	97.3	57.9	48.3
Assam	11612	9806	11252	14019	20.7	25.6	11.5	3.2	2.2	8.4	2.9	0.7	780	1970	1288	-1105	220.7	67.6	70.3	-101.5
Bihar	12559	11747	12733	13267	11.6	16.0	6.6	6.9	0.7	4.7	-2.1	-0.1	988	1507	966	437	154.8	66.3	91.3	510.4
Haryana	23466	18630	22793	29587	20.9	11.0	7.1	8.1	2.1	-1.6	0.2	2.0	6056	4263	5563	8597	91.9	97.9	74.6	83.0
Karnataka	24144	16678	28233	27897	29.6	21.0	24.0	8.6	2.4	3.5	-4.2	-2.3	7237	7886	7656	6051	47.5	30.9	48.9	61.3
Madhya Pradesh	10769	8276	11809	12385	21.2	19.0	8.1	12.4	2.2	4.7	1.3	2.2	741	1567	425	174	292.9	78.7	330.6	1822.2
Orissa	14426	11678	14261	17662	20.4	21.6	8.8	3.0	2.3	6.5	1.7	0.8	1335	2318	1839	-319	132.5	27.3	79.0	-559.0
Punjab	25076	23779	22653	29210	13.2	5.6	4.7	8.7	0.9	0.1	-0.4	2.1	6935	4435	5729	11054	53.8	26.6	28.9	34.7
Tamil Nadu	30297	29947	30623	30326	13.7	22.3	6.4	4.0	0.4	7.7	-2.1	-0.2	1528	1495	2048	989	128.2	52.1	56.4	313.7
Uttar Pradesh	15544	12899	15414	18629	18.4	17.5	7.9	7.1	1.9	4.9	-1.6	2.3	1856	1877	2067	1597	91.6	84.5	45.5	145.6
West Bengal	20874	18111	21103	23690	16.8	22.0	9.9	3.4	1.6	6.5	-0.1	0.4	1077	2698	1854	-1589	237.0	41.3	113.7	-125.2
India	19530	16667	19808	22403	18.4	18.1	9.1	6.7	1.5	4.6	-0.5	0.9	2813	2834	2853	2747	139.4	61.0	90.7	203.0

Source: Computed from various reports of *Cost of Cultivation of Principle Crops in India*, DES, MoA, GoI.

Table A-14.3.3: Paddy: Revenue Details

State	*Revenue Details*											
	Revenue per Hectare (₹) at 2004-05 Prices				*CV (%)*				*Growth Rate (%)*			
	1981-2009	*1981-1990*	*1991-2000*	*2001-2009*	*1981-2009*	*1981-1990*	*1991-2000*	*2001-2009*	*1981-2009*	*1981-1990*	*1991-2000*	*2001-2009*
Andhra Pradesh	28479	22949	28959	34091	20.3	18.2	7.2	12.5	2.1	5.2	-2.0	3.9
Assam	12393	11776	12540	12914	14.4	20.5	9.6	9.1	0.8	5.3	2.2	1.4
Bihar	13547	13254	13699	13704	12.3	13.0	8.2	14.7	0.3	3.4	-2.5	3.1
Haryana	29522	22893	28356	38183	27.7	14.8	12.8	20.6	2.5	4.6	-3.0	6.5
Karnataka	31381	24563	35889	33948	27.2	20.7	29.0	7.0	1.5	2.5	-5.5	1.7
Madhya Pradesh	11510	9843	12234	12559	27.7	19.3	13.3	36.5	1.3	3.7	-1.7	8.0
Orissa	15760	13996	16100	17343	15.7	18.5	7.4	12.3	1.3	5.7	-1.3	3.7
Punjab	32011	28214	28382	40264	19.9	4.3	7.5	12.4	1.7	0.2	-0.2	4.0
Tamil Nadu	31826	31442	32670	31314	14.3	22.1	5.4	9.3	0.4	7.4	-1.6	1.2
Uttar Pradesh	17400	14776	17480	20227	18.7	15.7	8.8	15.5	1.7	4.7	-1.9	5.5
West Bengal	21951	20809	22958	22101	15.2	19.7	13.0	9.8	0.6	6.0	-2.2	3.6
India	22344	19501	22661	25150	19.4	17.0	11.1	14.5	1.3	4.4	-1.8	3.9

Source: Computed from various reports of *Cost of Cultivation of Principle Crops in India*, DES, MoA, GoI.

Table A-14.3.4: Paddy: Yield Details

State	Yield Details											
	Yield per Hectare (Quintal)				CV (%)				Growth Rate (%)			
	1981-2009	1981-1990	1991-2000	2001-2009	1981-2009	1981-1990	1991-2000	2001-2009	1981-2009	1981-1990	1991-2000	2001-2009
Andhra Pradesh	45	37	46	52	15.3	9.7	5.7	5.5	1.8	3.3	1.6	1.6
Assam	22	19	22	24	14.5	9.0	7.5	11.9	1.3	2.0	1.7	-0.2
Bihar	22	19	22	25	14.0	9.3	5.5	10.8	1.3	1.2	1.6	-0.4
Haryana	40	36	39	45	11.9	6.2	5.7	9.3	1.1	1.1	-0.4	1.6
Karnataka	40	33	41	47	16.7	15.4	5.1	6.5	1.7	-0.3	1.0	1.7
Madhya Pradesh	17	14	18	18	22.4	13.6	17.7	23.7	1.3	1.8	1.8	3.9
Orissa	26	21	27	31	18.3	14.4	6.8	6.3	2.1	4.4	-0.5	0.9
Punjab	56	52	52	64	12.0	5.9	7.2	6.0	0.9	-1.2	0.4	1.2
Tamil Nadu	44	42	45	46	6.2	2.8	4.2	6.2	0.5	0.9	1.3	0.6
Uttar Pradesh	30	25	31	34	15.4	13.6	5.9	7.8	1.7	4.6	1.4	1.5
West Bengal	32	26	33	37	17.2	14.6	9.0	3.4	1.9	4.2	0.5	1.2
India	34	29	34	38	14.9	10.4	7.3	8.9	1.4	2.0	0.9	1.2

Source: Computed from various reports of *Cost of Cultivation of Principle Crops in India*, DES, MoA, GoI.

Table A-14.3.5: Wheat: A2-Cost Details

State	*A2-Cost Details*												*A2-Profit Details*							
	A2-Cost per Hectare (₹) at 2004-05 Prices				*CV (%)*				*Growth Rate (%)*				*A2-Profit per Hectare (₹) at 2004-05 Prices*				*CV (%)*			
	1981-2009	*1981-1990*	*1991-2000*	*2001-2009*	*1981-2009*	*1981-1990*	*1991-2000*	*2001-2009*	*1981-2009*	*1981-1990*	*1991-2000*	*2001-2009*	*1981-2009*	*1981-1990*	*1991-2000*	*2001-2009*	*1981-2009*	*1981-1990*	*1991-2000*	*2001-2009*
Bihar	8727	8838	8926	8384	16.2	23.4	10.8	8.0	0.0	8.2	-3.2	-2.7	11026	14048	10406	8356.2	26.0	11.1	15.8	20.8
Gujarat	9246	6953	10502	10399	28.3	46.8	6.4	9.0	3.0	17.7	-1.2	-2.9	10530	4898	12377	14735.5	45.8	57.2	17.1	14.3
Haryana	10443	9972	9959	11502	9.8	4.2	7.4	8.5	0.5	0.1	-0.7	-2.1	18434	14059	20460	21044.7	30.2	49.6	6.3	16.1
Himachal Pradesh	4879	4558	4539	5615	23.6	24.2	19.0	20.5	0.9	2.6	-4.2	0.4	6917	6320	6787	7723.0	33.0	25.4	25.2	40.1
Madhya Pradesh	6571	5414	6929	7458	17.1	17.7	8.7	6.3	1.6	5.2	-2.8	-1.2	9204	7238	9002	11612	29.6	20.5	17.5	25.2
Punjab	12318	12059	12416	12497	6.9	5.3	7.0	7.6	0.1	1.5	-1.0	-2.4	16471	12049	18057	19622	25.4	16.1	16.2	14.6
Rajasthan	9253	7971	10035	9809	21.0	33.7	9.2	6.8	1.8	7.0	-2.9	-1.7	18421	15196	19570	20726	20.9	25.2	12.2	12.4
Uttar Pradesh	10385	9968	10523	10694	11.3	11.1	13.0	7.8	0.3	2.3	-4.2	0.6	12061	7539	14899	13932	37.9	60.2	7.6	20.1
India	8978	8217	9229	9545	16.8	20.8	10.2	9.3	1.0	5.6	-2.5	-1.5	12883	10168	13945	14719	31.1	33.2	14.7	20.4

Source: Computed from various reports of *Cost of Cultivation of Principle Crops in India*, DES, MoA, GoI.

Table A-14.3.6: Wheat: C2-Cost Details

State	C2-Cost Details												C2-Profit Details							
	C2-Cost per Hectare (₹) at 2004-05 Prices				CV (%)				Growth Rate (%)				C2-Profit per Hectare (₹) at 2004-05 Prices				CV (%)			
	1981-2009	1981-1990	1991-2000	2001-2009	1981-2009	1981-1990	1991-2000	2001-2009	1981-2009	1981-1990	1991-2000	2001-2009	1981-2009	1981-1990	1991-2000	2001-2009	1981-2009	1981-1990	1991-2000	2001-2009
Bihar	16169	18006	16080	14227	16.4	16.6	11.1	9.2	-1.0	5.4	-3.5	-3.0	3584	4880	3252	2512	55.3	32.7	31.4	93.5
Gujarat	14818	9965	17601	17119	39.1	77.1	5.7	7.1	3.0	17.7	-0.9	-2.3	4958	1886	5278	8016	102.7	367.0	42.6	27.3
Haryana	20703	16865	21281	24325	16.1	6.1	8.7	4.0	1.8	0.4	1.6	0.6	8174	7166	9138	8222	54.3	94.7	19.0	31.8
Himachal Pradesh	11733	10216	11734	13419	23.5	26.8	13.9	20.8	1.3	5.4	-3.8	-0.8	63	663	-408	-81	1990.2	114.1	-242.3	-1998.1
Madhya Pradesh	12941	10378	13589	15071	17.9	16.7	6.9	5.2	1.9	4.7	-1.7	0.9	2833	2274	2342	4000	64.9	58.5	49.6	58.3
Punjab	22383	20049	23270	23992	9.4	8.0	4.7	3.1	0.9	2.2	0.3	-0.1	6406	4060	7203	8127	40.6	31.7	31.4	26.2
Rajasthan	18881	16018	20502	20262	15.6	18.2	8.4	5.9	1.2	3.9	-2.6	-1.1	8792	7149	9103	10272	28.4	31.9	17.3	24.4
Uttar Pradesh	18751	16709	19751	19908	13.1	13.5	10.6	6.7	0.9	3.2	-3.4	1.4	3695	798	5671	4717	97.5	507.0	21.5	57.0
India	17048	14776	17976	18540	18.9	22.9	8.8	7.8	1.3	5.3	-1.7	-0.6	4813	3609	5198	5723	304.2	154.7	-3.7	-209.9

Source: Computed from various reports of *Cost of Cultivation of Principle Crops in India*, DES, MoA, GoI.

Table A-14.3.7: Wheat: Revenue Details

State	Revenue Details											
	Revenue per Hectare (₹) at 2004-05 Prices				CV (%)				Growth Rate (%)			
	1981-2009	1981-1990	1991-2000	2001-2009	1981-2009	1981-1990	1991-2000	2001-2009	1981-2009	1981-1990	1991-2000	2001-2009
Bihar	19753	22886	19332	16740	16.8	11.2	12.6	7.1	-1.4	2.6	-4.0	1.9
Gujarat	19776	11851	22879	25134	31.2	21.3	7.8	6.7	3.8	3.5	1.5	0.9
Haryana	28877	24031	30419	32547	19.6	28.4	4.9	8.4	1.6	0.0	0.2	2.3
Himachal Pradesh	11796	10878	11326	13338	27.7	23.6	20.4	31.4	0.7	3.7	-6.5	-0.3
Madhya Pradesh	15775	12652	15931	19071	21.6	14.9	11.4	15.0	2.0	2.2	-2.1	5.2
Punjab	28789	24108	30473	32119	14.8	10.4	8.3	7.4	1.5	2.4	1.3	1.0
Rajasthan	27674	23167	29605	30535	17.9	20.7	10.8	8.6	1.5	4.8	-3.3	2.1
Uttar Pradesh	22446	17507	25422	24626	22.4	28.8	7.5	11.2	2.1	3.9	-2.2	3.4
India	21861	18385	23173	24264	21.5	19.9	10.5	12.0	1.5	2.9	-1.9	2.0

Source: Computed from various reports of *Cost of Cultivation of Principle Crops in India*, DES, MoA, GoI.

Table A-14.3.8: Wheat: Yield Details

State	*Yield Details*											
	Yield per Hectare (Quintal)				*CV (%)*				*Growth Rate (%)*			
	1981-2009	*1981-1990*	*1991-2000*	*2001-2009*	*1981-2009*	*1981-1990*	*1991-2000*	*2001-2009*	*1981-2009*	*1981-1990*	*1991-2000*	*2001-2009*
Bihar	22	22	22	23	7.1	4.8	6.7	8.2	0.3	0.5	1.0	0.5
Gujarat	33	32	33	34	17.0	1.2	23.7	16.2	0.3	0.3	1.4	-0.8
Haryana	33	30	33	37	16.3	12.1	11.7	15.0	1.2	3.7	-2.9	5.4
Himachal Pradesh	12	12	12	14	22.2	11.0	16.6	27.5	0.4	0.0	-4.3	-1.0
Madhya Pradesh	26	14	34	30	54.4	19.1	46.9	32.8	3.8	-1.2	14.9	-8.4
Punjab	39	33	41	42	13.1	8.6	10.7	5.1	1.3	0.0	2.6	-0.5
Rajasthan	29	23	29	34	17.5	10.5	6.5	4.1	2.0	2.1	1.7	0.7
Uttar Pradesh	29	25	30	32	13.0	6.9	7.6	6.4	1.4	2.0	2.5	1.3
India	28	24	29	31	20.1	9.3	16.3	14.4	1.4	0.9	2.1	-0.4

Source: Computed from various reports of *Cost of Cultivation of Principle Crops in India*, DES, MoA, GoI.

Table A-14.3.9: Maize: A2-Cost Details

State	A2-Cost Details												A2-Profit Details							
	A2-Cost per Hectare (₹) at 2004-05 Prices				CV (%)				Growth Rate (%)				A2-Profit per Hectare (₹) at 2004-05 Prices				CV (%)			
	1981-2009	1981-1990	1991-2000	2001-2009	1981-2009	1981-1990	1991-2000	2001-2009	1981-2009	1981-1990	1991-2000	2001-2009	1981-2009	1981-1990	1991-2000	2001-2009	1981-2009	1981-1990	1991-2000	2001-2009
Andhra Pradesh	9619	10726	7357	10903	24.6	11.4	14.6	23.1	0.0	0.3	-2.5	6.9	6308	4910	5932	8277	40.9	43.8	39.1	24.4
Bihar	7037	5798	7211	8220	21.5	29.7	11.9	4.5	2.2	10.1	-0.1	1.3	9241	8589	7773	11596	40.9	36.7	28.7	39.5
Himachal Pradesh	3131	2972	2792	3686	22.5	17.0	19.0	20.1	0.9	1.8	-4.1	2.9	6708	6176	6664	7349	22.6	21.9	10.7	27.5
Karnataka	5834	3890	6043	7762	30.4	15.5	16.1	11.0	3.4	2.1	4.7	-1.3	4317	2081	3304	7926	82.6	36.0	100.1	36.6
Madhya Pradesh	3318	2317	3189	4573	34.9	25.9	22.6	17.4	3.6	5.3	1.2	2.4	2905	3227	3319	2089	32.1	22.1	27.2	28.8
Rajastan	4877	3699	4807	6263	26.1	20.4	15.0	11.9	2.7	4.8	0.7	1.5	5500	5608	5282	5623	27.8	18.8	19.0	40.2
Uttar Pradesh	4770	4111	4933	5321	28.5	17.0	32.3	25.1	1.4	4.3	-2.5	7.0	3186	2480	3159	4002	48.7	59.8	44.9	34.1
India	5512	4788	5190	6675	26.9	19.5	18.8	16.2	2.0	4.1	-0.4	3.0	5452	4724	5062	6695	42.2	34.2	38.5	33.0

Source: Computed from various reports of *Cost of Cultivation of Principle Crops in India*, DES, MoA, GoI.

Table A-14.3.10: Maize: C2-Cost Details

State	*C2-Cost Details*												*C2-Profit Details*							
	C2-Cost per Hectare (₹) at 2004-05 Prices				*CV (%)*				*Growth Rate (%)*				*C2-Profit per Hectare (₹) at 2004-05 Prices*				*CV (%)*			
	1981-2009	*1981-1990*	*1991-2000*	*2001-2009*	*1981-2009*	*1981-1990*	*1991-2000*	*2001-2009*	*1981-2009*	*1981-1990*	*1991-2000*	*2001-2009*	*1981-2009*	*1981-1990*	*1991-2000*	*2001-2009*	*1981-2009*	*1981-1990*	*1991-2000*	*2001-2009*
Andhra Pradesh	17633	20114	13282	19711	24.6	12.3	15.5	20.7	-0.2	-0.5	-1.8	6.4	-1706	-4477	7	-531	-189.6	-72.5	36898.2	-272.0
Bihar	13176	12545	12729	14373	14.7	17.9	13.7	7.1	0.9	5.5	-1.2	0.2	3102	1842	2255	5443	121.0	166.4	79.1	88.8
Himachal Pradesh	9787	8608	9649	11249	17.9	22.0	10.9	8.8	1.5	5.5	-1.3	0.0	53	539	-193	-214	1882.6	113.1	-159.9	-706.0
Karnataka	9235	5148	9759	13194	38.8	11.6	21.5	9.5	4.7	1.8	7.1	0.4	916	823	-412	2494	255.9	101.6	-550.5	104.1
Madhya Pradesh	6975	5057	7020	9055	29.4	21.2	21.3	12.9	3.1	4.2	2.5	2.5	-751	487	-512	-2393	-187.7	167.2	-180.9	-21.3
Rajastan	11054	8057	11780	13578	25.9	26.0	13.9	9.5	2.9	6.0	2.4	0.5	-677	1250	-1691	-1692	-295.5	88.0	-60.3	-119.6
Uttar Pradesh	10291	7740	10267	13151	28.6	34.2	8.1	15.0	3.2	10.3	0.3	3.4	-2335	-1150	-2176	-3828	-72.7	-120.5	-21.4	-46.0
India	11164	9610	10641	13473	25.7	20.7	15.0	12.0	2.3	4.7	1.2	1.9	-200	-98	-389	-103	216.3	63.3	5143.5	-138.8

Source: Computed from various reports of *Cost of Cultivation of Principle Crops in India*, DES, MoA, GoI.

Table A-14.3.11: Maize: Revenue Details

State	Revenue Details											
	Revenue per Hectare (₹) at 2004-05 Prices				*CV (%)*				*Growth Rate (%)*			
	1981-2009	*1981-1990*	*1991-2000*	*2001-2009*	*1981-2009*	*1981-1990*	*1991-2000*	*2001-2009*	*1981-2009*	*1981-1990*	*1991-2000*	*2001-2009*
Andhra Pradesh	15927	15637	13289	19180	24.8	10.7	24.0	21.8	0.9	3.2	-4.8	7.9
Bihar	16278	14387	14984	19816	27.2	24.4	17.9	24.2	1.6	2.4	-4.3	8.1
Himachal Pradesh	9840	9148	9456	11035	19.0	18.2	12.3	19.2	0.9	4.1	-1.9	-0.2
Karnataka	10151	5971	9347	15688	48.4	12.4	43.5	18.2	5.3	3.3	15.4	5.6
Madhya Pradesh	6223	5544	6508	6662	19.1	13.5	19.3	17.5	1.0	1.7	-0.1	2.7
Rajastan	10377	9308	10089	11885	21.1	16.2	13.6	22.5	1.4	4.2	0.2	5.0
Uttar Pradesh	7956	6591	8091	9323	23.1	28.1	7.7	17.7	2.1	5.1	0.4	4.5
India	10965	9512	10252	13370	26.1	17.6	19.8	20.2	1.9	3.4	0.7	4.8

Source: Computed from various reports of *Cost of Cultivation of Principle Crops in India*, DES, MoA, GoI.

Table A-14.3.12: Maize: Yield Details

State	Yield Details											
	Yield per Hectare (Quintal)				*CV (%)*				*Growth Rate (%)*			
	1981-2009	*1981-1990*	*1991-2000*	*2001-2009*	*1981-2009*	*1981-1990*	*1991-2000*	*2001-2009*	*1981-2009*	*1981-1990*	*1991-2000*	*2001-2009*
Andhra Pradesh	21	10	19	34	54.7	10.6	29.3	20.4	6.2	0.0	8.6	7.5
Bihar	26	22	24	35	25.6	7.8	10.3	15.5	2.2	0.3	-1.6	4.8
Himachal Pradesh	13	13	13	14	12.5	6.0	6.4	16.8	0.2	-0.8	-0.8	-1.5
Karnataka	22	17	24	26	25.1	9.3	19.1	21.8	2.3	3.2	4.9	6.4
Madhya Pradesh	10	9	10	10	19.4	12.2	16.3	24.0	0.6	0.3	-0.7	3.6
Rajastan	13	10	12	15	29.7	22.0	15.0	30.7	1.9	1.9	-0.6	5.0
Uttar Pradesh	14	13	14	16	16.5	9.9	9.8	20.2	0.9	2.1	0.1	3.5
India	17	13	16	22	26.2	11.1	15.2	21.3	2.1	1.0	1.4	4.2

Source: Computed from various reports of *Cost of Cultivation of Principle Crops in India*, DES, MoA, GoI.

Table A-14.3.13: Urad: A2-Cost Details

State	A2-Cost Details												A2-Profit Details							
	A2-Cost per Hectare (₹) at 2004-05 Prices				CV (%)				Growth Rate (%)				A2-Profit per Hectare (₹) at 2004-05 Prices				CV (%)			
	1981-2009	1981-1990	1991-2000	2001-2009	1981-2009	1981-1990	1991-2000	2001-2009	1981-2009	1981-1990	1991-2000	2001-2009	1981-2009	1981-1990	1991-2000	2001-2009	1981-2009	1981-1990	1991-2000	2001-2009
Andhra Pradesh	3982	4167	3741	4043	20.7	16.8	25.0	18.8	-0.2	1.6	-5.0	2.8	8390	6628	9009	9661	59.8	62.3	66.1	42.9
Madhya Pradesh	2825	2749	2710	3035	19.4	10.2	27.3	15.1	0.3	-0.4	-3.9	1.5	3583	4125	3361	3228	38.7	21.8	33.5	56.8
Maharashtra	4535	3947	4025	5755	23.7	9.1	23.3	11.7	1.7	-1.7	2.4	-0.4	2307	2467	2725	1666	56.6	40.2	33.4	101.2
Orissa	2070	1899	2147	2175	26.0	37.3	19.7	16.1	1.4	13.9	-2.7	-3.3	4660	5526	4475	3904	26.2	16.2	23.5	28.5
Rajasthan	3290	4348	3093	2334	29.9	7.0	22.9	22.0	-3.2	-0.5	-4.9	-2.6	2671	1193	2005	5051	102.6	168.4	81.6	56.4
Tamil Nadu	4459	5069	4464	3778	20.9	10.4	22.5	18.4	-1.6	2.3	-4.5	-2.9	5737	6324	6345	4409	26.5	20.3	19.6	26.1
Uttar Pradesh	2696	3196	2601	2245	29.1	27.7	26.6	13.0	-1.4	8.3	-7.5	0.9	4541	4458	4821	4321	37.3	43.3	30.2	37.4
India	3408	3625	3254	3338	24.3	16.9	23.9	16.4	-0.4	3.4	-3.7	-0.6	4556	4389	4677	4606	49.6	53.2	41.1	49.9

Source: Computed from various reports of *Cost of Cultivation of Principle Crops in India*, DES, MoA, GoI.

Table A-14.3.14: Urad: C2-Cost Details

State	*C2-Cost Details*												*C2-Profit Details*							
	C2-Cost per Hectare (₹) at 2004-05 Prices				*CV (%)*				*Growth Rate (%)*				*C2-Profit per Hectare (₹) at 2004-05 Prices*				*CV (%)*			
	1981-2009	*1981-1990*	*1991-2000*	*2001-2009*	*1981-2009*	*1981-1990*	*1991-2000*	*2001-2009*	*1981-2009*	*1981-1990*	*1991-2000*	*2001-2009*	*1981-2009*	*1981-1990*	*1991-2000*	*2001-2009*	*1981-2009*	*1981-1990*	*1991-2000*	*2001-2009*
Andhra Pradesh	7993	7658	7859	8514	19.2	15.6	22.8	16.8	0.5	2.5	-4.3	4.6	4379	3137	4891	5190	96.2	119.4	103.7	61.8
Madhya Pradesh	6131	5989	5897	6549	16.7	10.1	23.5	11.4	0.3	-0.8	-3.0	1.5	277	886	175	-286	424.4	88.4	425.4	-544.1
Maharashtra	6827	5926	6260	8459	21.6	9.5	22.1	9.1	1.7	-0.2	1.6	0.3	16	488	491	-1037	9092.5	210.1	178.0	-163.0
Orissa	5363	5540	5208	5338	13.5	9.6	19.0	9.0	-0.3	-1.3	-2.1	-0.9	1367	1885	1414	740	65.6	28.2	40.8	148.6
Rajasthan	5695	4544	6251	6356	21.3	16.0	15.5	14.6	1.5	-0.2	-0.1	-1.2	266	998	-1153	1030	803.5	147.8	-109.2	259.5
Tamil Nadu	8645	9780	8544	7497	23.2	10.8	23.3	28.5	-1.6	2.1	-4.9	-4.9	1551	1612	2264	689	97.9	48.2	36.6	313.6
Uttar Pradesh	6566	7486	6538	5576	24.6	24.2	22.3	12.3	-1.1	8.1	-5.5	-0.2	670	168	884	989	181.5	588.8	123.5	139.8
India	6746	6703	6651	6898	20.0	13.7	21.2	14.6	0.1	1.5	-2.6	-0.1	1218	1311	1281	1045	1537.4	175.9	114.1	30.9

Source: Computed from various reports of *Cost of Cultivation of Principle Crops in India*, DES, MoA, GoI.

Table A-14.3.15: Urad: Revenue Details

State	Revenue Details											
	Revenue per Hectare (₹) at 2004-05 Prices				CV (%)				Growth Rate (%)			
	1981-2009	1981-1990	1991-2000	2001-2009	1981-2009	1981-1990	1991-2000	2001-2009	1981-2009	1981-1990	1991-2000	2001-2009
Andhra Pradesh	16165	10795	23750	13704	128.6	40.3	141.3	31.6	1.3	7.2	-9.4	8.6
Madhya Pradesh	6408	6875	6072	6263	24.0	14.0	25.1	30.3	-0.6	-2.3	-1.3	6.5
Maharashtra	6843	6415	6750	7422	18.7	16.8	19.2	17.0	0.6	-1.8	-0.9	2.8
Orissa	6730	7425	6622	6079	16.9	6.0	21.6	14.0	-1.0	-0.9	-3.6	3.5
Rajasthan	5961	5542	5098	7385	40.8	36.3	36.9	37.3	1.2	-6.0	2.8	2.5
Tamil Nadu	10196	11392	10808	8186	21.1	14.4	19.5	12.2	-1.5	2.7	-4.0	1.6
Uttar Pradesh	7236	7654	7422	6565	29.9	33.6	26.8	24.8	-0.2	11.8	-5.9	-0.9
India	8506	8014	9503	7943	40.0	23.1	41.5	23.9	0.0	1.5	-3.2	3.5

Source: Computed from various reports of *Cost of Cultivation of Principle Crops in India*, DES, MoA, GoI.

Table A-14.3.16: Urad: Yield Details

State	Yield Details											
	Yield per Hectare (Quintal)				CV (%)				Growth Rate (%)			
	1981-2009	1981-1990	1991-2000	2001-2009	1981-2009	1981-1990	1991-2000	2001-2009	1981-2009	1981-1990	1991-2000	2001-2009
Andhra Pradesh	6	6	6	8	28.2	23.6	16.9	24.3	1.8	4.3	-0.2	3.5
Madhya Pradesh	4	4	4	4	23.9	13.1	19.7	33.1	0.0	-2.4	-1.6	7.8
Maharashtra	4	4	4	5	24.4	12.4	18.2	28.0	1.1	2.6	-2.3	-1.5
Orissa	4	4	4	4	13.3	11.1	11.2	14.6	-0.7	-2.7	-0.6	0.0
Rajasthan	2	1	2	4	58.3	35.0	20.4	49.8	1.4	-4.5	4.4	1.4
Tamil Nadu	5	5	5	4	11.7	5.9	0.0	18.7	-0.8	-0.8	0.0	-2.8
Uttar Pradesh	5	5	5	4	24.0	12.5	21.9	30.8	-1.4	-0.7	-4.5	-3.3
India	4	4	4	5	26.3	16.2	15.5	28.5	0.2	-0.6	-0.7	0.7

Source: Computed from various reports of *Cost of Cultivation of Principle Crops in India*, DES, MoA, GoI.

Table A-14.3.17: Groundnut: A2-Cost Details

State	A2-Cost Details												A2-Profit Details							
	A2-Cost per Hectare (₹) at 2004-05 Prices					CV (%)			Growth Rate (%)				A2-Profit per Hectare (₹) at 2004-05 Prices				CV (%)			
	1981-2009	1981-1990	1991-2000	2001-2009	1981-2009	1981-1990	1991-2000	2001-2009	1981-2009	1981-1990	1991-2000	2001-2009	1981-2009	1981-1990	1991-2000	2001-2009	1981-2009	1981-1990	1991-2000	2001-2009
Andhra Pradesh	9607	7864	9993	11115	18.9	16.6	10.6	11.6	1.9	4.2	2.0	2.4	5627	5173	5857	5876	41.5	44.8	35.1	43.4
Gujarat	9428	7875	9103	11514	19.9	18.5	7.6	10.0	2.0	0.4	2.1	2.1	10335	8361	9166	13826	48.3	42.6	47.7	37.6
Karnataka	7371	6349	7977	7834	15.3	16.2	8.0	10.6	1.2	4.1	2.1	-2.3	5184	7403	4528	3449	49.9	25.6	42.1	62.1
Maharashtra	10777	6369	11704	14644	36.9	27.7	18.4	14.8	4.5	9.7	4.4	0.0	6372	5211	8020	5831	59.1	69.0	39.8	67.0
Orissa	7697	5767	8143	9347	27.8	36.5	15.9	9.9	3.3	10.9	4.1	-2.5	13061	11820	13407	14054	18.9	24.1	17.1	10.0
Tamil Nadu	13515	12357	13951	14317	19.2	22.1	12.3	19.4	0.9	6.8	1.8	-6.1	8340	8040	9134	7792	24.3	13.4	15.5	37.6
India	9733	7764	10145	11462	23.0	22.9	12.1	12.7	2.3	6.0	2.7	-1.1	8153	7668	8352	8471	40.3	36.6	32.9	42.9

Source: Computed from various reports of *Cost of Cultivation of Principle Crops in India*, DES, MoA, GoI.

Table A-14.3.18: Groundnut: C2-Cost Details

State	C2-Cost Details												C2-Profit Details							
	C2-Cost per Hectare (₹) at 2004-05 Prices					CV (%)			Growth Rate (%)				C2-Profit per Hectare (₹) at 2004-05 Prices				CV (%)			
	1981-2009	1981-1990	1991-2000	2001-2009	1981-2009	1981-1990	1991-2000	2001-2009	1981-2009	1981-1990	1991-2000	2001-2009	1981-2009	1981-1990	1991-2000	2001-2009	1981-2009	1981-1990	1991-2000	2001-2009
Andhra Pradesh	16512	13089	17471	19250	21.1	19.4	11.3	12.9	2.2	5.1	2.4	3.9	-1278	-52	-1621	-2259	-153.1	-2765.3	-111.4	-84.8
Gujarat	15445	12640	15145	18897	19.9	16.1	8.9	9.5	2.1	1.5	2.4	0.6	4317	3596	3124	6444	92.0	80.1	120.5	67.9
Karnataka	11737	10516	12659	12070	13.3	16.7	6.7	7.8	0.9	4.3	1.3	-0.2	819	3236	-154	-787	295.4	41.4	-1169.3	-224.8
Maharashtra	17033	10181	18742	22748	35.7	25.4	18.7	14.0	4.2	7.4	4.8	-3.0	115	1399	982	-2273	3398.8	243.7	373.5	-161.1
Orissa	16404	11541	17467	20627	31.5	43.4	16.0	9.6	4.6	18.5	4.1	-2.3	4354	6046	4082	2774	48.7	25.9	55.8	18.6
Tamil Nadu	22799	20501	23738	24310	19.3	22.3	12.5	18.5	1.1	7.1	2.1	-5.1	-945	-105	-653	-2201	-292.9	-1039.4	-293.6	-186.2
India	16655	13078	17537	19650	23.5	23.9	12.3	12.1	2.5	7.3	2.8	-1.0	1230	2354	960	283	564.8	-568.9	-170.8	-95.1

Source: Computed from various reports of *Cost of Cultivation of Principle Crops in India*, DES, MoA, GoI.

Table A-14.3.19: Groundnut: Revenue Details

State	*Revenue Details*											
	Revenue per Hectare (₹) at 2004-05 Prices				*CV (%)*				*Growth Rate (%)*			
	1981-2009	*1981-1990*	*1991-2000*	*2001-2009*	*1981-2009*	*1981-1990*	*1991-2000*	*2001-2009*	*1981-2009*	*1981-1990*	*1991-2000*	*2001-2009*
Andhra Pradesh	15234	13037	15850	16991	21.8	24.5	14.7	17.0	1.7	5.9	1.0	5.3
Gujarat	19762	16236	18268	25340	30.3	25.2	24.6	20.5	2.2	4.4	0.7	-1.6
Karnataka	12556	13752	12505	11283	18.0	18.2	12.9	16.4	-0.8	3.8	-2.3	1.8
Maharashtra	17149	11580	19724	20475	33.6	41.9	14.4	20.6	3.9	14.0	1.5	-0.4
Orissa	20758	17587	21550	23401	20.1	25.9	13.2	9.9	1.7	8.0	0.4	-2.7
Tamil Nadu	21855	20396	23085	22109	14.8	18.0	11.2	14.6	0.7	5.3	2.3	-1.0
India	17886	15432	18497	19933	23.3	25.6	14.5	16.7	1.6	7.1	0.6	0.2

Source: Computed from various reports of *Cost of Cultivation of Principle Crops in India*, DES, MoA, GoI.

Table A-14.3.20: Groundnut: Yield Details

State	Yield Details											
	Yield per Hectare (Quintal)				CV (%)				Growth Rate (%)			
	1981-2009	1981-1990	1991-2000	2001-2009	1981-2009	1981-1990	1991-2000	2001-2009	1981-2009	1981-1990	1991-2000	2001-2009
Andhra Pradesh	9	8	9	10	21.4	17.1	14.9	20.2	1.7	3.7	1.5	6.7
Gujarat	11	10	10	12	30.4	24.2	33.5	25.9	0.9	4.1	-4.5	-2.5
Karnataka	7	8	6	6	23.7	16.6	24.5	20.9	-1.1	0.7	1.3	2.3
Maharashtra	10	8	10	12	26.6	16.9	14.4	23.4	2.1	0.5	1.5	1.3
Orissa	12	11	12	12	6.4	6.3	7.5	2.2	0.2	1.1	-1.5	-0.1
Tamil Nadu	14	13	15	15	13.1	9.6	8.1	15.0	0.6	2.2	1.4	-3.6
India	10	9	10	11	20.2	15.1	17.1	17.9	0.8	2.1	-0.1	0.7

Source: Computed from various reports of *Cost of Cultivation of Principle Crops in India*, DES, MoA, GoI.

Table A-14.3.21: Cotton: A2-Cost Details

State	A2-Cost Details												A2-Profit Details							
	A2-Cost per Hectare (₹) at 2004-05 Prices				CV (%)				Growth Rate (%)				A2-Profit per Hectare (₹) at 2004-05 Prices				CV (%)			
	1981-2009	1981-1990	1991-2000	2001-2009	1981-2009	1981-1990	1991-2000	2001-2009	1981-2009	1981-1990	1991-2000	2001-2009	1981-2009	1981-1990	1991-2000	2001-2009	1981-2009	1981-1990	1991-2000	2001-2009
Andhra Pradesh	13903	10960	13133	18029	31.1	40.8	17.1	12.4	3.1	8.6	-0.4	1.9	13806	9657	13008	19301	43.4	52.2	16.8	29.3
Gujarat	12088	12468	9373	14685	29.4	28.2	13.0	21.7	0.8	0.7	-2.6	6.4	13153	9360	11255	19475	52.9	54.0	34.3	36.7
Haryana	7503	5435	6033	11434	43.5	23.7	22.0	24.6	3.8	5.7	1.2	9.4	15280	11825	15668	18689	46.4	45.8	31.4	47.0
Karnataka	7880	8767	6624	8289	24.2	23.3	18.1	19.6	-0.2	-0.5	-1.7	5.6	8230	10321	5773	8637	42.7	28.4	36.1	42.9
Madhya Pradesh	7961	5980	6646	11622	44.9	37.3	26.1	30.6	3.3	11.5	-6.6	2.6	9807	9304	7838	12555	58.6	29.4	27.5	71.6
Maharashtra	10634	7700	9729	14901	35.2	41.7	18.4	8.4	3.9	11.1	0.5	1.5	6279	5170	6720	7020	40.9	42.7	25.3	46.3
Punjab	12709	9616	10672	18411	33.3	22.7	17.9	3.9	3.3	6.3	2.2	-0.5	15981	15709	11549	21207	54.7	40.9	63.1	45.2
Rajasthan	7546	8647	6366	7635	21.7	17.9	22.1	12.5	-0.4	3.1	-0.2	2.2	16863	13309	16884	20789	37.9	43.3	17.2	36.0
Tamil Nadu	13223	11099	12503	16383	29.7	44.2	14.0	12.2	3.0	16.7	-0.1	0.3	12067	11195	11767	13370	31.9	35.3	11.5	37.9
India	10383	8963	9009	13488	32.5	31.1	18.7	16.2	2.3	7.0	-0.9	3.3	12385	10650	11162	15672	45.5	41.3	29.2	43.7

Source: Computed from various reports of *Cost of Cultivation of Principle Crops in India*, DES, MoA, GoI.

Table A-14.3.22: Cotton: C2-Cost Details

State	C2-Cost Details												C2-Profit Details							
	C2-Cost per Hectare (₹) at 2004-05 Prices				CV (%)				Growth Rate (%)				C2-Profit per Hectare (₹) at 2004-05 Prices				CV (%)			
	1981-2009	1981-1990	1991-2000	2001-2009	1981-2009	1981-1990	1991-2000	2001-2009	1981-2009	1981-1990	1991-2000	2001-2009	1981-2009	1981-1990	1991-2000	2001-2009	1981-2009	1981-1990	1991-2000	2001-2009
Andhra Pradesh	24867	19525	23529	32289	30.6	41.6	16.3	8.8	3.0	8.2	-1.1	1.3	2841	1092	2612	5041	122.8	148.6	91.3	92.5
Gujarat	19981	18775	16059	25678	30.3	28.4	14.8	21.0	1.6	2.7	-3.2	6.7	5261	3053	4569	8482	84.9	104.0	66.7	60.6
Haryana	18421	11923	16896	27337	40.7	27.6	16.0	20.6	4.5	7.2	3.0	8.2	4362	5337	4805	2785	117.6	81.7	91.0	223.3
Karnataka	13268	14748	11255	13860	21.1	18.3	16.4	17.8	-0.2	0.7	-1.8	5.2	2842	4339	1142	3067	95.8	57.7	141.7	93.7
Madhya Pradesh	15786	11970	13253	22842	43.4	34.4	23.7	30.4	3.3	10.5	-5.1	4.3	1982	3314	1230	1336	199.2	61.3	106.1	477.8
Maharashtra	16193	11814	15090	22283	33.9	41.0	16.7	9.3	3.8	11.0	0.4	2.3	721	1056	1359	-362	290.2	119.1	142.9	-698.8
Punjab	24421	19012	20638	34635	31.8	20.8	16.9	9.5	3.1	6.2	-0.2	3.2	4269	6313	1582	4982	150.3	86.1	389.9	133.5
Rajasthan	16755	16548	14748	19214	23.5	26.6	16.1	17.7	1.2	6.6	-0.6	5.5	7655	5408	8501	9210	52.3	57.8	31.2	53.4
Tamil Nadu	27510	25448	26165	31296	22.7	32.2	14.5	11.8	1.6	11.8	0.8	-1.9	-2220	-3154	-1894	-1542	-176.7	-120.5	-136.6	-321.0
India	19689	16640	17515	25493	30.9	30.1	16.8	16.3	2.4	7.2	-0.9	3.9	3079	2973	2656	3667	104.1	66.2	102.7	12.8

Source: Computed from various reports of *Cost of Cultivation of Principle Crops in India*, DES, MoA, GoI.

Table A-14.3.23: Cotton: Revenue Details

State	Revenue Details											
	Revenue per Hectare (₹) at 2004-05 Prices				CV (%)				Growth Rate (%)			
	1981-2009	1981-1990	1991-2000	2001-2009	1981-2009	1981-1990	1991-2000	2001-2009	1981-2009	1981-1990	1991-2000	2001-2009
Andhra Pradesh	27709	20617	26141	37330	35.2	45.8	11.7	18.3	3.6	9.5	-1.8	1.8
Gujarat	25241	21828	20628	34160	37.8	34.5	21.6	28.0	2.4	4.8	-5.8	10.9
Haryana	22783	17259	21701	30123	40.5	37.1	21.0	35.9	3.3	12.1	-1.5	16.6
Karnataka	16110	19087	12397	16926	28.9	20.7	21.5	25.6	-0.7	0.3	-5.4	6.3
Madhya Pradesh	17768	15284	14484	24178	46.2	27.3	24.5	46.0	2.2	7.3	-5.7	13.3
Maharashtra	16913	12870	16450	21921	31.0	39.1	11.7	16.3	3.3	11.0	-1.4	5.5
Punjab	28690	25324	22220	39617	38.5	29.9	33.7	23.8	2.3	7.6	-8.3	9.7
Rajasthan	24410	21956	23249	28424	28.5	32.4	13.6	28.2	1.9	8.4	-1.2	9.6
Tamil Nadu	25290	22294	24270	29754	27.3	39.6	10.0	18.1	2.3	13.9	-0.9	3.2
India	22768	19613	20171	29159	34.9	34.0	18.8	26.7	2.3	8.3	-3.5	8.5

Source: Computed from various reports of *Cost of Cultivation of Principle Crops in India*, DES, MoA, GoI.

Table A-14.3.24: Cotton: Yield Details

State	Yield Details											
	Yield per Hectare (Quintal)				*CV (%)*				*Growth Rate (%)*			
	1981-2009	*1981-1990*	*1991-2000*	*2001-2009*	*1981-2009*	*1981-1990*	*1991-2000*	*2001-2009*	*1981-2009*	*1981-1990*	*1991-2000*	*2001-2009*
Andhra Pradesh	12	8	12	18	39.0	17.8	10.9	18.9	4.0	0.6	1.2	2.6
Gujarat	11	9	10	15	36.3	19.1	24.4	28.7	2.2	-1.0	-6.6	11.3
Haryana	11	9	11	14	36.3	21.4	21.5	38.0	2.3	6.1	-1.0	18.1
Karnataka	7	8	6	7	26.0	21.5	21.6	27.2	-0.6	-4.0	-7.1	9.3
Madhya Pradesh	7	6	6	11	45.1	21.2	25.8	40.3	2.5	1.5	-6.5	15.7
Maharashtra	7	5	7	10	35.7	16.0	10.8	17.2	4.1	4.4	1.8	6.4
Punjab	13	12	10	18	39.1	27.6	32.0	27.1	2.2	5.4	-8.0	10.8
Rajasthan	11	8	12	13	29.4	8.2	19.4	29.3	2.4	2.8	3.1	10.4
Tamil Nadu	12	10	11	14	22.3	8.7	11.4	22.5	1.7	2.5	-1.5	5.0
India	10	8	9	13	34.4	18.0	19.7	27.7	2.3	2.0	-2.7	9.9

Source: Computed from various reports of *Cost of Cultivation of Principle Crops in India*, DES, MoA, GoI.

Table A-14.3.25: Sugarcane: A2-Cost Details

State	*A2-Cost Details*												*A2-Profit Details*							
	A2-Cost per Hectare (₹) at 2004-05 Prices				*CV (%)*				*Growth Rate (%)*				*A2-Profit per Hectare (₹) at 2004-05 Prices*				*CV (%)*			
	1981-2009	*1981-1990*	*1991-2000*	*2001-2009*	*1981-2009*	*1981-1990*	*1991-2000*	*2001-2009*	*1981-2009*	*1981-1990*	*1991-2000*	*2001-2009*	*1981-2009*	*1981-1990*	*1991-2000*	*2001-2009*	*1981-2009*	*1981-1990*	*1991-2000*	*2001-2009*
Andhra Pradesh	37679	36523	39070	37419	16.0	9.0	10.9	24.0	0.2	2.0	-1.5	7.2	47358	46309	50205	45360	28.1	25.6	15.8	40.1
Haryana	19157	15793	18813	23278	28.3	29.0	26.5	16.1	2.1	8.6	-3.5	0.9	53982	41923	60318	60340	37.9	28.0	15.4	48.5
Karnataka	31097	26901	33073	33564	23.6	29.4	16.4	19.2	1.3	3.5	-3.3	2.8	76055	77039	76982	73933	28.2	17.6	18.7	43.6
Maharashtra	44696	41671	42926	50025	20.1	26.1	15.5	12.1	0.9	-0.9	-3.0	2.6	39717	38436	37221	43915	61.2	44.9	20.2	87.7
Tamil Nadu	50020	48684	54122	46946	16.3	19.2	10.0	16.0	0.1	4.7	-2.4	5.2	67408	69007	75314	56846	22.5	18.0	9.2	32.3
Uttar Pradesh	14470	12557	15033	15970	18.4	23.8	8.6	13.2	1.6	3.9	1.8	2.2	64685	42788	100316	49426	161.0	25.3	169.6	44.0
India	32853	30355	33840	34534	20.4	22.8	14.6	16.8	1.0	3.6	-2.0	3.5	58201	52584	66726	54970	56.5	26.6	41.5	49.4

Source: Computed from various reports of *Cost of Cultivation of Principle Crops in India*, DES, MoA, GoI.

Table A-14.3.26: Sugarcane: C2-Cost Details

State	C2-Cost Details												C2-Profit Details							
	C2-Cost per Hectare (₹) at 2004-05 Prices				CV (%)			Growth Rate (%)					C2-Profit per Hectare (₹) at 2004-05 Prices				CV (%)			
	1981-2009	1981-1990	1991-2000	2001-2009	1981-2009	1981-1990	1991-2000	2001-2009	1981-2009	1981-1990	1991-2000	2001-2009	1981-2009	1981-1990	1991-2000	2001-2009	1981-2009	1981-1990	1991-2000	2001-2009
Andhra Pradesh	70047	66220	74208	69674	16.0	10.7	7.9	23.8	0.4	2.2	-2.4	7.8	14990	16612	15066	13104	58.9	52.3	23.7	92.5
Haryana	48269	36117	53571	55880	28.4	32.8	16.5	18.6	2.6	8.4	-0.1	2.0	24870	21599	25561	27738	53.5	30.2	16.6	79.3
Karnataka	59897	52842	63346	63905	20.6	19.5	14.4	21.9	0.9	2.4	-3.5	4.3	47255	51099	46709	43592	35.5	22.6	24.0	55.6
Maharashtra	67048	57997	65084	79287	26.5	33.8	13.6	20.3	1.9	1.4	-2.2	4.9	17366	22110	15063	14653	98.8	37.8	45.5	190.7
Tamil Nadu	75105	65000	85078	75253	25.1	39.3	10.2	14.2	2.5	-3.1	-1.0	4.2	42322	52690	44358	28539	39.0	28.4	19.4	53.2
Uttar Pradesh	35382	27932	38612	40071	22.6	27.1	8.6	14.8	2.4	6.6	0.8	4.4	43773	27413	76737	25325	233.8	51.2	219.0	70.0
India	59291	51018	63317	64012	23.2	27.2	11.9	18.9	1.8	3.0	-1.4	4.6	31763	31920	37249	25492	86.6	37.1	58.0	90.2

Source: Computed from various reports of *Cost of Cultivation of Principle Crops in India*, DES, MoA, GoI.

Table A-14.3.27: Sugarcane: Revenue Details

State	*Revenue Details*											
	Revenue per Hectare (₹) at 2004-05 Prices				*CV (%)*				*Growth Rate (%)*			
	1981-2009	*1981-1990*	*1991-2000*	*2001-2009*	*1981-2009*	*1981-1990*	*1991-2000*	*2001-2009*	*1981-2009*	*1981-1990*	*1991-2000*	*2001-2009*
Andhra Pradesh	85037	82832	89275	82779	19.8	15.5	9.7	30.0	0.2	2.7	-3.1	9.4
Haryana	73139	57715	79132	83618	32.1	27.7	14.8	36.4	2.2	8.4	-0.5	6.1
Karnataka	107153	103941	110055	107497	22.4	13.3	15.9	33.3	0.1	1.5	-5.3	7.5
Maharashtra	84414	80107	80147	93940	35.8	34.5	15.1	45.1	1.0	1.1	-3.4	9.3
Tamil Nadu	117427	117690	129436	103792	18.9	17.9	8.3	24.1	-0.3	4.7	-2.1	7.0
Uttar Pradesh	79155	55345	115349	65396	131.7	16.3	147.4	34.9	0.8	1.9	-4.5	8.4
India	91054	82938	100566	89504	43.4	20.9	35.2	34.0	0.7	3.4	-3.2	8.0

Source: Computed from various reports of *Cost of Cultivation of Principle Crops in India*, DES, MoA, GoI.

Table A-14.3.28: Sugarcane: Yield Details

State	*Yield Details*											
	Yield per Hectare (Quintal)				*CV (%)*				*Growth Rate (%)*			
	1981-2009	*1981-1990*	*1991-2000*	*2001-2009*	*1981-2009*	*1981-1990*	*1991-2000*	*2001-2009*	*1981-2009*	*1981-1990*	*1991-2000*	*2001-2009*
Andhra Pradesh	746	758	752	727	9.9	13.5	4.9	8.7	-0.1	-0.1	-1.0	2.4
Haryana	489	368	513	596	21.9	9.7	9.4	11.5	2.4	3.0	2.2	-1.9
Karnataka	848	799	829	925	9.2	7.8	5.1	7.2	0.7	-0.2	-0.8	1.7
Maharashtra	809	791	790	851	8.9	6.5	5.9	11.0	0.4	1.2	0.1	1.3
Tamil Nadu	1001	975	1058	968	8.0	4.1	7.1	8.8	0.1	0.3	1.4	1.1
Uttar Pradesh	461	417	462	510	10.8	6.6	6.4	8.0	1.0	1.7	1.0	0.7
India	726	685	734	763	11.5	8.1	6.5	9.2	0.7	1.0	0.5	0.9

Source: Computed from various reports of *Cost of Cultivation of Principle Crops in India*, DES, MoA, GoI.

15 An Analysis of District-wise Farm Household Income

A Case Study of Bihar

Ghanshyam Kumar Pandey

I

Introduction

Bihar is an agrarian state, with the 12th largest in the term of geographical area and third most populous state in the country with about 8.5 per cent of the Indian population (GoI, 2012). Agriculture has been way of life and continues to be a single most important livelihood option of the rural people in Bihar. The bifurcation of Bihar has increase the importance of agriculture of its economy. Nearly 90 per cent of the population lives in the rural areas and depends directly or indirectly on agriculture. During the recent years, the economy has shown turnarounds, throwing up new issues and acquired considerable attentions throughout the country and even abroad for its outstanding performance on the development fronts. But, despite this progress, rural Bihar is far from benefiting fully from the opportunities generated by rapid growth. This can be easily observed from the present scenario of agriculture in the state. The support mechanism for enhancement of agriculture development has been further stretchered to ensure that growth of state economy is sustainable in the years ahead. Bihar's journey of accelerated agricultural development has been accompanied by few advantages and number of handicaps. At the national level, it is being recognised that the Bihar is a sleeping giant as far as the potential for agriculture growth is concerned. Even as Bihar is being talked of as the next big hope for the agriculture sector in the country, this sector remains the most crucial factor for the state economy (Pandey, 2012).

Fortunately, the state is well endowed with land resources. The soil is predominantly thick alluvial (Gangetic) deposits, and terail soil, rich in nutrients, such as nitrogen, potash and phosphates (World Bank, 2007).

Agriculture sector continues to be an important driver in the state economy; according to its quick estimate it has contributed more than one-third in the gross domestic product (GDP). Average size of farm holdings is one of the lowest in the state. As 77 per cent of states' main workforce is employed in the agriculture sector, average size of operational holdings in Bihar is only 0.68 hectare as against the national level of 1.69 hectare (Bansil, 2011). The per capita agricultural income of Bihar is about half that of India as a whole and about one-fifth that of Punjab. The productive employment in the non-agriculture sector has not grown as in other states (Sharma, 2005). Infact agricultural development through infrastructure facilities was carried out in the state under successive five year plan. The density of population in the state is much more than the national average level, thus resulting in adverse land-man ratio. Declining land-man ratio has major impact on agricultural worker productivity (Bhalla and Singh, 2010). Despite continues decline in the net sown areas, there has been steady growth of principle crops in Bihar during two last decades, but the increase in their productivity has not been impressive during the last 50 years. The productivity/yield of the major crops in the state is much lower as compared to the national level. For example the production of rice, wheat, maize and pulses in the state happened to be 1120 kg/ha, 2084 kg/ha, 2341 kg/ha and 836 kg/ha during the year 2009-10, while in another major state, Punjab the figures were 4010 kg/ha, 4307 kg/ha, 3417 kg/ha and 896 kg/ha. In the same year the production of rice, wheat, maize and pulses all over the country reached 2125 kg/ha, 2829 kg/ha, 2024 kg/ha and 630 kg/ha respectively (GoI, 2012). Agricultural growth leads to immiserisation of the working class, a positive association between the level of economic development and the level of per capita earning even the lowliest among the rural labour households (Chadha and Khurana, 1989).

Importance of the Study

The socioeconomic conditions of all the farmers are miserable, mostly of agricultural labourers, landless farmers, marginal farmers and small farmers. Illiteracy, low level of skill, low level of income, poverty and lack of awareness are the main factor influencing their miserable condition in the state.

Further, a number of studies have recently been conducted for India on methodological issues for estimating income and income inequality, and actual measurement of these variables. But most of these studies are either based on secondary data available from NSS or conducted for depicting picture at national level and does not provide any breakdown of household income and expenditure data by different household characteristics such as the size of operational holdings. Majority of researcher in their studies on estimating farmer's income of different strata of households ignore micro level analysis. Literature based on in-depth village study is limited particularly from Bihar. Such a study can provide important insight that cannot be derived from secondary data due to lack of relevant information. The present study is an attempt to take into account some of the deficiencies mentioned with the objective to examine the levels of living of farmers in rural Bihar through their income. It is hoped that this study may prove useful as a contribution to the regional demand studies.

Objective of the Study

The present paper examines the pattern of income and expenditure on farming of the selected farm household in rural Bihar and variations and disparities in income across various segments of the farmers in the state.

The paper has been divided in to five parts and sub-parts. Section I deals with the introduction, importance and objectives, section II deals with methodological and sampling design of the study, section III and its sub-parts reveal the empirical findings at the state level, section IV reveals the district-wise empirical findings of the study and finally section V deals with the conclusion and suggestions.

II

Data, Methodology and Sampling Design

The present study is empirical study based on multistage random sampling technique related to the year 2010-11. Keeping in view the differences in agro-climatic conditions, one district from each zone has been selected, i.e., Sheohar district from Zone-I, Purnia district from Zone-II, Lakhisarai district from zone-III (A) and Bhojpur district from Zone-III (B).

Figure 15.1

Sampling Design of the Study

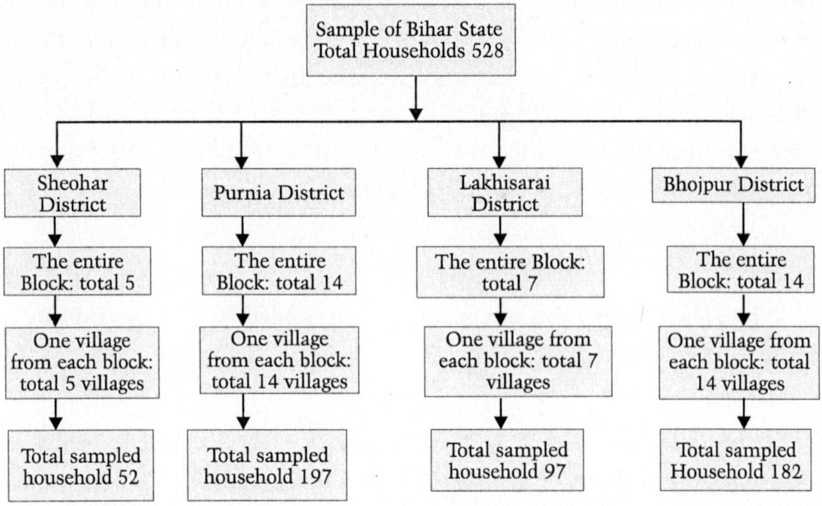

On the basis of random sampling method one village from each development block has been chosen. Thus, in all, 40 villages have been selected for survey. Five villages from Sheohar district, 14 from Purnia district, 7 from Lakhisarai district and 14 from Bhojpur district have been selected. Out of 40 villages 528 households have been selected for the study through the probability proportional technique. Out of total 528 households, 78 households are from landless farmers group, 257 from marginal farm-size group, 119 from small farm-size group, 52 from medium farm-size group and 22 from large farm-size group.

III

Empirical Findings

Levels and Composition of Farm Income

Household income is one of the alternative indicators to assess the level of living of the farmers in rural areas. The sources of income depict the opportunity enjoyed by the masses to generate income. Even in the rural areas, farmers try to diversify the risk and thereby avoid the loss if any. The agricultural activities are seasonal and their income is based on the vagaries of the monsoon. There are 10 major sources of income of the rural farm households surveyed in the different agro-climatic zone villages. Due to the non-availability of micro data, the government can easily access the per capita income and make decisions. In realising the importance of the per capita income, it is attempted here to compute and workout the distribution of per capita income to measure the inequality. An attempt has been made to discuss some of these results as they emerge from the primary field level survey undertaken during the course of present study.

The mean value of income per annum from the various sources by the different farm-size categories are given in Table 15.1. At the state-level farm households earns on an average ₹ 62,588 annually from all the sources of income. The income of farm households increased with the increase in the farm-size categories. The average family income of landless, marginal, small, medium and large farm-size categories of the sampled farmers worked out to be ₹ 32,615, ₹ 41,955, ₹ 74,071, ₹ 1,08,617 and ₹ 2,35,578 respectively per household per annum for the year 2010-11 for Bihar.

Farm business income is the most important source of their income followed by hiring out agricultural equipments and machinery, remittances, income from dairying and income from leased out land. The average incomes of farm households from these sources are ₹ 7,181, ₹ 3,864, ₹ 3,317 and ₹ 2,542 respectively. Table 15.1 depicts that the income earnings by remittances is more by landless farm households, marginal farm households and small farm households compared to the medium and large farm households, it is so because the unemployment rate is very high in these categories as compared to the medium and large farm households in rural Bihar. This suggests that work seekers of these categories of farm households face difficulties in finding stable position that provide high wage over the time within the state. Such increasing vulnerability of landless, marginal and small farm households generates large out migration to the other states thereby generating more remittances from these categories of farm households in rural Bihar.

Table 15.1

Levels of Income of Farmers in Rural Bihar: Category-wise

(Mean Values, in ₹ per Annum)

Sl. No.	Sources of Income	Landless Farmers	Marginal Farmers	Small Farmers	Medium Farmers	Large Farmers	All Sampled Farmers
1.	Farm Business Income	21413	26337	45358	65800	134714	38299
2.	(i) Poultry*	173	279	46	38	0	176
	(ii) Income from Dairying*	2991	2884	2629	5865	7227	3317
	(iii) Hiring out Agricultural Labour	3091	745	134	0	0	850
	(iv) Hiring out Agricultural Equipments	462	3135	10784	14298	41955	7181
	(v) Income from leased	0	39	1740	10769	25682	2542
	(vi) Salaries	769	650	3782	538	15455	1979
	(vii) Pensions	0	2214	3437	4173	6727	2544
	(viii) Remittances	3333	4078	4244	3038	3137	3864
	(ix) Others Sources**	383	1594	2548	4096	682	1839
	Sub Total	11203	15618	29343	42817	100864	24290
	Total Income	32615	41955	74701	108617	235578	62588

Note: *Net income is taken

**It includes income from hiring out labour in non-agricultural sector and income from small businesses like shop keeping etc.

Source: Field Survey, 2010-11.

Table 15.2 reveals that the relative shares of the income from various sources for different farm-size categories in rural Bihar. The table indicates that the main source of income in the case of an average sampled farm household is the farm business income. On an average, 61.19 per cent of total income of the farm household was dominated by farm business income. The table also reveals that there is an inverse relationship between the share of farm business income and farm-size categories in the total income of the farmers.

Table 15.2

Income Pattern of Farmers in Rural Bihar: Category-wise (Percentage of Total Income)

Sl. No.	*Sources of Income*	*Landless Farmers*	*Marginal Farmers*	*Small Farmers*	*Medium Farmers*	*Large Farmers*	*All Sampled Farmers*
1.	Farm Business Income	65.65	62.77	60.72	60.58	57.18	61.19
2.	(i) Poultry	0.53	0.66	0.06	0.04	0.00	0.28
	(ii) Income from Dairying	9.17	6.87	3.52	5.40	3.07	5.30
	(iii) Hiring out Agricultural Labour	9.48	1.78	0.18	0.00	0.00	1.36
	(iv) Hiring out Agricultural Equipments	1.42	7.47	14.44	13.16	17.81	11.47
	(v) Income from Leased	0.00	0.09	2.33	9.91	10.90	4.06
	(vi) Salaries	2.36	1.55	5.06	0.50	6.56	3.16
	(vii) Pensions	0.00	5.28	4.60	3.84	2.86	4.06
	(viii) Remittances	10.22	9.72	5.68	2.80	1.33	6.17
	(ix) Others Sources	1.18	3.80	3.41	3.77	0.29	2.94
	Sub Total	34.35	37.23	39.28	39.42	42.82	38.81
	Total Income	100.00	100.00	100.00	100.00	100.00	100.00

Source: Calculated from Table 15.1.

The landless farm households earns around 65.65 per cent of total income from the farm business income. The relative shares of farm business income in the total income are 62.77 per cent, 60.72 per cent, 60.58 per cent and 57.18 per cent respectively for marginal, small, medium and large farm households. The second most important source of income in the case of an average household size is income from hiring out agricultural equipments and machinery, it contribute near about 12 per cent in the total income. The share of this source in the total income by different farm-size categories is 1.42 per cent, 7.47 per cent, 14.44 per cent, 13.16 per cent and 17.81 per cent respectively for landless, marginal, small, medium and large farm-size categories. The relative share of this source shows the positive relationship with the farm-size categories. This is because as landholdings lead to increase in assets like machinery and agricultural equipments, it ultimately increases the income of farm households. The relative shares of the income from remintíes are 10.22 per cent, 9.72 per cent, 5.68 per cent, 2.80 per cent and 1.33 per cent respectively by the landless, marginal, small, medium and large farm-size categories.

Income from dairying is another major source of income among farm households in the state. There is found to be almost negative relationship between income from dairying and farm-size categories. This source contributes 5.30 per cent of total income of the farm households. The percentage share of income from dairying is 9.17 per cent, 6.87 per cent, 3.52 per cent, 5.40 per cent and 3.07 per cent respectively for landless, marginal, small,· medium and large farm-size categories. The next important source of income of the farm household income is income from leased out land and income from pension. These sources contribute 4.06 per cent respectively in the total income of the farm households in the selected districts of the state Bihar. Table 15.2 shows that there is positive relationship between the income from leased out land and farm-size categories. The table also shows that the percentage of income from pension is more for the marginal farmers in the state.

Per Capita Income

Per capita income is one of the important proxy indicators to assess the level of development. We have already analysed farm household income and pattern of income for different farm-size categories. Since the family-size across the sampled farming households varies, so it becomes relevant to look into the per capita income levels from this perspective. The information regarding the per capita income earnings by the different farm-size categories is given in Table 15.3.

An average annual per capita income of ₹ 8,118 is earned by farm-size categories in the rural Bihar. However, there are considerable variations in the per capita income earnings among farm-size categories. Table 15.3 highlights that during 2010-11, the per capita income earnings by the farm-size categories were ₹ 4,463, ₹ 5,714, ₹ 9,308, ₹ 13,196 and ₹ 22,436 respectively for the landless, marginal, small, medium and large farm-size categories. The table also shows that there is positive relationship between per capita income earned from all the sources of income and farm-size categories.

Table 15.3

Per Capita Income of Farmers in Rural Bihar: Category-wise

(In ₹ per Annum)

Sl. No.	*Sources of Income*	*Landless Farmers*	*Marginal Farmers*	*Small Farmers*	*Medium Farmers*	*Large Farmers*	*All Sampled Farmers*
1.	Farm Business Income	2930	3587	5652	7994	12830	4967
2.	(i) Poultry*	24	38	6	5	0	23
	(ii) Income from Dairying*	409	393	328	713	688	430
	(iii) Hiring out Agricultural Labour	423	101	17	0	0	110
	(iv) Hiring out Agricultural Equipments	63	427	1344	1737	3996	931
	(v) Income from Leased	0	5	217	1308	2446	330
	(vi) Salaries	105	89	471	65	1472	257
	(vii) Pensions	0	302	428	507	641	330
	(viii) Remittances	456	555	529	369	299	501
	(ix) Others Sources**	52	217	317	498	65	238
	Sub Total	1533	2127	3656	5202	9606	3150
	Total Income	4463	5714	9308	13197	22436	8118

Source: Calculated from Table 15.1.

Income from Agriculture and Allied Activities

Average Expenses and Income from Cultivation per Acre as well as from Dairying and Poultry

Table 15.4 describes the break-up of expenditure on cultivation per farm households by the farm-size categories. In the above analysis it has been shown that the total income increases with farm-size categories, but here the picture is different for the per acre gross returns; the relationship is inversely related with the farm-size. Total expenditure incurred on various inputs amount to ₹ 18,676/acre per annum. For the all farm-size taken together, the maximum expenditure is incurred on seeds, fertilisers and pesticides jointly. As against this, an average gross returns from cultivation per farm household per acre works out to be ₹ 10,521. This means the net income from cultivation is only ₹ 10,521 per acre. There are considerable variations in cost of inputs used by the farmers. The table indicates that there is negative relationship between the cost of inputs used and farm-size categories. The total cost of inputs per acre decreases with the increase in size holdings of the farmers; the medium and large farm-size categories enjoying the lower cost per acre compared to the other categories of farm households.

In Bihar the use of agricultural new technology is negligible among the small landholdings and even for majority of large farmers because of expensive cost and maintenance. The agriculture in the state is still based on traditional technology. There is inverse relationship found by the analysis of data between the net income from cultivation and farm-size categories. Table 15.4 indicates that per acre net income by the marginal and small farmers is highest followed by the landless, medium and large farmers. This also raises the frequently debated question of inverse relationship between farm-size and productivity: weather, marginal and small farmers are relatively efficient *vis-à-vis* larger class farmers? As reminder it should be borne in mind that, many studies have supported the existence of an inverse relationship between farm-size and productivity (Mazumdar, 1965; Rao, 2005; Khusro, 1964; Bardhan, 1973; Sen, 1964; Rudra, 1968a; Rosenzwenig and Biswanger, 1993; Saini, 1969; Krishna, 1995, Chattopadhyay and Sengupta, 1997; Dyer, 1998 among others). The present study also supports the evidence of inverse relationship between the farm-size and productivity in agriculture of Bihar.

Figure 15.2a and 15.2b shows the share expenses of different inputs on cultivation and per acre income from agriculture and allied activities for rural Bihar. The maximum expenditure incurred is on seeds, fertilisers, pesticides jointly, followed by lease rent for land, charges for machinery, labour charges and irrigation etc., for the all sampled. The above analysis also emphasises that the cost to maintain and feed animals and poultry as well as net income from it, shows positive relationship with the farm-size categories. The annual cost spent on animal and poultry stands at ₹ 13,650 while the net income from these sources annually stand at ₹ 3,492. The income from these sources increase with the increase of farm-size because they have more capacity/income to buy more and better quality of animals and poultry, and also maintain them well than others.

Income Inequalities in Bihar

Table 15.5 describes the income inequalities in Bihar among the different farm-size categories. Income inequality is affected by intra as well inter-sectoral changes. Income distribution refers to the percentage spread

Figure 15.2a

Expenses on Cultivation in Rural Bihar

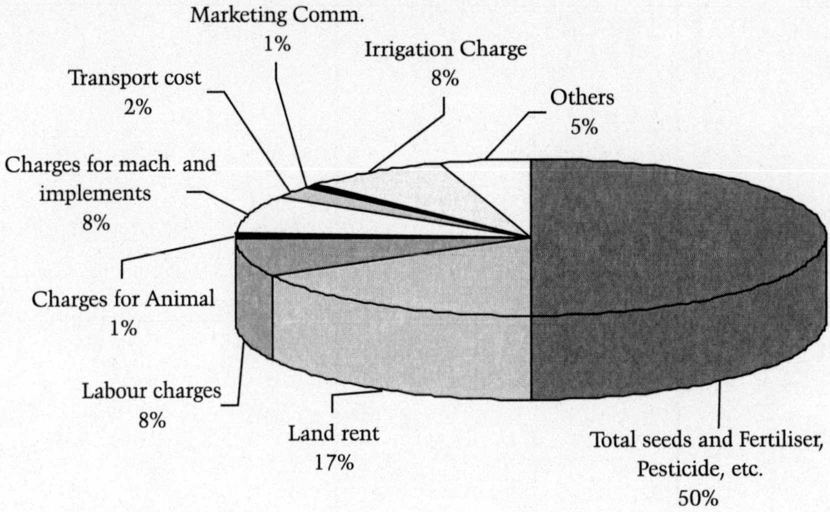

Figure 15.2b

Income from Agr./Acre and Allied Activities

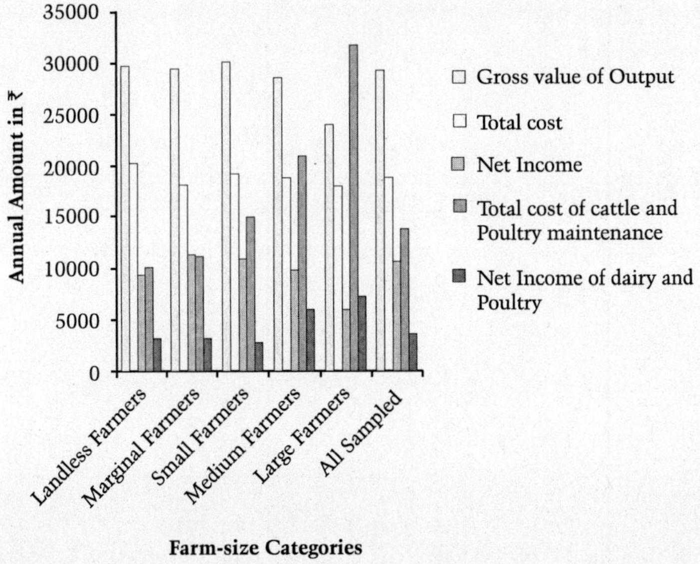

Table 15.4

Gross Return and Input Cost per Acre with Expenses on Cattle, Poultry and their Net Income by the Different Farm-Size Categories

(Amount in ₹ Per annum)

	Gross Value of Output	Total Seeds and Fertilisers, Pesticide etc.	Land Rent	Labour Charges	Charges for Animal	Charges for Mach. and Implements	Transport Cost	Marketing Comm.	Irrigation Charge	Others	Total Cost Per Acre	Net Income Per Acre	Total Cost of Cattle and Poultry Maintenance	Net Income from Dairy and Poultry
Landless Farmers	29588	6550	10360	174	98	1327	307	43	1316	61	20236	9351	10069	3164
Marginal Farmers	29325	9339	3032	1336	137	1618	419	47	1612	510	18047	11278	11154	3163
Small Farmers	29970	10701	793	2249	118	1777	464	178	1384	1469	19133	10836	14925	2675
Medium Farmers	28471	9983	353	2467	123	1710	495	262	1212	2133	18737	9733	20846	5904
Large Farmers	23858	8632	0	1913	77	961	391	194	878	4816	17861	5996	31591	7227
All Sampled	29197	9268	3220	1506	123	1592	419	103	1447	999	18676	10521	13650	3492

Sources: *Field Survey 2010-11.*

of state's income through its population and yields between the richest in a state to the poorest. When income is not proportionally distributed, it is called income inequality. Income inequality is most often measured using what is called a Gini coefficient. The Gini ratio lies between the scale 0 to 1, where 0 being a perfect distribution and 1 being perfect income inequality. Table 15.5 shows that the bottom of 10 per cent all farm household share only 1.53 per cent of the total income while the top 10 per cent of farm households attributes to near about 40 per cent of the total income earnings by the farm households. The bottom 10 per cent of large farm household income attributes just only 1.67 per cent of the total income, while the top 10 per cent of this category attributes 38.98 per cent of the total income. The Gini coefficient also indicates that the results for the large farm households. Over all Gini coefficient of the all farm-size categories is 0.49 per cent, which directly indicates that there is a large per capita income inequality among the selected farm households in the selected districts of the rural Bihar. This also shows that concentration of per capita income inequality is higher than per farm household income.

Table 15.5

Distribution of Per Capita Income of Farmers in Rural Bihar: Category-wise

Cumulative Percentage of Income	*Cumulative Percentage of Households*					
	Landless Farmers	*Marginal Farmers*	*Small Farmers*	*Medium Farmers*	*Large Farmers*	*All Sampled Farmers*
10	1.75	2.48	2.53	3.00	1.67	1.53
20	5.22	6.25	6.88	7.75	4.79	4.09
30	9.92	11.21	12.23	13.44	8.30	7.44
40	15.27	16.60	18.80	20.42	12.00	11.85
50	21.62	23.47	26.47	28.20	16.64	17.41
60	29.81	31.87	35.21	37.09	22.66	24.37
70	39.29	42.08	45.36	47.56	30.52	33.02
80	51.44	55.27	57.18	60.32	40.79	44.23
90	66.43	71.33	72.54	75.67	61.02	59.99
100	100.00	100.00	100.00	100.00	100.00	100.00
Gini coefficient	0.41	0.37	0.34	0.31	0.50	0.49

Source: *Field Survey 2010-11.*

IV

District-wise Per Capita Income

There are large inter-district differences in the per capita income received from the different sources. Table 15.6 shows the district-wise level of income earnings by different farm-size categories annually across the selected districts. An average sampled farm household earns ₹ 62,588 annually in the rural areas of the state. On an average, family size of sampled farm household is 7.71 for the state as a whole. Whereas the sizes of families are different for the selected districts i.e., 8.19, 7.60, 8.05 and 7.50 for Sheohar, Purnia, Lakhisarai and Bhojpur district respectively. The table analyses the significant inter-district per capita farm income in Bihar with Bhojpur having the highest district i.e., ₹ 9,116 and Purnia the lowest per capita income level i.e., ₹ 7,602.

The table also reveals that the per capita income earned by the farm business income is highest in the Bhojpur district with ₹ 5,595 annually followed by the Purnia district with ₹ 4, 810, Lakhisarai district with ₹ 4,533 and lowest in the Sheohar district with ₹ 4,305. There are also differences of per capita income among the farm-size categories within the district among the farm-size categories. It is clear from the table that marginal farm-size categories per capita income is highest in the Bhojpur district with ₹ 6,952 annually followed by the Lakhisarai district with ₹ 5,681, Sheohar district with ₹ 6,180 and lowest in the Purnia district with ₹ 4,545 in the rural Bihar. The similar trend is found in case of small farm households in the State, where Purnia district is at number three and Sheohar is at the lowest among the selected districts in the rural Bihar. For the large farm household per capita income earn by the Purnia district is highest with the ₹ 25,179 annually followed by Lakhisarai district with ₹ 22,595, Bhojpur district with ₹ 21,091 and lowest in the Sheohar district with ₹ 16,435 in the rural Bihar.

Composition of Income across the Districts

The district wise relative shares of income of farm household from the different sources are given in Table 15.7. It is clear from the table that the relative income of farm business income is highest in the Purnia district with 63.27 per cent followed by Sheohar district with 62.09 per cent, Bhojpur district with 61.38 per cent and Lakhisarai district with lowest 58.36 per cent among the farm-size categories.

The table also depicts that the share of income from dairying in the total income in the rural Bihar is the highest in Lakhisarai district with 8.37 per cent followed by Sheohar district with 7.64 per cent, Purnia district with 4.50 per cent and Bhojpur district with only 3.93 per cent. The income from the hiring out agricultural equipments and machinery is plays a crucial role to generate farm household income. Income from the this source in the total income is highest in the Bhojpur district with 12.75 per cent followed by Purnia district with 11.87 per cent, Lakhisarai district with 10.46 per cent and lowest in Sheohar district with just 6.70 per cent. The table also indicates that remittances share of income in total income is highest among the landless, marginal farm households compared to the small, medium and large farm households. It also differs from district to districts. On an average, income from the remittances of farm household in the state is highest in the Lakhisarai district with 10.20 per cent followed by Sheohar with 6.50 per cent, Purnia with 6.04 per cent and finally Bhojpur with 4.20 per cent. Contribution by remittances in the total farm household income for the large farm-a size category is very low; it is because there is less migration of people from the family.

Large farm-size categories have more occupational options than other farm-size categories. The share of income from the sources of pension and salaries also differs among the farm-size categories across the selected districts in the rural Bihar and intra-district within the state.

Table 15.6

Per Capita Income of Farmers in Rural Bihar: District-wise

(In ₹ per annum)

Sl. No.	Sources of Income	Sheohar District						Purnia District					
		Landless Farrmers	*Marginal Farmers*	*Small Farmers*	*Medium Farmers*	*Large Farmers*	*All Sampled Farmers*	*Landless Farrmers*	*Marginal Farmers*	*Small Farmers*	*Medium Farmers*	*Large Farmers*	*All Sampled Farmers*
1	Farm Business Income	2286	4075	4216	5119	10435	4306	3155	3035	5700	9731	13760	4810
2.	(i) Poultry	0	15	0	0	0	7	46	55	16	17	0	38
	(ii) Income from Dairying	321	585	533	813	0	530	300	244	281	1215	279	342
	(iii) Hiring out Agricultural labourers	319	69	0	0	0	79	463	135	0	0	0	140
	(iv) Hiring out Agricultural Equipments	129	121	361	1979	1652	465	89	364	1281	1963	4698	902
	(v) Income from leased	0	0	0	1042	4348	352	0	0	101	471	2674	214
	(vi) Salaries	0	118	222	0	0	103	0	71	0	231	2791	212
	(vii) Pensions	0	369	978	0	0	383	0	68	208	1190	0	175
	(viii) Remittances	500	399	633	479	0	451	113	389	662	769	803	459
	(ix) Others Sources	35	429	47	354	0	259	51	185	709	545	174	309
	Sub-Total	1305	2105	2774	4667	6000	2629	1062	1510	3258	6401	11419	2792
	Total Income	3591	6180	6991	9785	16435	6934	4217	4545	8959	16131	25179	7602

contd...

...contd...

Sl. No.	Sources of Income	Lakhisarai District						Bhojpur District					
		Landless Farrmers	*Marginal Farmers*	*Small Farmers*	*Medium Farmers*	*Large Farmers*	*All Sampled Farmers*	*Landless Farrmers*	*Marginal Farmers*	*Small Farmers*	*Medium Farmers*	*Large Farmers*	*All Sampled Farmers*
1	Farm Business Income	2700	3152	5360	6301	13186	4533	2960	4343	6134	8432	12309	5595
2.	(i) Poultry	0	80	0	0	0	38	12	0	0	0	0	1
	(ii) Income from Dairying	698	543	672	851	1341	670	447	409	136	262	974	359
	(iii) Hiring out Agricultural labourers	385	61	87	0	0	95	423	98	0	0	0	96
	(iv) Hiring out Agricultural Equipments	0	215	951	1885	5341	837	36	737	1870	1436	3154	1163
	(v) Income from leased	0	0	273	1034	1364	256	0	17	357	2110	2244	492
	(vi) Salaries	659	120	393	0	0	227	0	80	1038	0	1282	371
	(vii) Pensions	0	223	847	0	1364	383	0	602	278	424	1128	453
	(viii) Remittances	813	1056	683	483	0	817	751	491	290	0	0	382
	(ix) Others Sources	0	231	87	172	0	151	89	175	128	669	0	204
	Sub-Total	2555	2529	3995	4425	9409	3474	1757	2609	4096	4901	8782	3521
	Total Income	5255	5681	9355	10726	22595	8007	4718	6952	10230	13333	21091	9116

Source: *Field Survey 2010-11.*

Table 15.7

Income Pattern of Farmers in Rural Bihar: District-wise

(Percentage of Total Income)

Sl. No.	Sources of Income	Sheohar District						Purnia District					
		Landless Farrmers	Marginal Farmers	Small Farmers	Medium Farmers	Large Farmers	All Sampled Farmers	Landless Farrmers	Marginal Farmers	Small Farmers	Medium Farmers	Large Farmers	All Sampled Farmers
1.	Farm Business Income	63.66	65.94	60.31	52.31	63.49	62.09	74.81	66.78	63.63	60.32	54.65	63.27
2.	(i) Poultry	0.00	0.24	0.00	0.00	0.00	0.10	1.10	1.20	0.18	0.10	0.00	0.51
	(ii) Income from Dairying	8.94	9.47	7.63	8.30	0.00	7.64	7.11	5.36	3.14	7.53	1.11	4.50
	(iii) Hiring out agricultural Labour	8.89	1.12	0.00	0.00	0.00	1.14	10.98	2.97	0.00	0.00	0.00	1.85
	(iv) Hiring out Agricultural Equipments	3.59	1.95	5.17	20.23	10.05	6.70	2.10	8.01	14.30	12.17	18.66	11.87
	(v) Income from Leased	0.00	0.00	0.00	10.65	26.46	5.08	0.00	0.00	1.13	2.92	10.62	2.82
	(vi) Salaries	0.00	1.91	3.18	0.00	0.00	1.49	0.00	1.56	0.00	1.43	11.08	2.79
	(vii) Pensions	0.00	5.98	13.99	0.00	0.00	5.52	0.00	1.49	2.32	7.38	0.00	2.30
	(viii) Remittances	13.92	6.46	9.06	4.90	0.00	6.50	2.68	8.56	7.39	4.76	3.19	6.04
	(ix) Others Sources	0.99	6.93	0.67	3.62	0.00	3.74	1.21	4.07	7.92	3.38	0.69	4.07
	Sub-Total	36.34	34.06	39.69	47.69	36.51	37.91	25.19	33.22	36.37	39.68	45.35	36.73
	Total Income	100.00	100.00	100.00	100.00	100.00	100.00	100.00	100.00	100.00	100.00	100.00	100.00

contd...

...contd...

Sl. No.	Sources of Income	Lakhisarai District						Bhojpur District					
		Landless Farrmers	*Marginal Farmers*	*Small Farmers*	*Medium Farmers*	*Large Farmers*	*All Sampled Farmers*	*Landless Farrmers*	*Marginal Farmers*	*Small Farmers*	*Medium Farmers*	*Large Farmers*	*All Sampled Farmers*
1	Farm Business Income	51.38	55.48	57.30	58.74	58.36	56.61	62.75	62.47	59.96	63.24	58.36	61.38
2.	(i) Poultry	0.00	1.40	0.00	0.00	0.00	0.48	0.25	0.00	0.00	0.00	0.00	0.02
	(ii) Income from Dairying	13.28	9.55	7.18	7.93	5.93	8.37	9.47	5.89	1.33	1.96	4.62	3.93
	(iii) Hiring out agricultural Labour	7.32	1.08	0.93	0.00	0.00	1.18	8.97	1.41	0.00	0.00	0.00	1.05
	(iv) Hiring out Agricultural Equipments	0.00	3.79	10.16	17.57	23.64	10.46	0.75	10.60	18.28	10.77	14.95	12.75
	(v) Income from Leased	0.00	0.00	2.92	9.64	6.04	3.20	0.00	0.24	3.49	15.83	10.64	5.39
	(vi) Salaries	12.55	2.11	4.21	0.00	0.00	2.83	0.00	1.15	10.14	0.00	6.08	4.07
	(vii) Pensions	0.00	3.93	9.05	0.00	6.04	4.78	0.00	8.66	2.72	3.18	5.35	4.97
	(viii) Remittances	15.47	18.58	7.30	4.50	0.00	10.20	15.93	7.06	2.83	0.00	0.00	4.20
	(ix) Others Sources	0.00	4.07	0.93	1.61	0.00	1.89	1.88	2.51	1.25	5.01	0.00	2.24
	Sub-Total	48.62	44.52	42.70	41.26	41.64	43.39	37.25	37.53	40.04	36.76	41.64	38.62
	Total Income	100.00	100.00	100.00	100.00	100.00	100.00	100.00	100.00	100.00	100.00	100.00	100.00

Source: Field Survey 2010-11.

District-wise Distribution of Per Capita Income

Table 15.8 shows the per capita income disparities across the selected districts. Among the farm-size categories of the selected districts, there is more unequal distribution of income in the Bhojpur followed by Purnia; the bottom 10 per cent farm-size categories of these districts just attribute 2.35 per cent and 2.89 per cent respectively. While the top 10 per cent of farm household attribute 39.25 per cent and 28.18 per cent respectively. The data also reveal that there is an approximately similar tendency of income distribution among the farming community of Sheohar and Lakhisarai districts for the bottom 10 per cent of the total income.

Table 15.8

Distribution of Per Capita Income of Farmers in Rural Bihar: District-wise

Cumulative Percentage of Households	*Cumulative Percentage of Per capita Income of Sampled Farmers in*			
	Sheohar District	*Purnia District*	*Lakhisarai District*	*Bhojpur District*
10	2.47	2.89	2.71	2.35
20	7.38	7.32	7.03	5.85
30	13.75	12.78	12.98	10.17
40	21.81	19.08	20.70	15.23
50	30.61	26.28	28.00	21.23
60	40.25	34.83	37.39	28.16
70	51.35	44.96	48.18	36.51
80	65.05	57.11	61.02	46.82
90	81.14	71.82	77.00	60.75
100	100.00	100.00	100.00	100.00
Gini Coefficient	0.27	0.34	0.31	0.44

Source: *Field Survey, 2010-11.*

The bottom 50 per cent of income by farm households is less than the top 20 per cent of the sampled farm households in rural Bihar. Bottom 50 per of the farm households just contribute 30.61 per cent, 26.28 per cent, 28.00 per cent and 21.23 per cent of the total income in Sheohar, Purnia, Lakhisarai and Bhojpur district respectively. Even in Bhojpur and Purnia districts, bottom 50 per cent of farm household income is less than top 10 per cent of sampled farm household in rural Bihar. The table also shows the Gini coefficient for the each selected district of the state. An inter-district analysis shows that Sheohar and Lakhisarai districts are less skewed as compared as Bhojpur and Purnia districts in the term of income inequalities.

V

Conclusion

An analysis of district-wise farm household income brings out several important and interesting facts about the levels of living of farm households in the rural Bihar. Income of farmer households varied with the farm-size categories. First, the share of farm business income in total income is inversely correlated to the farm-size categories while with increase in the farm-size categories the income from other sources increases i.e., hiring out agricultural machinery and equipments, salaries and pension etc. The study also found that per capita income from all the sources is positively correlated with the farm-size categories. Cultivation remained the main source of income for the farm households in Bihar and the annual average net income from cultivation for farmer household was ₹ 10,521 per acre. The study finds that while the small farmers in Bihar are superior in terms of production performance; they are weak in terms of generating adequate income and sustaining livelihood. It found inverse relationship between farm-size and productivity, however for income from agriculture and allied activities like poultry and dairying, it was found that large farm household earn more from these activities *vis-à-vis* smaller farmers. Per acre cost of cultivation is also inversely correlated with the farm-size categories.

Further district-wise analysis shows that in all the selected districts farm business income is predominating in the total income. Income from this source is more than 60 per cent in all the district except in the Lakhisarai district. An inter-districts analysis shows that household income distribution among the farm-size categories and across the selected districts is shown by the Gini coefficient, which shows that there is more skewed situation in the Bhojpur district in term of income inequalities followed by the Lakhisarai, Purnia and Sheohar district of rural Bihar.

Serious steps should be taken to create employment avenues for small holders outside the agriculture. Small as well as marginal farmer's supplementary sources of income i.e., poultry and dairying etc., can directly increase their income level. In theory, titling is supposed to improve farm productivity by providing secure security for inputs loans for small land holder farmers. Better credit and subsidies directly to small land holders means more and better inputs and thus better productivity. Like government of Punjab, Bihar should also provide facilities like electricity free of cost to the farmers, which can directly reduce the cost of irrigation for the small and marginal farmers. Rural electrification is one state initiative which can provide major boost to agriculture in Bihar. It can bring down the cost of irrigation and improve the working capital situation of the small farmers. This seems to be the better way in achieving higher productivity and sustaining agricultural growth together with augmenting the income of small holders for improving their socioeconomic conditions.

References

Bansil, P.C. (2011). *Bihar Agriculture: A Perspective*. New Delhi: Concept Publishing Company Pvt. Ltd.

Bardhan (1973). "Size, Productivity and Returns to Scale: An Analysis of Farm-level Data in Indian Agriculture", *Journal of Political Economy* 81(6).

Bhall, G.S and G. Singh (2010). *Growth of Indian Agriculture: A District-level Study*. Project Submitted to Planning Commission, Govt. of India.

Bhalla, G.S. (2008). "Income, Consumption and Expenditure of Farmers", in S.S. Bhaumik (ed.), *Reforming Indian Agriculture*. Sage Publication Pvt. Ltd.

Chadha and Khurana (1989). "Backward Agriculture, Unrewarded Labour and Economic Deprivation: Bihar's Contrast with Punjab", *Economic and Political Weekly* XXIV(47).

Chattopadhyay, M. and A. Sengupta (1997). "Farm Size and Productivity: A New Look at the Old Debate", *Economic and Political Weekly* 32(52).

Dyer, G. (1998) "Farm Size and Productivity: A New Look at the Old Debate", *Economic and Political Weekly* 33(26).

GoI (2012). *Census of India, 2011*. New Delhi: Census of Population.

Khusro, A.M. (1964). "Returns to Scale in Indian Agriculture", *Indian Journal of Agricultural Economics* 19(3/4).

Krishna, R. (1995). "Toward a Theory of Optimum Land Reform for a Dualistic Agriculture", in Vijay Krishna (ed.), *Raj Krishna's selected Writings*. New Delhi: Oxford University Press.

Mazumdar, D. (1965). "Size of Farm and Productivity: A Problem of Indian Peasant Agriculture", *Economica* 32 (126).

Pandey, G.K. (2012). "Backbone of Bihar Economy on Backfoot" *Mainstream* 50(34).

Rao, C.H.H. (2005). *Agricultural Growth, Farm Size, and Rural Poverty Alleviation in India*. New Delhi: Academic Foundation.

Rosenzweig, M.R. and H.P. Binswanger (1993). "Wealth, Weather Risk and the Composition and Profitability of Agricultural Investments", *Economic Journal* 103 (416).

Rudra, A. (1968). "Farm Size and Yield per Acre", *Economic and Political Weekly* 3 (26/28).

Sharma, A.N. (2005). "Agrarian Relation and Socio-economic Changes in Bihar", *Economic and Political Weekly* XL(10).

Saini, G.R. (1969). "Farm Size, Productivity and Returns to Scale", *Economic and Political Weekly* 4(26).

Sen, A.K. (1964). "Size of Holdings and Productivity", *The Economic Weekly* 16(5/7).

World Bank (2007). *Bihar: Towards the Development Strategy*. A World Bank Report. Available at *http://siteresources.worldbank.org/INTINDIA/Resources/Bihar_report_final_June2005.pdf*

Economic Reforms, Agricultural Credit and Indebtedness

16 Wither Farm Profitability?

A Cause of Agrarian Distress in India

Elumalai Kannan

Background

India's agricultural sector has undergone considerable changes since the introduction of high yielding varieties in mid-1960s. It has been established that green revolution technology has helped to improve the rural livelihood pattern, nutrition and education of children at farm household level (Foster and Rosenzweig, 1996; Baker and Jewitt, 2007). But, the technology has brought some negative aspects as well. Since it proved successful only in irrigated areas, dry land regions and crops grown therein were left out of the process and hence had created regional disparity in rural income (Rao, 1996; Vaidyanathan, 1988; Bhalla and Singh, 2009). Further, the yield improving technology was considered to be scale neutral, but not resource neutral. Hence, the capital intensive nature of the technology forced the poor peasants to depend on rich peasants and village money lenders for taking credit (Ramachandran and Swaminathan, 2002). The yield improving technology had not only hastened the differentiation within the peasantry (Byres, 1981: 425), it also created disparity in growth across crops and regions. Notwithstanding, a massive public investment made in irrigation infrastructures and public extension system helped to spread the high yielding varieties (HYVs) of rice and wheat during 1980s.

However, by the end of 1980s the productivity growth seemed to have exhausted and yield of major crops slowed down. This along with neglect of agriculture in policy priority and marginalisation of farmers' interests had begun at the central and state government level. A complacency was set at the policymaking level that the country achieved self-sufficiency in food grains production and that agriculture reached a stage where removal of market restrictions would help to maintain and achieve higher growth trajectory by responding to domestic market and international prices (Reddy and Mishra, 2009: 18). Economic reforms launched in 1990s also emphasised to reorient the agricultural production towards meeting the exports demand without touching upon the existing production constraints like land ownership and inequitable access to resources. The neo-liberal policies also led to reduction in public investment in agriculture, withdrawal of subsidies, weakening of institutional support including institutional credit and extension, and lowering of tariff protection on agricultural trade (Mahendra Dev, 2000; Rao, 2003). All these developments forced the farmers to depend more on markets to purchase costly inputs like seed and fertilisers; more particularly usurious money lenders for credit, unscrupulous traders for sale of crop output and input dealers for crop advice. With inadequate government support service systems, marginal and small farmers experience severe distress in the event of crop failure and ultimately commit suicide. Various scholars pointed out that suicides were concentrated among emerging capitalist farmers with small holdings (Reddy and Mishra, 2009; Deshpande and Arora, 2010; Lerche, 2011) who are willing to take risk and invest in new high yielding technology.

Literature on India's agrarian distress and farmers' suicides started increasing considerably since the first reporting of farmers' suicides in the late 1990s. Various studies have pointed out different reasons for agrarian distress. Among others, they included failure of institutional credit delivery system and indebtedness (Assadi, 1998; Mohanty, 2005; Mishra, 2009), crop loss due to pests-diseases and natural disasters (Assadi, 1998; Jeromi, 2007; Mishra, 2009), high input costs and low profitability from cultivation (Mohanty and Shroff, 2004; Nair and Ramakumar, 2007; Ramachandran and Rawal, 2010), family and social problems (Deshpande, 2002; Mohanty and Shroff, 2004; Vasavi, 2012) and commercialisation and adverse trade reforms (Sridhar, 2006; Jeromi, 2007; Nair and Menon, 2007). There seems to be a broader consensus on different causes although the reported degree of a particular cause responsible for distress and farmers' suicide vary. It is because that it is not a single factor, but the constellation of different factors that put the peasants in distress condition and pushes them to the brink of committing suicide (Deshpande and Arora, 2010: 11-12).

The issue of declining profitability from cultivation is something serious and it can be attributed to the intensified neo-liberal policies leading to withdrawal of input subsidies, cut in public investment and defunct public extension system. According to the Situation Assessment Survey of Farmers conducted by the National Sample Survey (NSS) in 2003, about 27 per cent of farmers did not like farming as it was not profitable and 40 per cent would quit if they were given a choice for some other occupation. In spite of such concerns little empirical evidences are available on changes in farm profitability and its implications for peasantry. The present study makes an attempt to analyse trends in real income of major crops in select states in India and discusses underlying reasons for changes in crop income.

The rest of the paper is organised as follows. The next section provides data source for the study. The third section discusses changes in agricultural income at national level. Trends in crop income at the select state level is presented in the fourth section. The fifth section discusses the factors affecting crop income and the final section provides concluding remarks.

Data Sources

The study relies on secondary data compiled from the National Accounts Statistics and the Cost of Cultivation of Principal Crops in India for the period 1981-82 to 2007-08. While data from the National Accounts Statistics are used to analyse the changes in agricultural income at macro level, the cost of cultivation surveys are used for the state level analysis. To compute per hectare farm business income, both inputs and output data were deflated by relevant price deflators. Cost A2 has been used to calculate the farm business income, which is the difference between paid out cost (Cost A2) and gross value of main output. The inputs considered include: human labour, bullock labour, seed, fertiliser, insecticides, irrigation, interest on working capital, rent paid for leased-in land, land revenue, cesses and taxes and depreciation on implements and farm buildings. While agricultural labour wages were deflated by consumer price index (CPI) for agricultural labourers, material inputs and other items were deflated by respective wholesale price indices (WPI). Various crop outputs were also deflated by using respective WPI. A weighted state level income series was constructed by using area share of crops in total cropped area as weight. The methodology of collection of farm level data, definitional issues of cost concepts and consistency in data are elaborated in Sen and Bhatia (2004). However, it is important to mention here that a major limitation of the cost of cultivation surveys is that data are not available by farm size categories and hence worked out figures are averages for the states or country.

Changes in Agricultural Income at National Level

The trend in annual growth in gross value added (GVA) and CPI for agricultural labourers (CPIAL) is plotted in Figure 16.1. GVA indicates agricultural income, while CPIAL shows the price that rural persons pay for purchase of consumer products. It is evident that the percentage change in CPIAL was higher than agricultural income during most years under study. The years which marked with high agricultural income over CPI, were affected by drought and hence higher base value has resulted in higher growth. After discounting the drought years, however, trend in movement of agricultural income clearly indicates that the purchasing power of farmers remained low and has worsened during recent years.

Figure 16.1

Annual Growth in Agricultural GVA and CPIAL

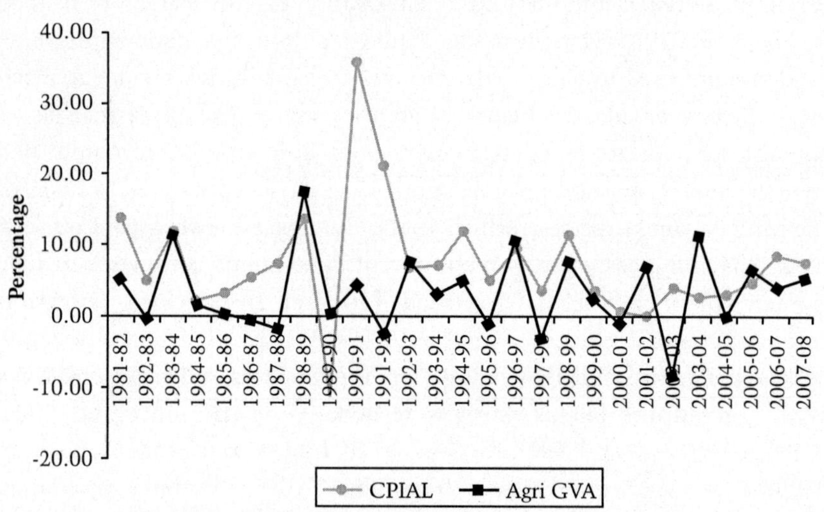

Source: Central Statistical Organisation, Government of India.

The impact of fall in income will be severe for marginal and small producers when compared to rich peasants who are better positioned in terms of relative resource endowments within agriculture and outside agriculture. Rich peasants are in dominant position in the extraction of surplus value from agricultural production, profit from money lending and trading of agricultural products. It has been established that landlords and rich peasants have considerably diversified their source of income particularly through non-farm sector, although they continue to retain a hold on land in order to assert their influence in village affair (Harriss, 2011), and therefore any fall in agricultural income will not affect them as much as poor peasants. Further, increase in the real price of consumer goods affects the food consumption of poor peasants and agricultural labourers, who happen to be the net food buyers in the market.

Notwithstanding, serious concerns had emerged since the mid-1990s on slow down in agricultural technology frontiers. This was marked with decline in crop productivity growth, lack of new technology and collapse of institutional support (Kumar *et al.*, 2004; Chand *et al.*, 2011). In fact, there was an increase in agricultural output to input ratio till mid-1980s, but it remained almost constant thereafter (Table 16.1). The output-input ratio has hovered around 4.5 during entire 1990s and post 2000s. Further, WPI of food articles was lower than that of agricultural inputs till 1988-99 indicating that the farmers received lower market price for agricultural commodities than the price paid for purchase of inputs. However, the prices of food articles have risen faster from 1989-2000 to 1992-93 and thereafter it increased much lower than input prices. These trends broadly imply that despite increase in overall agricultural output at aggregate level, proportionate rise in input costs due to progressive withdrawal of subsidies has led to sluggish output-input ratio over time (Kannan *et al.*, 2012).

In this context, an attempt has been made to look into the changes in composition of intermediate inputs consumption, whose value is actually used in the estimation of agricultural income. Among total input costs, feed of livestock has constituted the highest proportion followed by chemical fertilisers (Table 16.2). Though in percentage terms the livestock feed cost has declined, but in absolute values they have increased substantially. Since the use of purchased feed has been increasing, rise in feed cost is likely to affect the profitability of animal rearing, which is an important livelihood activity for marginal and small farmers, and landless agricultural labourers.

Table 16.1

Trends in Terms of Trade (TOT) and Agricultural Output-Input Ratio

Year	*Agricultural Output-Intermediate Input Ratio*	*Wholesale Price Index (Base: 1999-2000 =100)*				
		TOT	*Food Articles*	*Fertilisers*	*Diesel*	*Electricity (Irrigation)*
1981-82	3.98	86.4	20.5	37.53	23.54	26.80
1982-83	3.94	89.0	23.6	39.40	24.72	29.77
1983-84	4.19	89.2	26.2	36.55	25.48	31.00
1984-85	4.17	91.4	27.1	37.02	26.07	30.42
1985-86	4.15	91.1	28.1	38.49	27.78	30.20
1986-87	4.08	93.2	30.7	40.47	28.35	36.35
1987-88	4.09	94.8	33.8	39.47	28.65	40.21
1988-89	4.30	95.7	36.6	37.19	28.21	37.02
1989-90	4.27	96.8	37.4	37.19	29.49	35.50
1990-91	4.33	99.2	42.5	37.19	38.36	37.95
1991-92	4.20	102.8	51.7	49.17	40.97	40.41
1992-93	4.39	101.2	56.0	65.43	48.27	46.73
1993-94	4.40	100.9	59.5	68.40	52.27	60.68
1994-95	4.47	103.8	69.0	83.09	55.36	73.29
1995-96	4.36	102.5	74.5	88.71	56.87	81.28
1996-97	4.68	100.4	82.8	89.58	65.34	84.49
1997-98	4.44	102.8	86.3	93.33	79.50	90.42
1998-99	4.49	102.4	96.2	95.26	76.99	94.77
1999-00	4.51	100.0	100.0	100.00	100.00	100.00
2000-01	4.55	98.2	101.7	108.79	125.37	125.21
2001-02	4.62	100.1	105.3	111.36	131.99	146.17
2002-03	4.39	100.9	108.4	115.70	160.67	163.81
2003-04	4.57	98.3	111.7	117.62	197.22	166.77
2004-05	4.49	97.7	118.0	119.76	233.57	182.52
2005-06	4.53	99.2	127.7	121.93	243.71	191.53
2006-07	4.53	99.3	134.0	126.27	240.12	191.02
2007-08	4.47	98.7	145.9	134.59	341.99	188.62

Source: TOT from *Agricultural Statistics at a Glance 2010*, Ministry of Agriculture; Agricultural output-input ratio computed (at 1999-2000 prices) from National Accounts Statistics, CSO and WPI for Food Articles from Office of the Economic Advisor, Ministry of Commerce and Industry.

Table 16.2

Composition of Intermediate Inputs Consumed (%)

Items	*1981-82*	*1990-91*	*2000-01*	*2007-08*
Seed	9.45	8.94	8.15	6.97
Organic manure	7.77	6.49	5.35	5.48
Chemical fertilisers	9.20	15.06	17.44	18.22
Current repairs, maintenance of fixed assets & other operational costs	1.12	1.27	1.69	1.45
Feed of livestock	58.49	51.52	47.76	45.27
Irrigation charges	0.47	0.55	0.53	0.44
Market charges	7.66	7.93	8.01	7.73
Electricity	0.62	1.48	2.69	2.07
Pesticides & insecticides	1.34	1.27	0.85	1.02
Diesel oil	2.43	3.71	4.98	6.19
Financial intermediation services indirectly measured	0.54	1.78	2.54	5.16
Total Intermediate Inputs	100.00	100.00	100.00	100.00

Source: *National Accounts Statistics* (various issues), Government of India.

The cost share of seed has declined due to operation of various input subsidy programmes, while that of electricity and diesel oil have increased. The increase in share of electricity cost is due to imposition of tariffs on electricity used for agriculture during recent years with the intention of cost recovery (Reddy and Mishra, 2009). But, by the imposition of tariffs, which is a reversal of policy on free power supply to agriculture, the government failed to provide quality supply due to inadequate investment in power generation, failure to curtail distribution and transmission losses, and corruption (Gulati, 2011). The severity of power crisis is felt more by the marginal and small famers as compared to rich capitalist farmers who have financial resources to invest in diesel engine. In fact, subsidy on diesel is mostly appropriated by the rich farmers whose asset ownership of tractors, combine harvesters and threshers is high. Similarly, despite several initiatives taken by the government to supply agricultural credit at low interest rates, the cost share has surprisingly increased from 1.78 per cent in 1990-91 to 5.16 per cent in 2007-08. It is because the direct lending to farmers by the institutional sources has declined while non-institutional lending has increased since 1991 (Ramakumar and Chavan, 2007). On the whole, the analysis at the macro level indicates that while value of agricultural output has increased over time, cost of inputs has also increased at the same or higher proportions.

Changes in Agricultural Income at State Level

The present section discusses trends in agricultural income in the selected states in India. Though information related to livestock and related activities are compiled through this survey, but they are not made available for use. Thus, the present analysis is restricted only to changes in crop income. For the purpose of present study four states *viz.*, Karnataka, Maharashtra, Punjab and West Bengal representing different regions and agrarian structures were selected. Some of the agrarian characteristics of these states are provided in Table 16.3. It can be noted that agriculture plays an important role in accelerating overall economic growth, though differentially across the selected states. Except Punjab, other states have registered overall growth in gross state domestic product (GSDP) at about 6 per cent during 1980-81 to 2007-08. Trend annual growth in agricultural GDP was higher in West Bengal (4.6 per cent) than the all India growth rate of 3.0 per cent. The share of agricultural income and employment are also quite different. Agriculture has contributed over a quarter of total state income in Punjab and one-fifth in West Bengal. However, the proportion of rural persons engaged in agriculture was substantially higher in Maharashtra and Karnataka implying that agrarian question of labour remains an important issue. The dependence of large work force on low agricultural income gives rise to lower rural per capita income; as a result prevalence of rural poverty and malnutrition is high in rural areas.

The extent of cultivated area under irrigation and favourable weather conditions determine the cropping pattern and cropping intensity. The cultivation of paddy and wheat in rotation is the prominent cropping pattern in Punjab. In fact, paddy was not a major crop in Punjab before 1970, but its area has increased tremendously over time occupying 33.2 per cent of total cropped area. The effective operation of incentive factors like market price support and regular government procurement seem to be the major forces behind cultivation of paddy by farmers in that state. However, in West Bengal, the cropping pattern is dominated by paddy with 58.7 per cent. With regard to Maharashtra, it is a rapidly industrialising state in Western India dominated by sugar and textile factories but its record

Table 16.3

Important Indicators of Agricultural Development in Select States

Items	*Karnataka*	*Maharashtra*	*Punjab*	*West Bengal*	*India*
Share of agriculture in total income (%, 2007-08)	18.3	13.3	29.0	21.3	17.8
Share of rural persons engaged in agriculture (%, 2009-10)	62.5	76.4	50.8	51.4	57.6
Trend growth in GDP (%, 1980-81 to 2007-08)	6.6	6.6	4.9	6.4	5.7
Trend growth in agricultural GDP (%, 1980-81 to 2007-08)	3.1	3.6	3.0	4.6	3.0
Net irrigated area (%, 2007-08)	30.1	18.9	97.2	59.2	44.2
Cropping intensity (%, 2007-08)	123.7	129.7	188.0	184.1	139.0
Food grains in GCA (%, 2007-08)	61.0	58.3	80.1	65.2	63.5
Two major crops (% of GCA)	Rice (11.0%) Jowar (10.7%)	Jowar (18.3%) Cotton (14.1%)	Wheat (44.3%) Rice (33.2%)	Rice (58.7%) Jute (6.3%)	Rice (22.4%) Wheat (14.3%)
Number of tractors per thousand hectare of net cropped area (2003)	59	60	714	35	168

Source: Income data from *National Accounts Statistics*, CSO; Employment from Government of India (2010); Crop and irrigated area from Ministry of Agriculture and Tractors from *Statistical Abstract of India 2007*.

on agricultural development is not commendable. The level of irrigation development is very low at 19 per cent. Sorghum (*jowar*) and cotton are the two major crops grown in the state. Similarly, Karnataka has vast stretch of un-irrigated area and is dominated by paddy and coarse cereals. Thus, existence of diverse crop production pattern and agrarian relations prevailing in these states makes the analysis highly complex and it is even more challenging at micro level. Nevertheless, aggregate analysis at state and crop level throws some meaningful results on changes in farm business income over time across different agrarian settings in India.

Table 16.4 to 16.7 provide average real cost, farm business income and wage rate in Karnataka, Maharashtra, Punjab and West Bengal. The number of crops for which real income and cost estimated in these selected states was determined by the availability of data from the cost of cultivation survey for the study period. Farm business income is estimated as the difference between paid out cost and gross value of output (only main product). In addition to cultivators' income, average earning to agricultural labour was also estimated from the cost of cultivation. For Karnataka, data were available for nine crops, which accounted for about 60 per cent of

Table 16.4

Average Real Crop Output Value, Cost, Wage and Income in Karnataka

Items	*Paid Out Cost (₹/ha)*	*Gross Value of Output (₹/ha)*	*Farm Business Income (₹/ha)*	*Wage Rate (₹/day)*
Paddy				
1981-82 to 1990-91	7747	21781	14033	33
1991-92 to 2000-01	10811	27677	16867	43
2001-02 to 2007-08	14902	29994	15092	51
Jowar				
1981-82 to 1990-91	2671	7373	4702	29
1991-92 to 2000-01	3070	6245	3176	37
2001-02 to 2007-08	3603	4936	1333	40
Ragi				
1984-85 to 1990-91	4376	6125	1750	27
1991-92 to 2000-01	5714	5856	142	41
2001-02 to 2007-08	7072	6946	-125	52
Arhar				
1981-82 to 1990-91	3056	10037	6981	30
1991-92 to 2000-01	4159	8936	4777	36
2001-02 to 2007-08	5624	10681	5057	41
Groundnut				
1981-82 to 1990-91	5547	10402	4855	27
1991-92 to 2000-01	6470	9154	2684	36
2001-02 to 2007-08	6272	7212	939	33
Sunflower				
1984-85 to 1990-91	3080	5593	2513	27
1991-92 to 2000-01	3643	5792	2149	38
2001-02 to 2007-08	4226	4813	587	38
Safflower				
1984-85 to 1990-91	2358	3895	1537	34
1991-92 to 2000-01	2402	5021	2619	38
2001-02 to 2007-08	3077	5751	2674	43
Cotton				
1981-82 to 1990-91	7143	23772	16630	28
1991-92 to 2000-01	6208	11695	5487	31
2001-02 to 2007-08	6276	13167	6891	35
Sugarcane				
1981-82 to 1990-91	14629	75288	60659	31
1991-92 to 2000-01	17163	74849	57686	18
2001-02 to 2007-08	24154	62687	38533	15
Overall (Weighted)				
1981-82 to 1990-91	4979	13788	8809	29
1991-92 to 2000-01	6167	13526	7359	37
2001-02 to 2007-08	7647	13318	5671	41

total cropped area. Among crops, average farm business income from *jowar*, finger millet (*ragi*), groundnut, sunflower and sugarcane has secularly declined across study periods (Table 16.4). The largest fall in business income was observed in case of *ragi* and it turns out to be completely non-remunerative during the recent period. The business income from paddy has declined marginally during recent period. Surprisingly, per hectare average income from two major oilseeds *viz.*, groundnut and sunflower has declined steeply from ₹ 4855 and ₹ 2513 in 1981-82 to 1990-91 to ₹ 2684 and ₹ 2149 in 1991-92 to 2000-01 and then to ₹ 939 and ₹ 587 during 2001-02 to 2007-08, respectively. However, business income from safflower was more or less stagnant during recent periods.

Although the average paid out cost has increased over time for pigeon pea (*arhar*), it registered higher value of output and thus resulted in higher income during recent period. This was due to rise in productivity resulted from the cultivation of improved varieties of *arhar* by farmers and also higher support price offered by the government as an incentive to increase the production. Similarly, cultivation of Bt cotton since early 2000s has helped to improve the income (Naik *et al.*, 2005) during 2001-02 to 2007-08, but it was much less than the level of income obtained during 1980s. For Karnataka as whole, average per hectare business income has fallen from ₹ 8809 to ₹ 7359 and then to ₹ 5671 during the sub-periods, respectively. The overall income has actually declined by 23 per cent between 1990s and 2000s; the period during which agrarian crisis and reporting of farmers' suicides was high in the state. With respect to real wage rates, there has been a significant increase from ₹ 29 to ₹ 41 per day between 1981-82 to 1990-91 and 2001-02 to 2007-08. However, the rate of increase in real wages is observed to be lower during recent period. It is for the reason that incremental gross value of output (agricultural surplus) was itself low during this period.

For the estimation of farm business income in Maharashtra, cost of cultivation data for eight crops which accounted for about 50 per cent of gross cropped were compiled. Farm business income from *jowar*, sunflower, cotton and sugarcane has declined across the periods under study (Table 16.5). Despite increase in value of output from these crops, more than proportionate rise in cost of inputs has led to decline in income. Sugarcane is the most profitable crop, whose area under cultivation had increased

Table 16.5

Average Real Crop Output Value, Cost, Wage and Income in Maharashtra

Items	*Paid Out Cost (₹/ha)*	*Gross Value of Output (₹/ha)*	*Farm Business Income (₹/ha)*	*Wage Rate (₹/day)*
Jowar				
1981-82 to 1990-91	3592	6925	3333	26
1991-92 to 2000-01	4707	7811	3105	37
2001-02 to 2007-08	7126	8067	940	46
Bajra				
1986-87 to 1990-91	3509	5989	2480	29
1991-92 to 2000-01	3963	5662	1699	36
2001-02 to 2007-08	7241	9170	1929	50
Urad				
1986-87 to 1990-91	3147	6308	3161	27
1991-92 to 2000-01	3963	7829	3866	33
2001-02 to 2007-08	6045	8521	2476	41
Groundnut				
1986-87 to 1990-91	7551	10699	3147	29
1991-92 to 2000-01	9645	14549	4904	37
2001-02 to 2007-08	11592	13602	2010	49
Sunflower				
1981-82 to 1990-91	2634	4735	2101	25
1991-92 to 2000-01	4469	5606	1138	33
2001-02 to 2007-08	6242	6256	14	46
Safflower				
1986-87 to 1990-91	2357	4122	1765	30
1991-92 to 2000-01	2294	4622	2329	33
2001-02 to 2007-08	3667	5796	2129	47
Cotton				
1986-87 to 1990-91	6257	13903	7645	26
1991-92 to 2000-01	7970	14564	6594	38
2001-02 to 2007-08	12191	17029	4838	40
Sugarcane				
1981-82 to 1990-91	32819	61343	28524	36
1991-92 to 2000-01	25172	46362	21190	42
2001-02 to 2007-08	34092	50136	16044	58
Overall (Weighted)				
1981-82 to 1990-91	4740	8933	4194	25
1991-92 to 2000-01	6278	10671	4392	37
2001-02 to 2007-08	10002	13049	3047	45

tremendously from 0.8 per cent of gross cropped area (GCA) in 1987-88 to 3.0 per cent in 2000-01. Extension of sugarcane cultivation to non-conventional areas, which is largely driven by expansion of sugar mills, poses threat to sustainable use of groundwater.

However, farm business income from the cultivation of sunflower and *jowar* had registered a steep fall during the study periods. Though there is increase in per hectare income from pearl millet (*bajra*) during 2001-02 to 2007-08, but it is much lower than the income level registered during 1981-82 to 1990-91. As for cotton, it is claimed in literature that cost of cultivation of Bt cotton (officially introduced in 2002 in India) is lower than that of conventional cotton varieties (Naik *et al.*, 2005; Subramanian and Qaim, 2010), but as per the official cost of cultivation data the average paid out cost has increased by more than 50 per cent during 2001-02 to 2007-08 as compared to the previous periods. The increase in input costs has brought down its income from ₹ 6594 to ₹ 4838 per hectare. The agrarian crisis leading to farmers' suicides in Maharashtra was highly connected with rise in the cost of production of commercial crops like cotton in the dry land regions, where the incidences of monsoon failure was common. Mohanty and Shroff (2004) and Mohanty (2005) showed that a large number of small farmers who committed suicides in the Vidarbha region were the cotton growers.

Despite a poor performance of agriculture in Maharashtra, rise in real wages for agricultural labourers is commendable. The real wage has increased by 12 percentage points between 1980s and 1990s as compared to only 8 percentage points during 2000s. This has actually happened due to a higher 'reservation wage' set in agriculture by Employment Guarantee Scheme (Gaiha, 1997). Gaiha had established that employment guarantee scheme helped to increase the agricultural wages significantly in Maharashtra.

In Punjab, four crops *viz.*, paddy, wheat, cotton and rapeseed and mustard dominate the cropping pattern accounting for 86 per cent of total cropped area. Interestingly, farm business income from paddy and wheat has showed increasing trend, while it fluctuated for cotton and rapeseed and mustard (Table 16.6). Among crops, the per hectare gross value of paddy output has increased tremendously from ₹ 28,452 to ₹ 37,899

Table 16.6

Average Real Crop Output Value, Cost, Wage and Income in Punjab

Items	*Paid Out cost (₹/ha)*	*Gross Value of Output (₹/ha)*	*Farm Business Income (₹/ha)*	*Wage Rate (₹/day)*
Paddy				
1981-82 to 1990-91	13377	29030	15653	50
1991-92 to 2000-01	11453	28452	16998	65
2001-02 to 2007-08	12058	37899	25842	74
Wheat				
1981-82 to 1990-91	10177	20631	10454	53
1991-92 to 2000-01	10581	25404	14823	73
2001-02 to 2007-08	10031	26990	16959	86
Cotton				
1981-82 to 1990-91	8864	24768	15904	45
1991-92 to 2000-01	11201	18798	7597	62
2001-02 to 2007-08	14893	30472	15579	69
Rapeseed & Mustard				
1986-87 to 1990-91	5098	11524	6427	43
1991-92 to 2000-01	5700	9816	4116	62
2001-02 to 2007-08	5270	13906	8636	64
Overall (Weighted)				
1981-82 to 1990-91	10914	23473	12559	51
1991-92 to 2000-01	10878	25681	14803	69
2001-02 to 2007-08	11179	31425	20247	81

between 1991-92 to 2000-01 and 2001-02 to 2007-08 which has resulted in higher business income. The rate of increase in value of paddy output is much higher than rate of increase in the paid out cost. Although income from cotton and rapeseed and mustard has declined during 1991-92 to 2000-01, it has risen significantly during the recent period. As compared to other states under study, average real wage rate is relatively high in Punjab and it has increased considerably over time. Although higher degree of farm mechanisation or capitalist development of agriculture has affected the wage employment, it has led to increase in real wages. In a way, it can also be argued that the rise in productivity has not only benefited the cultivators but also the agricultural labourers. However, rate of increase in real wages

(17%) was much less than rate of rise in farm business income (37%) between 1990s and 2000s.

In West Bengal, paddy and jute dominate the cropping pattern accounting for about 65 per cent of total cropped area in the state. It can be seen from the Table 16.7 that the overall farm business income increased between 1980s and 1990s, but declined substantially during the recent period. In case of paddy, during 2001-02 to 2007-08 value of its output declined, while paid out cost increased. The productivity of paddy is the highest in West Bengal, but surprising fall in output value has reduced the per hectare income from ₹ 9402 to ₹ 7073 between 1990s and 2000s. In spite of increase in value of output of jute, more than proportionate raise in cost has led to decline in farm business income. The fall in crop income has also slowed down the rise in real wages for agricultural labourers.

Table 16.7

Average Real Crop Output Value, Cost, Wage and Income in West Bengal

Items	*Paid Out Cost (₹/ha)*	*Gross Value of Output (₹/ha)*	*Farm Business Income (₹/ha)*	*Wage Rate (₹/day)*
Paddy				
1981-82 to 1990-91	7398	16206	8808	36
1991-92 to 2000-01	10176	19578	9402	49
2001-02 to 2007-08	10862	17935	7073	59
Jute				
1981-82 to 1990-91	7454	12752	5298	33
1991-92 to 2000-01	8535	13258	4723	47
2001-02 to 2007-08	11065	15614	4550	56
Overall (Weighted)				
1981-82 to 1990-91	7409	15918	8509	35
1991-92 to 2000-01	10036	19044	9009	49
2001-02 to 2007-08	10880	17720	6840	59

The above discussion clearly brings out that agricultural income has fallen at national, select states and crop level. The decline in income was more conspicuous after 1991; the year during which economic reforms were introduced. However, there is some improvement in farmers' income since 2006-07. Economic reforms were largely focused on industries, but it was considered that removal of industrial protection would help agricultural sector indirectly through increased availability of cheap material inputs. But, analysis showed a substantial increase in materials input costs, which has actually led to decline in crop income. The fall in crop profitability is worrisome and it causes distress among cultivators, which is already manifested in the form of suicides (Deshpande and Arora, 2010; Reddy and Mishra, 2009). However, average real wage rate has increased in all the states implying that there is improvement in the welfare of agricultural labours, but such improvement appears to be uneven across the states.

Discussion of Important Factors Affecting Crop Income

There are several factors operating within and outside the farm that influence the level of income obtained from various crops. Among others, productivity of crops, soil and water management, prices, procurement support, market infrastructure and government policies assume great significance in affecting the farm income. Although some of these factors have already been dealt in the paper, a further explanation of a few of them would be reinforcing in better understanding of the changes in average crop income in the selected states.

The green revolution technology has changed the landscape of Indian agriculture registering respectable growth rate of about 3.0 per cent per annum during the late 1960s to 1980s (Bhalla and Singh, 2001; 2009). But, the slump in agricultural growth rate during 1990s and early 2000s was attributed to falling or stagnating crop yields. The stagnation in yield could be traced to declining public investment in agriculture research and education since 1990s (Singh and Kumar, 2011). The period of decline in public spending was coincided with promotion of private sector investment in India's agricultural research sector. But, unfortunately private sector investment focused only on the segments like hybrid seed development through which higher profit could be extracted from capitalist small and medium farmers who are forced to buy the hybrid seeds every season (Ramasundaram *et al.*, 2011). Further, private investment in agricultural research was narrowly focused on commercial crops largely suitable for cultivation in irrigated regions and it failed to develop technologies suitable for crop varieties grown for dry land regions. Therefore, it can be argued that sluggish public investment in agricultural research and narrow interests of private sector investment did not bring any significant breakthrough in the yield of major crops in the country.

The role of public extension system in achieving higher agricultural growth during the 1980s is well recognised. At the same time, it has also been documented that slow down in agricultural growth during 1990s was attributed to among others, ineffective functioning of public extension system in transfer of knowledge to farmers (Rao, 2003; Vyas, 2008). The Working Group on Agricultural Extension for the Eleventh Plan noted a virtual collapse of public extension system in most states partly due to constraints on government expenditure. This had led to a situation of knowledge and technology deficit particularly for marginal and small farmers as compared to large farmers who have easy access to experts in the state agriculture department and agricultural universities. In spite of a serious concern about the prevailing weak public extension system, initiatives taken by the government were inadequate and thus situation has not improved much on the ground. But, taking advantage of the weak public extension system and its low accessibility to marginal and small farmers, input dealers have emerged to be major exploiters of peasantry in rural areas. Given the exploitative nature of the "input dealer led extension system", it may not help to foster higher productivity and hence achieve higher income for farmers, but it would rather be a source of distress for the peasantry.

There is a debate among researchers that the minimum support price (MSP) announced by the government for agricultural commodities have increased substantially over time distorting the free operation of agricultural markets. It may be true that the MSP in nominal terms has increased over time, but in real value (deflated by respective commodity specific WPI) it has actually declined. Further, available empirical evidences show that high proportion of farmers are not aware of MSP itself (Deshpande and Naika, 2002). Further, government regulated markets are located far away from the villages (Acharya, 2006) and farmers with small marketed surplus will be discouraged to carry them due to high transportation cost. The recent Agricultural Produce Marketing Committee (APMC) reforms do not seem to have any impact on the ground benefiting the farmers.

Since the beginning of agricultural development in 1960s, dual pricing policy of output price support and input subsidisation were followed with an objective of encouraging farmers to produce sufficient food domestically. There is greater discussion in the literature that policy of supplying important inputs like fertilisers, credit, electricity and water at subsidised rate helped to increase agricultural production in the country during 1970s and 1980s (Rao, 1994; Bhalla and Singh, 2001; Acharya, 2006). However, neo-liberal policies introduced in 1991 increasingly focused on withdrawal of the input subsidies and hence forced the farmers to pay higher price for inputs. The main thrust behind these policy changes was to allow the market forces to determine the prices of different type of fertilisers. But, consequence is a steep rise in the price of certain most commonly used fertilisers like urea and di-ammonium phosphate (DAP). As per the *Fertiliser Statistics of India*, the per tonne average price of DAP was ₹ 4860 in 1991-92, which had increased to ₹ 9350 in 2008-09 with a rise of about 92.4 per cent. Increase in the price of Muriate and Potash during the corresponding period was ₹ 1700 to ₹ 4455 with a growth of 162.1 per cent. In case of urea, the average price increased by 67 per cent from ₹ 3180 in 1991-92 to ₹ 5310 in 2009-10.

Rising cost of cultivation increases cash requirement for meeting the production and family consumption needs. As regards availability of credit, medium and large farmers have better access to institutional sources than marginal and small farmers because the former have assets to offer as collateral to take loans (NCEUS, 2008: 14). Land ownership and land size continues to be major criteria for accessing institutional loans in most states, which actually keeps away marginal farmers, small farmers and tenant cultivator from taking formal loans. In this context, the government policy of increasing flow of agricultural credit at subsidised interest rate since 2004 has not helped much on the ground because those who earlier had access to institutional sources continue to get loans at low interest rate. For marginal and small farmers, village money lenders continued to be important source of loans. Therefore, increase in the price of crucial input like fertiliser along with increased dependence on informal sources of credit, which comes with the condition of selling produce to credit agencies results in low income realisation to poor peasants.

Conclusions

The present study has analysed the changes in agricultural income at the national and select state level in India. At macro level, annual growth in agricultural income and CPI for agricultural labours indicate that the purchasing power of farmers has remained low and it has worsened during recent years. Analysis of crop income at state level showed that average real farm business income has fallen in all the four states under study. The fall in income was observed to be sharp for coarse cereals and certain oilseeds. The decline in income was more conspicuous after 1991. However, there is some improvement in farmers' income since 2006-07. Interestingly, real wage rate has increased in the study states. The study results therefore, broadly indicate that the real income of cultivators has declined, while that of agricultural labours has increased.

These results have policy implications for improving the conditions of poor farmers in India. This calls for intervention in critical areas such as technology, public service delivery system and domestic market reforms. There are discussions among Indian policymakers for launching of second green revolution, for which the role of public agricultural research assumes greater significance in terms of developing varieties that not only produce higher yield but also better nutritional qualities. Government extension system needs to be revived and also change in extension approach from individual to group based approach needs to be effected to include all farm size groups. Conducive policy environment should be created to realise better price for the agricultural produce by marginal and small farmers through overcoming existing bias in the operation of input and output markets. This will go a long way in generating higher income and employment to large rural populace.

References

Acharya, S.S. (2006). "Agricultural Marketing and Rural Credit for Strengthening Indian Agriculture", *INRM Policy Brief No3*. New Delhi: Asian Development Bank.

Assadi, M. (1998). "Farmers' Suicides Signs of Distress in Rural Economy", *Economic and Political Weekly* 33(14): 747-48.

Baker, K. and S. Jewitt (2007). "Evaluating 35 Years of Green Revolution Technology in Villages of Bulandshahr District, Western UP, North India", *Journal of Development Studies* 43(2): 312-39.

Bhalla, G.S. and G. Singh (2001). *Indian Agriculture: Four Decades of Development*. New Delhi: Sage Publications.

———. (2009). "Economic Liberalisation and Indian Agriculture: A State-Wise Analysis", *Economic and Political Weekly* 44(52): 34-52.

Byres, T.J. (1981). "The New Technology, Class Formation and Class Action in the Indian Countryside", *The Journal of Peasant Studies* 8(4): 405-54.

Chand, R., P. Kumar and S. Kumar (2011). "Total Factor Productivity and Contribution of Research Investment to Agricultural Growth in India", *Policy Paper* 25. New Delhi: National Centre for Agricultural Economics and Policy Research.

Deshpande, R.S. (2002). "Suicide by Farmers in Karnataka Agrarian Distress and Possible Alleviatory Steps", *Economic and Political Weekly* 37(26/29):2607-10.

Deshpande, R.S. and S. Arora (eds.) (2010). *Agrarian Crisis and Farmer Suicides*. New Delhi: Sage Publications.

Deshpande, R.S. and T. Naika (2002). "Impact of Minimum Support Prices on Agricultural Economy: A Study in Karnataka", *ADRTC Research Report*. Bangalore: Institute for Social and Economic Change.

Foster, A.D. and M.R. Rosenzweig (1996). "Technical Change, Human-Capital Returns and Investments: Evidence from the Green Revolution", *The American Economic Review* 86(4): 931-53.

Gaiha, R. (1997). "Do Rural Public Works Influence Agricultural Wages? The Case of the Employment Guarantee Scheme in India", *Oxford Development Studies* 25(3): 301-14.

Gulati, M. (2011). "How Free or Subsidised Power to Agriculture can be Delivered", *The Economic Times*, 2 April.

Harriss, J. (2011). "Reflections on Agrarian Change in Indian Since Independence: Does 'Landlordism' Still Matter", *Simons Papers in Security and Development*, No.14/2011. Vancouver: Simon Fraser University.

Jeromi, P.D. (2007). "Impact of Agricultural Trade Liberalisation: Farmers' Indebtedness and Suicides in Kerala", *Indian Journal of Agricultural Economics* 62(2): 159-75.

Kannan, E., R.S. Deshpande and K. Vishnu (2012). *Agrarian Change and Farm Sector Distress: An Exploratory Study.* Research Report. Bangalore: Institute for Social and Economic Change.

Kumar, P., A. Kumar and S. Mittal (2004). "Total Factor Productivity of Crop Sector in the Indo-Gangetic Plain of India: Sustainability Issues Revisited", *Indian Economic Review* 39(1): 169-201.

Lerche, J. (2011). "Agrarian Crisis and Agrarian Questions in India", *Journal of Agrarian Change* 11(1): 104-18.

Mahendra Dev, S. (2000). "Economic Reforms, Poverty, Income Distribution and Employment", *Economic and Political Weekly* 35(10): 823-35.

Mishra, S. (2009). "Agrarian Distress and Farmers' Suicides in Maharashtra", in N. Reddy and S. Mishra (ed.), *Agrarian Crisis in India*. New Delhi: Oxford University Press. pp.126-63

Mohanty, B.B. and S. Shroff (2004). "Farmers' Suicides in Maharashtra", *Economic and Political Weekly* 39(52): 5599-06.

Mohanty, B.B. (2005). "We are Like the Living Dead': Farmer Suicides in Maharashtra, Western India", *The Journal of Peasant Studies* 32(2): 243-76.

Naik, G., M. Qaim, A. Subramanian and D. Zilberman (2005). "Bt Cotton Controversy, Some Paradoxes Explained", *Economic and Political Weekly* 40(15): 1514-17.

Nair, K.N. and R. Ramakumar (2007). "Agrarian Distress and Rural Livelihoods: A Study in Upputhara Panchayat, Idukki District, Kerala", *CDS Working Paper No*.392. Thiruvananthapuram: Centre for Development Studies.

Nair, K.N. and V. Menon (2007). "Distress Debt and Suicides among Agrarian Households: Findings from Three Village Studies in Kerala", *CDS Working Paper* No. 397. Thiruvananthapuram: Centre for Development Studies.

NCEUS (2008). *A Special Programme for Marginal and Small Farmers*. New Delhi: National Commission for Enterprises in the Unorganised Sector, Government of India.

Ramachandran, V.K. and V. Rawal (2010). "The Impact of Liberalisation and Globalisation on India's Agrarian Economy", *Global Labour Journal* 1(1): 56-91.

Ramachandran, V.K. and M. Swaminathan (2002). "Rural Banking and Landless Labour Households: Institutional Reform and Rural Credit Markets in India", *Journal of Agrarian Change* 2(4): 502-44.

Ramakumar, R. and P. Chavan (2007). "Revival of Agricultural Credit in the 2000s: An Explanation", *Economic and Political Weekly* 42(52): 57-63.

Ramasundaram, P., S.A. Kurup and R. Chand (2011). "Manoeuvring Technology for Surplus Extraction: The Case of Bt Cotton in India", *Policy Brief* 37. New Delhi: NCAP.

Rao, C.H.H. (1994). *Agricultural Growth, Rural Poverty and Environmental Degradation in India*. New Delhi: Oxford University Press.

———. (2003). "Reform Agenda for Agriculture", *Economic and Political Weekly* 38(7): 615-20.

Rao, V.M. (1996). "Agricultural Development with A Human Face: Experiences and Prospects", *Economic and Political Weekly* 31(26): A50-A62.

Reddy, D.N. and S. Mishra (eds.) (2009). *Agrarian Crisis in India*. New Delhi: Oxford University Press.

Sen, A. and M.S. Bhatia (2004). "Cost of Cultivation and Farm Income", *State of the Indian Farmer A Millennium Study*. Volume 14. New Delhi: Academic Foundation.

Singh, A. and S. Kumar (2011). "Trends in Public and Private Investment in Agricultural R&D", in S. Pal. (ed.), *Agricultural for Inclusive Growth*. New Delhi: IARI.

Sridhar, S. (2006). "Why Do Farmers Commit Suicide? The Case of Andhra Pradesh", *Economic and Political Weekly* 41(16): 1559-65.

Subramanian, A. and M. Qaim (2010). "Impact of Bt Cotton on Poor Households in Rural India", *Journal of Development Studies* 46(2): 295-311.

Vaidyanathan, A. (1988). *India's Agricultural Development in a Regional Perspective*. Hyderabad: Orient Longman.

Vasavi, A.R. (2012). *Shadow Space Suicides and the Predicament of Rural India*. New Delhi: Three Essays Collective.

Vyas, V.S. (2008). "Market Reforms in Indian Agriculture: One Step Forward, Two Steps Back", in A. Gulati and S. Fan (eds.), *The Dragon and the Elephant Agricultural and Rural Reforms in China and India*. New Delhi: Oxford University Press.

17

Agricultural Risk and Efficacy of Crop Insurance Schemes in Odisha

Mamata Swain

Introduction

Agriculture is inherently a risky venture due to uncertainty in production and volatility in price. Agricultural production depends on various climatic variables such as rainfall, humidity, sunshine and temperature. Variations in these weather indicators from the requisite quantity needed for the plant at its various growth stages affect crop yield adversely. The devastation caused to the agrarian economy due to occurrence of weather induced natural disasters like drought, flood, hurricane, cyclone, hailstorm are well known. Apart from weather risk, unforeseen contingencies like infestation of plant disease and pest attack very often inflict huge crop loss. In addition to production risk, there is also substantial market risk in agriculture due to wide variations in input and output prices. Crop prices are more volatile because of difficulties in storage owing to bulkiness and perishability of the products. Demand for agricultural products is relatively inelastic with respect to variations in price and income. As a result, the supply shocks are manifested in wide variations in price. However, of all the risk factors affecting crop production, weather is typically the most significant (Miranda and Vedenov, 2001).

The government takes many measures to reduce risk and impart greater resilience to agriculture. These include promoting diversification of cropping pattern, inter-cropping, flood-control, drought-proofing, watershed development for reducing production risk, and price support through market intervention and futures trading for stabilising prices and hedging risks. However, in spite of preventive measures being taken, when there is crop failure, insurance is considered the most effective mechanism to compensate the farmers for their losses. Insurance involves the substitution of a small known cost (premium rate) for the possibility of a larger but uncertain loss which will be compensated through indemnification. Though crop insurance does not reduce the risk of crop failure and thus, is not preventive in nature, but certainly it is curative by helping the farmer to recoup the losses through indemnity payment. Crop insurance is a risk transfer mechanism that transfers the production risk from the insured to the insurer and reinsurer. In areas where production risk is high, farmers are required to insure their crops, so that in the event of crop failure due to occurrence of natural calamities like drought, flood, cyclone, hailstorm, hurricane, plant disease, pest attack etc., they will be adequately compensated through indemnity payment. This will provide them economic support, stabilise their farm income, induce them to invest in agriculture, reduce indebtedness and decrease the need for relief measures.

In recent times climate change and globalisation have emerged as two major threats to agriculture. Due to climate change accompanied with gradual increase in temperature, greater variability in rainfall, rise in sea level and increased frequency, intensity and duration of extreme weather events (IPCC, 2007), production risk in agriculture has increased appreciably. In the context of globalisation, with the integration of domestic market with world food market, price risk has also increased substantially. In India large scale farmers' suicide in states like Maharashtra, Karnataka, Andhra Pradesh and Odisha during the last few years speaks of the failure of public action in handling the problem of agricultural risk in the changed context of globalisation and climate change. Hence, there is a need to assess the efficacy of current crop insurance schemes as a risk management tool and to suggest new innovative insurance products to cater to the needs of the farmers in the changed environment.

In the above context, there has been an efflorescent of research studies on agricultural insurance assessing the feasibility and evaluating the performance of different crop insurance schemes. Divergent views have emerged on appropriate modality and mechanism of crop insurance with respect to risk coverage (single/multiple peril), level of application (area/group/individual), basis (yield/rainfall/weather index/revenue), mode of application (voluntary/compulsory), implementing agency (public/private/public and private) and premium setting (subsidy/no subsidy). The major problems in implementing crop insurance schemes are asymmetric information between insurer and the insured regarding the cause of crop failure which results in moral hazard and adverse selection inflicting high administrative and transaction costs. To overcome these problems, credit linked and area based index insurance schemes are usually implemented in developing countries like India with high subsidy.

Many evaluation studies of performance of crop insurance schemes around the world reveal that the schemes have become fiscally burdensome and unsustainable (Skees *et al.*, 1999). The main reasons adduced by the economists for disappointing performance of crop insurance in both developing and developed countries are that: (i) it attempts to provide multi-peril coverage for the correlated and covariate risks which in fact cannot be pooled and thus not insurable; (ii) it has caused moral hazard problem among insurers, who have relied on government bailouts with subsidised premium-setting (Yaron *et al.*, 1997); (iii) it has created moral hazard in farmers/regional level i.e., risky farmers/states/regions benefiting more from the scheme.

A number of research studies have been undertaken to identify the determinants of participation in crop insurance schemes. The factors that influence the decision to buy crop insurance product may be studied under three heads: socioeconomic characteristics, farm characteristics and risk attitude. Gine *et al.* (2008) have examined the purchase behaviour of Indian farmers of rainfall insurance and found that participation is decreasing with basis risk between insurance payouts and income fluctuations, increasing in household wealth and decreasing in the extent to which credit constraints bind. Rydant (1979) has analysed several socioeconomic and psycho-dynamic factors that might be expected to affect the decision to adopt crop-hail insurance in Illinois agriculture. He observed that the insurance adopter was a repeat buyer, with a higher income and a greater awareness of hail-hazard than a non-adopter. Sherrick *et al.* (2003) with data from Midwest in US find that demand for revenue insurance is greater by larger, younger farmers and farm in more separate location. Marcel *et al.* (2002) with empirical data from Netherlands show that the participation decision in insurance programmes is negatively and significantly associated with the producer's belief about the availability of disaster relief in the future.

Many vouch for increased government subsidy for agricultural insurance in the context of increased agricultural risk. Duncan and Myres (2000) develop an insurance model and show that catastrophic risk increases premium, reduces farmer coverage levels, and, under some conditions, lead to a complete breakdown of the crop insurance market. Reinsurance can increase participation, if it is subsidised. On the other hand, Mcleman and Smit (2006) are of the view that government subsidisation of insurance against risks associated with adverse climatic conditions and extreme weather events such as flood damage, may lead to individual decisions that actually increase the susceptibility of people, property and economic activities to those risks. With examples from New Zealand they illustrate how the removal of subsidy in crop insurance reduced the moral hazard and farmers took adaptive actions to reduce the likelihood of crop losses. Also, removal of subsidy reduced physical hazard, as farmers stopped cultivating marginal lands where production risk is more. Considerable research and location specific empirical studies are required to suggest appropriate agricultural insurance schemes in the context of increased agricultural risk keeping in view the interest of both the insurer and insured.

Objectives

In the above backdrop, this study attempts to critically examine the performance of current National Agricultural Insurance Scheme (NAIS) and Weather Based Crop Insurance Scheme (WBCIS) scheme under implementation in Odisha in Eastern India by using secondary data collected from the regional office of Agriculture Insurance Company of India Limited (AIC), Odisha, Bhubaneswar. NAIS is operative in all the 30 districts of Odisha, whereas WBCIS is under experimentation on pilot basis only in 3 backward rainfed districts.

The principal objectives of the study are

- To make a comparative assessment of performance of area based crop yield scheme (NAIS) and rainfall insurance scheme (WBCIS) under implementation in Odisha with respect to their coverage, financial performance and operational efficiency as a risk management strategy.

- To suggest measures to improve the crop insurance schemes so as to make them financially viable, administratively implementable, socially acceptable and more efficient and effective in managing agricultural risk in the context of increased vulnerability due to climate change.

Modality of Crop Insurance Schemes in Odisha

Realising the need for crop insurance for management of agricultural risk, Government of India and Odisha introduced different crop insurance schemes from time to time (AIC, 2007). Odisha is located in the eastern coast of India along the Bay of Bengal. The state has the dubious distinction of the poorest state in India. The state's economy is extremely vulnerable to natural disasters like drought, flood and cyclone because of its tropical climate, monsoon rain, long coast line, high dependence on agriculture, mass poverty, low irrigation coverage and rampant deforestation. Production risk is quite high in such fragile environment. There is a great need for crop insurance, which guarantees compensation in case of crop failure due to occurrence of natural disasters and other unforeseen contingencies.

During 1985 an area based crop yield insurance scheme known as Comprehensive Crop Insurance Scheme (CCIS) was launched in Odisha covering all the districts of Odisha. In this scheme the indemnification was on area basis and if there was a shortfall in actual average yield per hectare of the insured crop from the threshold yield, each of the insured farmers growing that crop in the defined area was eligible for indemnity. This scheme covered only the loanee farmers taking credit from institutional agencies. This scheme was vehemently criticised as a loan insurance scheme as it covered only the loanee farmers. Also, the financial performance of the scheme was very poor. To remove these shortcomings since 1999 *rabi* season the NAIS has been implemented in Odisha following the Government of India guidelines. NAIS replaced CCIS.

NAIS covered all food crops (cereals, millets and pulses), cotton, sugarcane and potato in the first year and other annual commercial/ horticultural crops in a period of three years. All loanee farmers were compulsorily covered under the scheme. The non-loanee farmers growing insurable crops could also opt for the scheme. The scheme provided comprehensive risk insurance against yield losses due to natural fire, lightening, storm, hailstorm, cyclone, typhoon, tempest, hurricane, tornado, flood, inundation and landslide, drought, dry spells, pests, diseases etc. The sum insured extends to the value of the threshold yield of the crop, with an option to cover up to 150 per cent of average yield of the crop on payment of extra premium. The premium rate for *kharif* crops *bajra* and oilseeds is 3.5 per cent of sum insured and 2.0 per cent for other food crops. In the *rabi* season, the premium rate is 1.5 per cent for wheat and 2.0 per cent for other food crops and oilseeds. Also, 50 per cent subsidy in premium is allowed to small and marginal farmers which will be shared equally by the Government of India and State Government/Union Territory. The subsidy shall be phased out in a period of five years. In Odisha during 2009, the subsidy has been reduced to only 10 per cent. The scheme operates on the basis of area approach. However, there is provision that the scheme would operate on individual basis for localised calamities such as hailstorm, landslides, cyclone and flood. NAIS was certainly an improvement over CCIS by extending insurance facility to non-loanee farmers, providing three levels of indemnity payment, setting higher premium and including horticultural and commercial crops. However, NAIS failed to achieve its objectives owing to its low coverage, poor financial performance and less effectiveness (Sinha, 2004; Vyas and Singh, 2006; Kalavakonda and Mahul, 2005).

Since 2008, an area based rainfall insurance scheme christened as WBCIS has been introduced in few drought prone blocks in Odisha on pilot basis. Currently the scheme is in operation only in eight blocks of three drought prone districts namely, Bolangir, Nuapada and Baragarh located in Western Odisha for *kharif* paddy only. Since *kharif* 2009, the scheme has been extended to both loanee and non-loanee farmers for paddy crop. In these pilot areas, for loanee farmers, WBCIS is compulsory and NAIS is not available to them. But for non-loanee farmers both NAIS and WBCIS are available. In case of WBCIS, if there is deficient or excess rainfall in the defined area, all the farmers cultivating the notified crop within the defined area will be eligible for compensation. The premium rate for *kharif* paddy is 2.5 per cent of sum insured. WBCIS is considered as an improvement over NAIS because, rainfall as an index is easy to measure objectively and transparently. As a result, the administrative cost is low and this facilitates quicker payment of indemnity to the buyers of insurance.

Coverage of Crop Insurance Scheme

To evaluate the performance of NAIS in Odisha, the trend in the coverage of the scheme has been shown in Table 17.1 and Figure 17.1. Total area under NAIS which includes both *kharif* area and *rabi* area shows an increasing trend over the period 2000-2009 excepting 2002 and 2008. The total area increased from 9.2 lakh ha in 2000 to 11.3 lakh ha in 2009. However, percentage of gross cropped area (GCA) under NAIS has marginally increased from 10.7 per cent in 2000 to 12.4 per cent in 2009 (Table 17.1). Thus, the penetration of NAIS is very low and slow, as more than 85 per cent of GCA is not yet covered under NAIS.

Table 17.1

Percentage of GCA under NAIS 2000-2009

in Ha.

Year	*GCA*	*Kharif Area under NAIS*	*Rabi Area under NAIS*	*Total Area under NAIS*	*Percentage of GCA under NAIS*
2000	8526000	751595	164703	916298	10.7
2001	7877000	625098	108810	733908	9.3
2002	8798000	1377756	174899	1552655	17.6
2003	7852000	633977	123475	757452	9.6
2004	8638000	943212	178181	1121393	13.0
2005	8718000	922854	198026	1120880	12.9
2006	8928000	890122	216780	1106902	12.4
2007	8960000	905934	199725	1105659	12.3
2008	9014000	590932	138534	729466	8.1
2009	9071000	981287	144564	1125851	12.4

Source: Agricultural Insurance Company of India Limited, Bhubaneswar Regional Office, Odisha.

To examine the adoption rate of NAIS, we have considered the adoption behaviour of non-loanee farmers only. For loanee farmers, i.e., the farmers taking loans from commercial banks, cooperatives and regional rural banks (RRBs), NAIS is compulsory, but for non-loanee farmers, it is voluntary. Therefore, a break-up analysis of area and farmers covered according to loanee and non-loanee categories has been made to examine the extent of adoption of NAIS by farmers voluntarily. In the year 2009-10 *kharif* season, only 3 per cent of farmers were non-loanees and they accounted for 5.4 per cent of total area under NAIS (Table 17.2). Thus, the adoption rate of NAIS by non-loanee farmers is abysmally low and a matter for great concern. A region-wise analysis reveals that the participation of non-loanee farmers in NAIS is higher for agriculturally backward regions such as northern plateau and central table land area, where agricultural risk is high. The percentage of area covered by non-loanee farmers is only 0.3 in the most agriculturally advanced coastal region, whereas it is as high as 12 per cent for the most agriculturally backward region of northern plateau. This indicates that in risky areas, the non-loanee farmers are coming forward to insure their crops.

Figure 17.1

Kharif and Rabi area under NAIS from 2000-2010

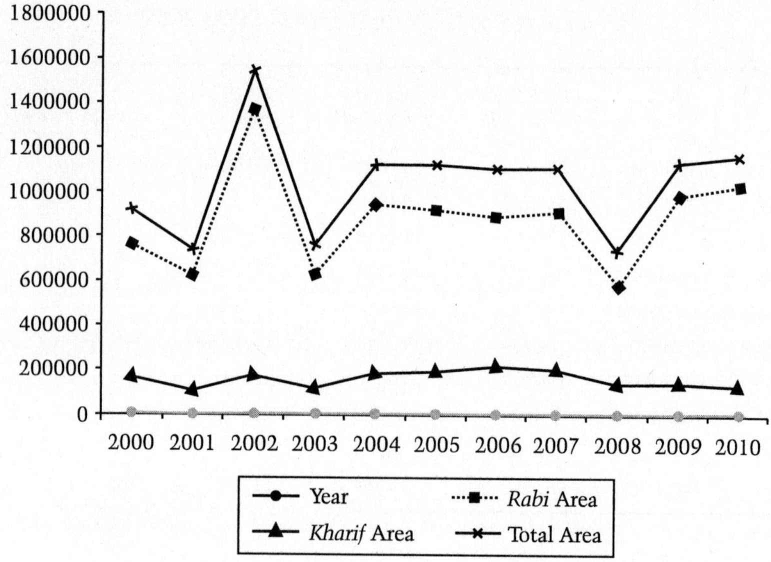

Source: Computed from data collected from Agricultural Insurance Company of India Limited, Bhubaneswar Regional Office, Odisha.

Table 17.2

Percentage Distribution of Farmers and Area Covered according to Loanee and Non-Loanee Category NAIS Kharif 2009-2010

Region	Rank in Agricultural Development	No. of Farmers			Area in Hectare		
		Loanee Farmers	Non-Loanee Farmers	Total	Loanee Farmers	Non-Loanee Farmers	Total
Northern Plateau	4	92.1	7.9	100.0	87.2	12.8	100.0
Central Table Land	2	93.5	6.5	100.0	89.5	10.5	100.0
Eastern Ghat	3	99.0	1.0	100.0	98.4	1.6	100.0
Coastal Plain	1	99.8	0.2	100.0	99.7	0.3	100.0
State Total		97.0	3.0	100.0	94.6	5.4	100.0

Source: Computed from data collected from Agricultural Insurance Company of India Limited, Bhubaneswar Regional Office, Odisha.

It is necessary to examine the trend in adoption rate of non-loanee farmers to ascertain whether over the period the coverage of the scheme is increasing and farmers are realising the importance of insurance as a risk management strategy. During the period 2000-2010 *kharif* seasons, it is observed that the percentage of non-loanee farmers availing NAIS has declined substantially from 11.9 per cent in 2000 to only 3 per cent in 2009 (Table 17.3). Likewise the percentage of area covered by non-loanee farmers has declined from 10.5 per cent in 2010 to only 5.4 per cent in 2009 (Table 17.3). Thus the insurance scheme has not received wide acceptance by the non-loanee farmers, for whom insurance is voluntary. In case of *rabi* season, the participation of non-loanee farmers is quite negligible and during 2009 it was nil (Table 17.4). Over the period 2000-2009, the percentage of non-loanee farmers and percentage of area show a decline from 0.7 per cent in 2000 to zero per cent in 2009 (Table 17.4).

Through focused group discussion (FGD) with the farmers and implementing agency personnel, it was discerned that the cooperatives, RRBs and State Bank of India (SBI) extend insurance facility to loanee farmers, for whom insurance is compulsory. But they are reluctant to provide insurance service to non-loanee farmers due to additional work burden and shortage of manpower. In spite of the provision for payment of service charges to the banks for providing crop insurance, the bankers complained that they do not get any service charges from the AIC. The farmers also expressed that the compensation should be paid on the basis of individual crop loss and not on the basis of crop loss as the block level.

Table 17.3

Percentage Distribution of Farmers and Area Covered according to Loanee and Non-Loanee category NAIS Kharif 2000-2010

Season	No. of Farmers			Area in Hectare		
	Loanee Farmers	Non-Loanee Farmers	Total	Loanee Farmers	Non-Loanee Farmers	Total
Kharif-2000	88.1	11.9	100.0	89.5	10.5	100.0
Kharif-2001	89.6	10.4	100.0	90.6	9.4	100.0
Kharif-2002	54.4	45.6	100.0	53.2	46.8	100.0
Kharif-2003	96.2	3.8	100.0	94.8	5.2	100.0
Kharif-2004	88.7	11.3	100.0	85.3	14.7	100.0
Kharif-2005	99.1	0.9	100.0	98.4	1.6	100.0
Kharif-2006	98.4	1.6	100.0	97.4	2.6	100.0
Kharif-2007	99.5	0.5	100.0	99.2	0.8	100.0
Kharif-2008	97.4	2.6	100.0	95.5	4.5	100.0
Kharif-2009	97.0	3.0	100.0	94.6	5.4	100.0
Kharif-2010	99.0	1.0	100.0	98.1	1.9	100.0

Source: Same as Table 17.1.

Table 17.4

Percentage Distribution of Farmers and Area Covered according to Loanee and Non-Loanee category NAIS Rabi 2000-2009

Season	No. of Farmers			Area in Hectare		
	Loanee Farmers	Non-Loanee Farmers	Total	Loanee Farmers	Non-Loanee Farmers	Total
Rabi-2000	99.3	0.7	100.0	99.3	0.7	100.0
Rabi-2001	99.7	0.3	100.0	99.8	0.2	100.0
Rabi-2002	99.7	0.3	100.0	99.9	0.1	100.0
Rabi-2003	100.0	0.0	100.0	100.0	0.0	100.0
Rabi-2004	100.0	0.0	100.0	100.0	0.0	100.0
Rabi-2005	100.0	0.0	100.0	100.0	0.0	100.0
Rabi-2006	100.0	0.0	100.0	100.0	0.0	100.0
Rabi-2007	99.9	0.1	100.0	100.0	0.0	100.0
Rabi-2008	99.9	0.1	100.0	99.8	0.2	100.0
Rabi-2009	100.0	0.0	100.0	100.0	0.0	100.0

Source: Same as Table 17.1.

An analysis of participation of farmers in NAIS according to farmer category reveals that majority of buyers of crop insurance (more than 80%) belong to small and marginal farmer category. However, in case of non-loanee farmers, the percentage of medium and large farmers buying crop insurance to total farmers is greater than that of loanee farmers. It indicates that the large farmers are aware of the benefits of insurance and therefore voluntarily insure their crop.

Benefits from Crop Insurance

During 2000-2009 *kharif* seasons, it is observed that the percentage of farmers who benefitted from crop insurance in the way of receipt of compensation varies substantially ranging from 1.7 per cent in 2001 to 69.7 per cent in 2002 (Table 17.5). During abnormal years of flood and drought, the number of farmers who benefitted is very large. During *rabi* seasons, the percentage of farmers who benefitted ranges from zero per cent in the year 2000 to 27.4 per cent in 2009 (Table 17.5).

Table 17.5

Percentage of Farmers Benefitted under NAIS Kharif and Rabi from 2000-2009

Year	*Kharif*			*Rabi*		
	Farmers Covered	*Farmers Benefitted*	*Percentage of Farmers Benefitted*	*Farmers Covered*	*Farmers Benefitted*	*Percentage of Farmers Benefitted*
2000	681010	349406	51.3	232836	15	0.0
2001	627568	10854	1.7	123964	25759	20.8
2002	1204849	839345	69.7	212162	18541	8.7
2003	638303	38188	6.0	142871	16806	11.8
2004	872551	45657	5.2	202699	1335	0.7
2005	900022	19352	2.2	210853	7451	3.5
2006	880330	68532	7.8	230039	7039	3.1
2007	840727	65885	7.8	199886	19541	9.8
2008	611477	55928	9.1	132418	964	0.7
2009	1068687	99184	9.3	161720	44323	27.4

Source: Same as Table 17.1.

A region-wise analysis of percentage of farmers who benefitted from NAIS during 2009 *kharif* season in case of loanee farmers reveals that it is the highest i.e., 31.3 per cent for the agriculturally most backward region of Northern Plateau and the least i.e., 0.07 per cent for the agriculturally advanced coastal region (Table 17.6). Similar finding is obtained in case of non-loanee farmers. However, in case of non-loanee farmers the percentage of farmers who benefitted is higher than that of loanee farmers for the state as a whole and also in northern plateau and central table land area. This shows that more of risky farmers voluntarily come forward to insure their crops and thereby get compensation in the event of crop failure.

Table 17.6

Farmers Who Benefitted under NAIS: Region-wise Kharif, 2009-10

Region	*Loanee*			*Non-Loanee*		
	No. of Loanee Farmers	*No.of Farmers Benefitted*	*% of Total*	*No. of Non-Loanee Farmers*	*No.of Farmers Benefitted*	*% of Total*
Northern Plateau	156252	48903	31.3	13408	10555	78.72
Central Table Land	245639	29196	11.89	16951	3309	19.52
Eastern Ghat	139211	6812	4.89	1391	35	2.51
Coastal Plain	494997	374	0.07	838	0	0
State Total	1036099	85285	8.23	32588	13899	42.65

Source: Agriculture Insurance Company of India Limited, Bhubaneswar Regional Office, Odisha.

Financial Performance of NAIS

To assess the performance of NAIS in Odisha, the claim-premium ratio was computed by dividing the indemnity claim or compensation payment by insurance premium collected. If the claim-premium ratio exceeds one, it indicates financial loss on the part of the insurer in insurance business. During the 2000-2009 *kharif* seasons, it is observed that the claim-premium ratio is unfavourable i.e., greater than 1 for six years and favourable for only four years (Table 17.7). The claim-premium ratio ranges from 0.15 in 2005 *kharif* season to 9.4 in 2000 *kharif* season. But for 2000-2009 *rabi* seasons, claim premium is favourable for nine years and unfavourable only for one year.

A region-wise analysis of claim-premium ratio unfolds that it is higher in case of agriculturally backward region and for non-loanee farmers

(Table 17.8). There is negative association between extent of agricultural development of the region and claim-premium ratio.

Table 17.7

Claim-Premium Ratio under NAIS Kharif and Rabi from 2000-2009

Year	*Kharif*			*Rabi*		
	Premium (₹)	*Claims (₹)*	*Claim/ Premium*	*Premium (₹)*	*Claims (₹)*	*Claim/ Premium*
2000	112392562	1054716278	9.38	22844642	16556	0.01
2001	102218804	23401452	0.23	18561159	14494434	0.78
2002	298560969	2440291965	8.17	33184437	10819387	0.33
2003	138160518	181844628	1.32	25437944	11741081	0.46
2004	254954481	146885827	0.58	33678681	981629	0.03
2005	243334143	37382700	0.15	49858874	3593957	0.07
2006	275039791	274771560	1.00	57848102	22079443	0.38
2007	282415471	240172446	0.85	52583352	41629009	0.79
2008	217694400	303489566	1.39	38119451	1703855	0.04
2009	396960881	469004449	1.18	49783305	83697450	1.68

Source: Same as Table 17.1.

Table 17.8

Claim Premium Ratio under NAIS: Region-wise Kharif, 2009-10

Region	*Loanee*			*Non-Loanee*		
	Premium (₹)	*Claims (₹)*	*Claim-Premium Ratio*	*Premium (₹)*	*Claims (₹)*	*Claim-Premium Ratio*
Northern Plateau	49925083	226124293	4.53	4073856	47700491	11.71
Central Table Land	105381168	159329369	1.51	13085837	23156777	1.77
Eastern Ghat	55077045	12010752	0.21	727688	1180	0.002
Coastal Plain	168389711	681586	0.004	28228034	46315914	1.04
State Total	657695848	344043414	0.52	28228034	46315914	1.64

Source: Same as Table 17.2.

Crops Covered

Crops covered under NAIS during *kharif* and *rabi* 2009 are shown in Table 17.9. Out of total insured area of 9.8 lakh ha, paddy coverage is as high as 9.5 lakh ha during *kharif* season. Other crops covered are maize (15276 ha), cotton (8457 ha), ginger (1039 ha), turmeric (767ha) and groundnut (140 ha). During *rabi* season also paddy is the most important insured crop followed by groundnut, potato and mustard. WBCIS covers only paddy during *kharif* season and does not cover any crop during *rabi* season.

Table 17.9

Crop Wise Business Coverage/Claims Statistics: 2009-10

Name of the Crops	*No. of Farmers Covered*	*Area in Hect.*	*Sum Insured (₹)*	*Premium (₹)*	*Claims (₹)*	*Farmers Benefitted*
			Season: Kharif, 2009 (NAIS)			
Paddy	1045969	955607	15378583880	386168749	467248604	98634
Groundnut	80	140	724024	9050		
Maize	14301	15276	243949867	6098747	1755845	550
Cotton	6335	8457	106105540	4032011		
Turmeric	910	767	9079370	272381		
Ginger	1092	1039	15197720	379943		
Total	1068687	981286	15753640401	396960881	469004449	99184
			Weather Based Crop Insurance Scheme (WBCIS)			
Paddy	81429	113266	1359188376	135918838	75014420	54811
			Season: Rabi, 2009-10 (NAIS)			
Paddy	112593	117862	2240829329	40334928	59265615	18131
Groundnut	13611	9149	190698630	3813973	5722330	3025
Mustard	1	1	2000	40		
Potato	8467	4490	156933915	2354009		
Total	134672	131502	2588463874	46502950	64987945	21156

Source: Agricultural Insurance Company of India Ltd, R.O.Bhubaneswar.

Comparison of NAIS and WBCIS

WBCIS has been implemented in Odisha on pilot basis since 2008 and covers only *kharif* paddy crop. A comparison of performance of NAIS and WBCIS reveals that percentage of non-loanee farmers and area covered by non-loanee farmers are higher in case of WBCIS, which implies that WBCIS is more popular than NAIS (Table 17.10). The percentage of farmers benefitted is higher in case of WBCIS than that of NAIS (Table 17.11). The claim-premium ratio in case of NAIS is higher than that of WBCIS and unfavourable exceeding one (Table 17.12). But in case of WBCIS the claim premium ratio is less than one indicating its financial viability. In case of WBCIS sum assured per farmer and per ha are smaller than that of NAIS (Table 17.13).

Table 17.10

Distribution of Farmers and Area Covered according to Loanee and Non-Loanee category from 2008-2010

Season	No. of Loanee Farmers	% of Total	No. of non-Loanee Farmers	% of Total	Total	Area of Loanee Farmers (ha)	% of Total	Area of Non-Loanee Farmers (ha)	% of Total	Total
					NAIS					
Kharif 2008	595858	97.4	15619	2.6	611477	564402	95.5	26530	4.5	590932
Kharif 2009	1036099	97.0	32588	3.0	1068687	928301	94.6	52986	5.4	981287
Kharif 2010	1096604	99.0	11106	1.0	1107710	1011782	98.1	19403	1.9	1031185
					WBCIS					
Kharif 2008	0	0.0	13289	100.0	13289	0	0.0	22278	100.0	22278
Kharif 2009	74283	91.2	7146	8.8	81429	97332	85.9	15934	14.1	113266
Kharif 2010	72557	97.1	2177	2.9	74734	95877	94.3	5841	5.7	101718

Source: Same as Table 17.9.

Table 17.11

Farmers Benefitted under NAIS/WBCIS Kharif, 2008 and 2009

Season	Farmers Covered	Farmers Benefitted	Percentage of Farmers Benefitted
		NAIS	
Kharif 2008	611477	55928	9.1
Kharif 2009	1068687	99184	9.3
		WBCIS	
Kharif 2008	13289	13289	100.0
Kharif 2009	81429	54811	67.3

Source: Same as Table 17.9.

Table 17.12

Claim-Premium Ratio of NAIS/WBCIS Kharif, 2008 and 2009

Season	Premium (₹)	Claims (₹)	Claim/Premium
		NAIS	
Kharif 2008	217694400	303489566	1.39
Kharif 2009	396960881	469004449	1.18
		WBCIS	
Kharif 2008	44556294	41480741	0.9
Kharif 2009	135918838	75014420	0.6

Source: Same as Table 17.9.

Table 17.13

Sum Assured/Farmer and Sum Assured/Hectare
NAIS /WBCIS Kharif 2009

in ₹

Year	*Sum Assured/Farmer*	*Sum Assured/ha*
NAIS 2009	16693	16227
WBCIS 2009	14848	12000

Source: Same as Table 17.2.

Several rounds of FGD with insurance users and implementing agencies of NAIS and WBCIS in drought prone Bolangir and Kalahandi district in western Odisha unfolded many interesting facts on the performance of NAIS and WBCIS, which are as follows:

- In case of NAIS, indemnity is higher whereas in case of WBCIS, indemnity is very low.
- Since last eight years, most of the farmers in the region did not get any indemnity under NAIS, whereas most of the farmers under WBCIS, received compensation twice since 2009.
- Bank officials promote NAIS but they do not show much interest for WBCIS, as they are not paid any service charge for implementing the scheme.
- Farmers grumbled that under NAIS, there is a great deal of discrepancy between the crop cutting experiment report and the actual crop loss.
- In case of WBCIS, the compensation is being paid within 45 days of obtaining rainfall data. In case of NAIS, there is unusual delay in payment of compensation, even more than six months elapse after the crop loss to get compensation.
- NAIS covers all types of crop loss. But WBCIS covers crop loss due to rainfall variations.
- The cooperatives, RRBs and SBI extend insurance facility to loanee farmers, for whom insurance is compulsory. But they are reluctant to provide insurance service to non-loanee farmers due to additional burden and shortage of manpower. In spite of there

being a provision for service charges, banks do not get any service charges.

- For loanees, the last date for applying for WBCIS is 31st September while it is 31st July for non-loanees. A short period of time is available to non-loanees for applying for crop insurance. They can apply for NAIS from 15th July to 31st July only. It becomes very difficult for farmers to apply. They have to arrange the necessary documents on time. The village agricultural worker (VAW) has to recommend their name and a sowing certificate has to be issued by the agricultural officer.
- The amount of indemnity is adjusted towards the loan taken through transfer voucher about which the farmer is usually unaware of.
- There is no awareness campaign on insurance schemes.
- In case of normal rainfall, farmers do not go for crop insurance. Usually they wait till there is a shortfall in rain. Only after this, they become interested to insure their crops. But by then, the deadline to apply for crop insurance for non-loanees is over. So they are unable to avail crop insurance.
- Due to Indravati Project, there is frequent flood in Tel and Haathi Rivers. Almost 50 acres of land is now under 6-10 ft of sand and unfit for cultivation. No insurance is provided against such risks.
- Due to heavy rain during harvest time or post harvest period, harvested paddy gets destroyed. Such type of loss is not covered under NAIS/WBCIS.

Conclusion and Suggestions

From the secondary data analysis and interaction with the farmers, it is inferred that the performance of NAIS in Odisha is not satisfactory due to low coverage, poor financial performance and delay in compensation payment. There is a need to create awareness about the benefits of insurance schemes. A comparison of NAIS and WBCIS reveals that the operational modality of WBCIS is better understood by the farmers and it is a more popular scheme. The indemnity is paid quickly. It is more transparent and there is less scope for manipulation as rainfall is measured by rain gauge stations in each block.

There is a need to take Gram Panchayat instead of Block as the reference unit both in case of NAIS and WBCIS. The procedure for applying for crop insurance should be made simple for non-loanee farmers, for whom insurance is voluntary. Insurance agents should be appointed like Life Insurance Companies to provide insurance service at the door step of the farmers, thereby saving the precious time of farmers which can be utilised for land preparation and sowing. Quick payment of compensation should be made to stabilise farm income.

As both NAIS and WBCIS are area based agricultural insurance schemes, they do not cover independent idiosyncratic individual risk. In a frequently disaster affected state like Odisha, where reasons for crop failure are many, there is need for multi-peril crop insurance scheme like NAIS. The public sector may address catastrophic risk and provide multi-peril insurance where subsidy requirement is high, but allow private sector to provide insurance products for less severe events and for individual independent idiosyncratic and localised risk. A conducive atmosphere for promotion of private sector participation in agricultural insurance needs to be created. The government can encourage private participation by providing subsidy, guarantee and reinsurance facility. For building capacity of the private sector, technical assistance needs to be provided. Appropriate regulatory framework may be developed to monitor private participation in agricultural insurance.

Varied insurance products like crop yield insurance, weather index insurance, single peril insurance and revenue insurance should be offered to meet the needs of different farmers in different areas, setting premium rates commensurate with risk, and the judicious use of premium subsidies. Private participation should be encouraged to provide varied products with actuarial premium rates for commercial farmers on individual basis. At the same time, micro-insurance i.e., insurance for the poor should be encouraged through participation of banks, non-governmental organisations and microfinance institutions. Self help groups (SHGs) and micro-finance institutions (MFIs) can play an important role in educating people about the need for and benefits of agricultural insurance. It will be wise to integrate micro-insurance with microfinance to manage the risk of default and enterprise loss in a more proactive manner. Farmers should be encouraged to insure their crop, as insurance acts as a guarantee for investment and can serve as collateral for obtaining credit from institutional sources.

References

Agricultural Insurance Company of India Limited (AIC). (2007). *Crop Insurance in India*.

Chambers, Robert G. and John Quiggin (2002). "Optimal Producer Behavior in the Presence of Area-Yield Crop Insurance", *American Journal of Agricultural Economics* 84(2): 320-34.

Coble Keith H., Thomas O. Knight, George F. Patrick and Alan E. Baquet (n.d.). "Understanding the Economic Factors Influencing Farm Policy Preferences", *Review of Agricultural Economics* 24(2): 309-21.

Duncan, John and Robert J. Myers (2000). "Crop Insurance under Catastrophic Risk", *American Journal of Agricultural Economics* 82(4): 842-55.

Gine Xavier, Roberts Townsend and James Vickery (2008). "Patterns of Rainfall Insurance Participation in Rural India", *The World Bank Economic Review* 1-28, October 22.

Goodwin, Barry K. (2001). "Problems with Market Insurance in Agriculture", *American Journal of Agricultural Economics* 83(3): 643-49.

Hazell, P.B.R., C. Pomareda and A. Valdes (1986). *Crop Insurance for Agricultural Development: Issues and Experience*. Baltimore: The John Hopkins University Press.

Hess, Ulrich (2003). "Innovative Financial Services for Rural India: Monsoon-Indexed Lending and Insurance for Smallholders", *Agricultural & Rural Development Working Paper 9*, World bank.

IPCC (2007). *Climate Change, Impacts, Adaptation and Vulnerability*. Contribution of Working Group III to the Fourth Assessment Report of the Intergovernmental Panel on Climate Change. Cambridge University Press.

Kalavakonda Vijay and Olivier Mahul (2005). "Crop Insurance in Karnataka", *World Bank Policy Research Working Paper* 3654.

Marcel A.P.M., Van Asseldonk, Miranda P.M. Meuwissen and Ruud B.M. Huirne (2002). "Belief in Disaster Relief and the Demand for a Public-Private Insurance Program", *Review of Agricultural Economics* 24(1): 196-207.

Mcleman, Robert and Barry Smit (2006). "Vulnerability to Climate Chnage Hazards and Risks: Crop and Flood Insurance", *The Canadian Geographer* 50(2): 217-26.

Miranda, Mario and Dmitry V. Vedenov. (2001). "Innovations in Agricultural and Natural Disaster Insurance", *American Journal of Agricultural Economics* 83(3): 650-55.

Raju, S.S. and Ramesh Chand (2008). "Agricultural Insurance in India: Problems and Prospects", *National Centre for Agricultural Economics and Policy Research Working Paper* No.8.

Rydant, A.L. (1979). "Adjustments to Natural Hazards: Factors Affecting the Adoption of Crop-Hail Insurance", *Professional Geographer* 31(3): 312-20.

Sherrick, Bruce J., Peter J. Barry, Gary D. Schnitkey, Paul N. Ellinger and Brian Wansink (2003). "Farmers' Preferences for Crop Insurance Attributes", *Review of Agricultural Economics* 25: 415-29.

Sinha, Sidharth (2004). "Agriculture Insurance in India: Scope for Participation of Private Insurers", *Economic and Political Weekly* 39(25): 2605-11.

Skees Jerry, P. Hazell and M. Miranda (1999). "New Approaches to Crop Yield Insurance in Developing Countries", *Environment and Production Technology Division Discussion Paper* No. 55. Washington, D.C.: International Food Policy Research Institute.

Smit, B. and O. Pilisova (2003). "From Adaptation to Adaptive Capacity and Vulnerability Reduction" in J.B. Smith, R.J.T. Klein and S. Huq (eds.), *Climate Change, Adaptive Capacity and Development*. London Imperial College Press.

Swain, Mamata (2008). "Agricultural Risk and Crop Insurance in a Globalised Economy", *Man & Development* 30(3): 107-28.

Swain, Mamata (2009). "Impact of Climate Change on Agriculture in Orissa: Adaptation and Mitigation", in R.K. Panda (ed.), *Orissa's Economic Development: Experiences and Expectations*. New Delhi: Serials Publisher.

Yaron Jacob, McDonald P. Nenjamin and Jr. Gerda L. Piprek (1997). *Rural Finance: Issues, Design, and Best Practices*. Washington, D.C.: The World Bank, Enviromentally and Socially Sustainable Development.

Vyas, V.S. and S. Singh (2006). *Crop Insurance in India: Scope for Improvement* 41(43&44): 4585-94.

This work has been undertaken with the financial support of the South Asian Network for Development and Environmental Economics (SANDEE) and its sponsors. Technical support and guidance has been provided by several SANDEE advisors and peers during the course of this research.

18 Neo-liberal Transformation of the Interventionist Economy

A Study of West Bengal

Sudipta Bhattacharyya

Objective and Problem

Based on primary field surveys conducted in 1993-94 and 2004-05 on same set of households, this study would like to analyse the changes that took place during the decade of neo-liberal reform. The prime research question is that whether the interventionist economy developed during the first 10-15 years of Left Front rule in West Bengal had eroded owing to economic reform and what was the form of peasant class differentiation. In other words whether the inequality prevailed in land and non-land assets, irrigation, credit, output and product marketed had expanded. Further whether the participation rate of family members increased and whether the Chayanovian notion that the ascending class/size status is related to higher participation rate, and therefore, lower well being can be accepted or rejected. Finally what happened to the interregional inequality—whether the same had increased or decreased? Finally, whether neo-liberal reform had affected only a class of farmers or peasantry as a whole?

Introduction

In spite of having a successful experiment of a limited land reform and pro-poor state intervention, the agrarian economy of West Bengal had been class differentiated even during the peak period of state intervention in eighties. The neo-liberal reform in 1990s has made this differentiation more intensified. A consequence has been a greater ownership concentration in the upper echelons of the rural class structure, combined with an equally pronounced tendency towards proletarianisation and a loss of ownership of the means of production affecting those at the lower ends of the same hierarchy. It is important to note, however, that West Bengal constitutes an unusual case, in that class differentiation has been accompanied by pro-poor state intervention particularly during 1980s. By moderating the effects of the market, the limited agrarian reform in West Bengal has had a positive impact on the economic condition of the rural poor. Simply put, pro-poor intervention has helped to stabilise petty commodity production by smallholding peasants (Sen and Sengupta, 1995). It has also helped to stop the market-led process of immiseration, and has made differentiation itself more broad-based—giving an upward impetus to more people (Ibid). But even during 1980s, these positive changes had not been sufficiently strong to stop the process of differentiation altogether, or to alter the concentration of means of production in the hands of a few. Since the early 1990s, along with other Indian states West Bengal too was exposed to market reforms that dampened the positive impact of interventionist regime to a great extent. Since the early 1990s agricultural production, yield, consumption of fertiliser, proportion of high yielding varieties (HYV) area, cropping intensity and institutional credit experienced a negative trend break (Bhattacharyya and Bhattacharyya, 2007). As the neo-liberal reforms involved withdrawal of state from pro-poor interventionist policies one can expect, therefore, to encounter a familiar and consistent pattern of maldistribution, whereby economic indicators (assets, irrigation, output, marketed surplus, etc.) are likely to be biased towards the upper echelons in the agrarian class structure.

This has been reflected in Table 18.1 which shows the value of Gini coefiicient measuring inequality over time. The increased inequality in the non-land asset market in West Bengal during 1980s was interconnected with declined inequality in the land market following land reform. A bulk of the erstwhile *zamindars* invested in the market of non-land assets. An inequality developed as a result in the non-land asset market. At the India as whole, almost similar tendencies were found except the fact that the concentration for owned land declined marginally from 1992 to 2002-03. Another scenario at the all-India level is the growing concentration of the borrowing during the last decade from 0.392 to 0.474 (we don't have data for West Bengal), which may be a reflection of increased unequal access to credit during the neo-liberal era.

Table 18.1

*Value of Gini Coefficients for Variables Representing Productive Capacity: West Bengal and India**

Variables	*1972*		*1982*		*1992*		*2002-03*	
	West Bengal	*India*	*West Bengal*	*India*	*West Bengal*	*India*	*West Bengal*	*India*
Asset (Non-land)	0.660**	0.585	0.547	0.585	0.570	0.621	0.438	0.561
Owned Land	0.572	0.665	0.593	0.665	0.503	0.652	0.517	0.636
Operated Land	0.608	0.671	0.501	0.588	0.434	0.588	0.525	0.694
Irrigation	0.452	0.501	0.469	0.508	0.472	0.543	0.540	0.659
Outstanding Credit	NA	0.387	0.361	0.406	0.403	0.398	0.377	0.381
Borrowing	NA	NA	0.353	0.396	0.352	0.392	NA	0.474

Note: *Cultivator Households where otherwise not mentioned.

**Value for All Rural Households.

NA = Not Available.

Methodology

The analysis which follows is based on a field survey conducted among 210 sample households in two blocks of Bankura district (one is advanced, Kotulpur and another backward Hirbandh) over 1993-94 and 2004-05. As a matter of fact we had initially surveyed these households in 1993-94. In 2004-05, we have re-surveyed same households once more. The 210 sample households (of which 100 households belong to the backward region) were classified on the basis of land owned/controlled and of Patnaik's (1987) E-criterion (see Appendix A-18.1). The latter is particularly relevant to an understanding of changes occurring in West Bengal agriculture, where there is a tremendous concentration of households below 2.5 acres. If we consider NSSO classification of acreage groups we find an increase in the concentration of households from 138 to 153 out of 210 total households (66% and 73% of total respectively) during 1993-94 to 2004-05 in the 'marginal holding' or 0.01-2.5 acres. However, as we will see later the households belonging to this category are internally differentiated among different economic classes. Patnaik's E-criteria would like to explore this dynamics.

Household and Class

The distribution of sample households in the combined region is presented in Table 18.2. As expected, while a close link exists between farm size and labour use, this is not absolute. The number of landless households is relatively small: only 5 per cent of the sample was wholly landless in 1993-94, neither owning nor operating any holdings, which appears to reflect the low incidence of immiseration in the villages concerned. However in 2004-05, the total number of landless households increased from 10 to 13. The swelling of the number of the landless households are owing to the many factors. The erstwhile 5 landless households in 1993-94 remained landless in 2004-05 as well and 6 households have added as new. Among the 6 new landless households, 3 households sold their land in distress and shifted to low paid non-farm activities, but they earn a major income from selling their labour as agricultural labour, particularly in the busy agricultural season. The non-agricultural works are not at all remunerative (Ghosh, 2006). In fact the wage rate for such non-agricultural work is less than the agricultural work in busy season. Such a conversion from agriculture to non-agriculture is primarily due to non-availability of agricultural job apart from busy season. The increase in non-agricultural job is therefore nothing but a situation of disguised proletarianisation (Murray, 2006; Kay, 2008). Two households mortgaged their land and become landless at least temporarily. One household lost its land in family feud. In general it might be presumed that in the post-liberalisation period these households lost their land due to deterioration in the living conditions and non-availability of the formal credit, for which they are compelled to mortgage or sell their land, which in effect reflects the process of the intense differentiation of the peasantry.

Table 18.2

Cross Classification of the Number of Households

	Land-less		*Poor Peasant*		*Small Peasant*		*Middle Peasant*		*Rich Peasant*		*Land-lord*		*All*	
	Pd1	*Pd2*	*Pd1*	*Pd2*	*Pd1*	*Pd2*	*Pd1*	*Pd2*	*Pd1*	*Pd2*	*Pd1*	*Pd2*	*Pd1*	*Pd2*
Landless	10	13	0	0	0	0	0	0	0	0	0	0	10	13
0.01-2.5 acres	0	0	71	69	27	27	13	14	12	16	15	27	138	153
2.5-5 acres	0	0	1	1	11	5	6	3	10	12	8	8	36	29
5-10 acres	0	0	1	0	4	2	5	2	2	5	9	5	21	14
10-15 acres	0	0	0	0	0	0	0	0	1	0	2	1	3	1
15 acres and above	0	0	0	0	0	0	0	0	2	0	0	0	2	0
All	10	13	73	70	42	34	24	19	27	33	34	41	210	210

Secondly, total households belonging to the size group 0.01-2.5 acre increased from 138 to 153. The backward region was dominated previously by landlord class who occupied a very large size landed property. However, over the decade households belonging to the landlord class had lost its character and adopted the same pattern of labour process, i.e., they cultivate purely on the basis of hired-in labour. At the same time, their land size declined drastically primarily because of government measure of land reform and natural inheritance rule. There are also changing composition of the households belonging to this size group. First of all, the proportion of small peasant belonging to this size group declined. The proportion of the middle peasant either increased marginally. The self-employed classes as a whole witnessed almost a stagnation for this size group. On the other hand, poor peasant within this size group increased from 27 to 34 in the backward region and declined from 44 to 35 in the advanced region. The exploiting classes (particularly landlords) actually swelled from 27 to 43. The increasing concentration of landlord households in this size group reveals the fact that the erstwhile owners of big sizes of land is losing their land either following the inheritance rule or the legal implementation of the land reform, but they maintain the same labour process by alienating themselves from all kinds of direct production on tiny plots of land. In other words, they remain the exploiting classes—though their ownership of land holding declined.

Household Structure and Workforce

Chayanov (1966) argued that the higher the consumer/worker ratio, the higher will be the number of days of work performed by family workers. Our findings, which employ a household, members-to-worker ratio that is the same as Chayanov's concept of a consumer/worker ratio, indicate a broad relation between the latter, farm size and economic class. However, the relationship that Chayanov identified—namely, rising consumer/worker ratios accompanied by a higher family labour input—was not found. We have not found Chayanov proposed inverse relationship between farm size and economic well being (Table 18.3). As in the farm management studies (FMS), a clear direct relationship (as opposed to Chayanovian inverse one) connects farm size and family size, and—more broadly—the latter and economic status. The work participation rate (WPR) is substantially higher in the backward region when compared to the advanced region, a not unexpected finding since rural families in the former context are likely to have a lower income per worker, and more family members must work to obtain a minimum family income. This is confirmed when the members to workers ratio is linked to economic class and land size. As the economic position of the household improves, therefore, it becomes possible even for adult members of working age to withdraw from work, a factor reflected in lower participation rates. By farm size, too, a higher than average participation rate occurs for holdings below 5 acres, and a lower than average rate on holdings above 5 acres. The only deviation from this pattern is the case of landless households in the backward region in 1993-94, which have a below average work-participation ratio, though in 2004-05 it was more than doubled; magnitude for which was far above the average.

Table 18.3 has also summarised the family labour, hired in labour and hired out labour. It is observed that landlords hire in all the labour they use per farm, not working a single day in the fields. The rich peasant on the other hand rely on hiring in for 77 per cent (1993-94)/68 per cent (2004-05) of their labour input with the remaining share contributed by family labour. The middle peasants hire in 30.25 per cent (1993-94)/38.60 per cent (2004-05) and the small peasant hire in a meagre proportion 5 per cent-6 per cent in both years. Poor peasants hire out 84 per cent to 82 per cent of total labour days and only 16 per cent to 18 per cent of total labour days they work on their own farm. These households are better thought as agricultural labourer with some land. We have seen in Table 18.2 that these households belong largely to the lowest landed size group (0.01-2.5 acres). In Chayanov-Sen (Chayanov, 1966; Sen, 1966) formulation, however, these farms are projected as 'family labour farm' as if these farms survive on 'self-exploitation' and devoid from any kind of off farm activities. The acreage grouping also shows a broad direct association between the size of farms and percentage of total labour hired in and a broad inverse relation between size of farm and the proportion of family labour days hired out on wage.

Table 18.3

Class and Size-Group-wise Participation Rate, Hired-in and Hired-out

	Participation Rate=(Worker/Member)×100		*% of HI to Total*		*% of HO to Total*	
	Pd1	*Pd2*	*Pd1*	*Pd2*	*Pd1*	*Pd2*
Landless	43.9	49.12	0.00	0.00	100.00	100.00
Poor Peasant	50.93	51.36	4.66	4.84	83.71	81.71
Small Peasant	55.21	49.12	5.27	6.13	21.77	23.07
Middle Peasant	42.35	44.78	30.25	38.60	1.40	10.23
Rich Peasant	25.78	39.83	77.01	68.07	0.00	1.08
Landlord	21.91	37.98	100.00	99.81	0.00	0.00
Landless	43.9	49.12	0.00	0.00	100.00	100.00
0.01-2.5 acres	44.76	46.23	41.94	49.21	70.64	67.00
2.5-5 acres	44.76	43.23	61.49	67.79	18.68	14.66
5-10 acres	33.64	43.70	78.01	70.36	16.12	5.05
10-15 acres	13.51	31.82	96.99	100.00	0.00	0.00
15 acres and above	26.83	0.00	63.89	0.00	0.00	0.00
Total	41.58	45.35	63.81	62.14	57.49	55.06

Land and Other Means of Production

Data on land tenure—owned/operated, as well as that leased-in and leased-out—by economic class and farm size are presented in Table 18.4 and 18.5. Nearly three quarters of all rural households in the two regions compared own/operate holdings under 2.5 acres, which account for only 28 per cent of total land; by contrast, only 2 per cent of households own/operate land of 10 acres and above, holdings which account for some 17 per cent of total land. This element of inequality is more pronounced in the advanced region, where labour-purchasing households comprise 39 per cent of households owning/controlling 82 per cent of the total area.

Table 18.4

Class and Size-Group-wise Participation Rate, Hired-in and Hired-out

	All		*Advanced*		*Backward*	
	Pd1	*Pd2*	*Pd1*	*Pd2*	*Pd1*	*Pd2*
1. Per-holding owned area	2.86	2.01	1.88	1.57	3.94	2.48
2. Per-holding operated area	2.50	1.83	1.81	1.45	3.25	2.25
3. % of leased in to operated area	6.57	4.23	12.53	6.90	3.19	2.54
4. Per-holding irrigated	1.29	0.56	1.13	0.88	1.47	0.20
5. % of irrigated to total operated area	51.47	31.53	65.01	67.17	43.79	8.95

Table 18.5

Economic Class-wise Percentage Distribution of Owned, Operated and Irrigated Land

	Households		*Owned Land*		*Operated Land*		*Irrigated Land*	
	Pd1	*Pd2*	*Pd1*	*Pd2*	*Pd1*	*Pd2*	*Pd1*	*Pd2*
All								
Exploited	39.52	39.52	11.33	11.25	11.14	12.94	9.13	7.26
Self-employed	31.43	25.24	36.52	28.36	35.93	28.86	25.98	21.81
Exploiter	29.05	35.23	52.15	60.39	52.93	58.19	64.88	70.93
Adv								
Exploited	44.55	41.82	8.89	11.25	10.05	7.05	9.40	6.74
Self-employed	31.43	25.24	9.59	28.36	11.11	19	14.07	20.07
Exploiter	29.05	35.23	81.52	60.39	78.84	73.96	76.53	73.20
Back								
Exploited	34	37	12.60	15.5	11.77	16.68	8.91	9.71
Self-employed	48	35	50.65	37.64	50	35.11	36.09	30.11
Exploiter	18	28	36.75	46.86	38.23	48.21	55.07	60.17

There was a decline in per holding owned (from 2.86 acres to 2.01 acres) and operated area (from 2.5 acres to 1.83 acres) over decade of neo-liberal reform. The decline was sharp in the backward region and also prominent in the advanced region. There was a decline of land holding in all classes but the same was sharp for the higher classes as revealed by the lower value of Gini coefficient for ownership and operated area. Taking both regions together, the extent of leasing-in and leasing-out is small: only seven to four per cent of operated area is leased-in and five to seven per cent of owned area is leased-out. The single digit low percentage figure for leasing-in/out have suggested that share-tenancy is on the decline following the impementation of Operation Barga in West Bengal (Dasgupta, 1984). In the wake of Operation Barga, no new tenancy contracts have been signed and after the liberalisation the incidence of tenancy declined.

A significant tendency to be noted is that percentage of irrigated to total operated area declined massively from around 44 per cent to 9 per cent in the backward region during two periods when neo-liberal policy of state withdrawal happened. Because of the state withdrawal-public investment in irrigation was neglected and at the same time due to the withdrawal of fertiliser and credit subsidies there was a decline in private investment in irrigation. It is a well-known fact that public and private investment are complimentary to each other particularly in agriculture. The agricultural extension system had also come to a collapse during neo-liberal regime. The outcome is nothing but the rise and growth of 'ultra Left terrorism' in backward region during the same period.

Non-Land Assets, Capital Investment and Credit

The data presented in Tables 18.6 indicate that, across both regions, about two-third (1993-94) and over 60 per cent (2004-05) of total non-land assets and owned/controlled by rich peasants and landlords. Categorised in terms of farm size, however, no consistent pattern of non-land assets emerges. Thus ownership/control of machinery and irrigation is concentrated in farms belonging to the 5-10 acre category, which account for 35 per cent (1993-94) and 46 per cent (2004-05) of these particular means of production; by contrast, holdings below 5 acres and above 10 acres account for a much smaller amount of such productive assets. Much the same is true of livestock, ownership of which is concentrated both among the landless, marginal (0.01-2.5 acres) and also among holdings in excess of 15 acres. The reason why farm size cannot explain the way means of production are distributed is that the majority among the total holdings in the 5-10 acre category, of which two (1993-94) and five (2004-05) holdings belong to rich peasants and nine (1993-94) and five (2004-05) to landlords, are cultivated intensively (see Table 18.2). No other farm size group contains such a high proportion of capitalist farms purchasing labour-power. One of the big impacts of neo-liberal reform was to remove fertiliser subsidies and as a result we have witnessed unprecedented increase in fertiliser price. At the same time there was increased prevalence of monetisation and market dependence. As a result purchased input as a percentage of total cost increased from 26 per cent to 72 per cent for all classes during the decades of neo-liberal reform. The increase was phenomenal in the backward region from 8.84 per cent to 70 per cent in comparison to the advanced (from 33% to 73%), which marks the increased dependence on market in the backward region.

The above data must be linked in turn to the availability and source of agricultural credit as shown in Table 18.6. That non-institutional credit increased from 62 per cent to 66 per cent of borrowing over the decade of neo-liberal reform underlines the degree to which moneylenders tightened their domination over agricultural production in West Bengal. By contrast, institutional credit had been more and more insignificant, owing to the state withdrawal policy of reform. The impact was disastrous for the backward region. Earlier this region was the big beneficiary of the IRDP policies that contributed to the ratio of institutional credit was as high as 51 per cent in 1993-94. Over the decade the percentage drastically declined to 27. The domination of moneylenders emerged to fill the farmers demand for credit in the environment of withdrawal of bank finance particularly the rural finance as a part of anti-poverty programme like Integrated Rural Development Programme (IRDP). Government even lowered its finance through priority sector and the latter was redefined such that even multinational companies located in rural areas can borrow under priority sector scheme. This has also contributed to rise and growth of moneylending. There was a broadly negative association between ascending class status and the proportion of institutional credit received during 1993-94 which has vanished in 2004-05. Rather the association became positive after a decade. Accordingly in 1993-94 the landless class had the highest access to this source of borrowing (78%), while the rich peasant had the lowest (around 28%). However, over the decade of neo-liberal reform the corresponding percentages were 26 per cent and 23 per cent. Both the regions also broadly follow this transformation, though it was much sharp in the advanced region. The farm-size-wise distribution showed that advanced and combined region strictly follow this trend.

Table 18.6

Economic Class-wise Percentage Distribution of Asset, Cost and Institutional Credit

	Households		*Asset*		*Cost of Prod*		*Inst. Credit*	
	Pd1	*Pd2*	*Pd1*	*Pd2*	*Pd1*	*Pd2*	*Pd1*	*Pd2*
All								
Exploited	39.52	39.52	9.35	12.71	10.57	8.30	27.30	20.36
Self-employed	31.43	25.24	24.23	26.46	28.18	21.96	32.89	17.93
Exploiter	29.05	35.23	66.42	60.84	61.26	69.74	39.81	61.71
Adv								
Exploited	44.55	41.82	4.79	10.33	10.63	7.03	32.18	16.6
Self-employed	31.43	25.24	8.88	11.88	14.37	16.42	11.04	14.00
Exploiter	29.05	35.23	86.33	77.79	75.01	79.55	56.77	69.4
Back								
Exploited	34	37	19.12	15.69	10.48	13.11	17.91	31.58
Self-employed	48	35	57.14	44.79	47.80	42.94	74.88	29.72
Exploiter	18	28	23.74	39.52	41.72	43.96	7.20	38.69

Cropping Patterns Output and Marketed Surplus

In terms of aggregate sown area and yield in the sample villages, paddy grown in different seasons (*aman* and *boro*) constitutes the main crop in each region, occupying some 73 per cent and 61 per cent of the gross output in 1993-94 and 2004-05 respectively. Although a substantial difference exists between the two regions in terms of the overall share of paddy cultivated—it accounts for nearly 90 per cent of total output in the backward region in both years, compared with only 56 per cent (1993-94) and 41 per cent (2004-05) in the advanced one. When output per holding and per acre is considered, this difference is not so acute. In this connection it should be noted that evidence of increased output in a backward region is not specific to our village level findings: other economically backward areas of West Bengal, such as Purulia, registered a very high level growth rate during the past (Saha and Swaminathan, 1994; Sen and Sengupta, 1995). The relatively high level of production in our sample villages in Bankura district—even by the poorer categories of farmer—was due mainly to the expansion both of irrigation facilities and thus also of *boro* cultivation, an effect again of the land reforms carried out in West Bengal.

A crucial indicator of the extent to which agriculture is commercialised, marketed surplus is linked to the purchase of both inputs and of manufactured consumption goods. Table 18.7 shows output and marketed product in percentage term in comparison with the household distribution by each main class. The data indicates an acute kind of inequality in both periods where in the combined region 39 per cent of landless and poor peasant combined (exploited class) constitute 8 per cent of total output, while 29-35 per cent of households belong to rich peasant and landlord (exploiter class) register 67-68 per cent of output. The corresponding figure for product marketed by exploited and exploiter classes are around 4 per cent and 80 per cent respectively. The distribution of marketed product are far more unequal than the output distribution. The inequality is far more in the advanced region.

Table 18.7

Economic Class-wise Percentage Distribution of Output and Product Marketed

	Households		*Output*		*Product Marketed*	
	Pd1	*Pd2*	*Pd1*	*Pd2*	*Pd1*	*Pd2*
All						
Exploited	39.52	39.52	8.12	8.18	3.58	4.22
Self-employed	31.43	25.24	24.65	23.17	14.37	17.34
Exploiter	29.05	35.23	67.23	68.65	82.05	78.44
Adv						
Exploited	44.55	41.82	8.48	5.43	5.71	4.58
Self-employed	31.43	25.24	12.48	17.66	11.78	15.5
Exploiter	29.05	35.23	78.71	76.91	82.51	79.92
Back						
Exploited	34	37	7.69	13.63	0.43	1.27
Self-employed	48	35	38.73	34.09	18.2	32.33
Exploiter	18	28	53.58	52.28	81.37	66.39

In overall terms, average per holding output is positively associated with ascending class status up to the rich peasant category, but declines for the landlord group for both in 1993-94 and 2004-05 (Table 18.8). Because of higher cropping intensity on labour hiring farms, however, output per acre sown shows a less positive variation with ascending class status than output per acre operated. Accordingly, those units which purchase labour-power have a significantly higher yield than holdings belonging to petty-commodity producers or sellers of labour-power. Up to the 10 acre limit, farm size is inversely related to yield in 1993-94 (defined as output per acre operated as well as sown), but units of 10 acres and above possess a high value output. However, in 2004-05 no such inverse relationship was found (Table 18.8). In fact there was no clear tendency. Compared with holdings owned/operated by sellers of labour-power and petty commodity producers, therefore, those owned/operated by labour-purchasing categories register nearly double the yield level in both the year. Overall our findings do not support the Chayanov-Sen thesis that smaller farm sizes/lower classes are associated with higher level of yield.

Table 18.8

Class and Size-Group-wise Distribution of Yield (All Crops)

(in ₹)

	Per Holding		*Per Acre*	
	Pd1	*Pd2*	*Pd1*	*Pd2*
Poor Peasant	4095	5251	5455	7600
Small Peasant	15690	15082	6210	7780
Middle Peasant	19310	27808	5890	12449
Rich Peasant	54245	53101	10399	16727
Landlord	31693	32475	8041	11805
Total	27025	21393	11032	12004
0.01-2.5	7416	12605	7155	12562
2.5-5	24638	41071	6825	11298
5-10	43604	87779	6253	12249
10-15	97027	144000	8244	9606
15 & above	423200	0	14170	0
Total	27025	21393	11032	12004

Note: All crops includes mainly paddy (*aman* & *boro*), potato, oilseed, *til*, wheat.

Policy Implication

First the government intervention has to be strengthened in agriculture, otherwise the impasse of production and distribution cannot be cured. The government intervention may be in the form of food, fertiliser and credit subsidies. When European Union (EU) or United States (US) extend huge subsidies to their farmer, there is no harm if Indian government too extend subsidies to its farmers. Only then cost structure of agriculture will be rationalised in comparison to the selling prices. At the same time, support prices for the farmers can be guaranted. We can move from targeted to universal public distribution system (PDS) which will benefit a huge rural poor population without having below poverty line (BPL) card. The institutional credit network can effectively be extended to marginal and poor peasant. However, only government initiative is not enough, the farmers should be enough conscious about their well-being. As the second recommendation we can say cooperative movement must be initiated. Presently the cooperatives are in a bad shape which is being tied with bureaucracy. The true cooperative movement can join government initiative with people's participation. Third, agricultural extension service should be revived.

Conclusion

Based on surveys conducted in 1993-94 and 2004-05 on same set of households, this study shows during the decade of neo-liberal reform, the interventionist economy eroded in rural West Bengal and peasant differentiation was going on. This study showed that landlessness increased and distress selling of land has happened. Households belong to 0-2.5 acres increased from 138 to 153 during decade. This is reflected in swelling of landlord households from 15 to 27 and rich peasant from 12 to 16 during the decade. Refuting Chayanov this study showed that the ascending class / size status is related to lower participation rate, and therefore, greater well being. It is interesting to note that participation rate increased for upper classes during neo-liberal reforms. This means the well-being of higher peasant classes have also been curbed during neo-liberal reforms.

A big decline of irrigated area particularly in the backward region in West Bengal owing to neo-liberal policy of state withdrawal may give birth to 'ultra Left terrorism'. The households belonging to self-employed declined, exploited classes remained same and that for exploiter increased. The control of land by exploiter classes increased in liberalisation period (from around 50 to 60%). The self-employed are major land loser. The condition of exploited remains more or less unchanged who continue to own around 11-12 per cent of total land area. For exploiter classes the access to institutional credit jumped to a large extent from 40 to 62 per cent and the same declined for the rest. The output and product marketed showed extreme kind of differentiation. Particularly the product marketed showed a huge inequality, where the exploiter classes occupy 78 per cent and the exploited 4 per cent. The pattern of inequality remains same between pre- and post-liberalisation, which reminds us about Harriss-White's note that deregulated markets may become more or less imperfect than regulated market (Harriss-White 2008). Left Front pre-occupied with the reforms in the structures of production, and nothing for market (Harriss-White 2008). No inverse relationship between farm size/economic class and productivity is found. This thoroughly refutes populist proposition of Chayanov and Sen.

References

Bhattacharyya, Sudipta (2007). "Capitalist Development, Peasant Differentiation and the State: Survey Findings from West Bengal", *Journal of Pleasant Studies* 28(4): 95-126.

Bhattacharyya, Maumita and Sudipta Bhattacharyya (2007). "Agrarian Impasse in West Bengal in the Liberalisation Era", *Economic and Political Weekly* Dec 29.

Chayanov, A.V. (1966). "Theory of Peasant Economy", in D. Thorner, R. E. F. Smith and B. Kerblay (eds.), *Homewood*. IL: The American Economic Association.

Dasgupta, Biplab (1984). "Sharecropping in West Bengal: From Independence to Operation Barga", *Economic and Political Weekly* 19(26): A85-A96.

Ghosh, Jayati (2006). "Being your own Boss", *www.macroscan.org*, 18.12.2006.

Harriss-White, Barbara (2008). *Commercial Capital in West Bengal Agriculture*. Delhi: Oxford.

Kay, C. (2008). "Reflectios on Rural Development and Development Strategies: Exploring Synergies, Eradicating Poverty", Innaurial Address delivered on 4 December at the Institute of Social Studies. The Hague, The Netherlands. *http://www.iss.nl/layout/set/print/content/download/14008/133876/file/Inaugural%20Address%20Prof%20Kay.pdf*

Murray, W.E. (2006). "Neo-feudalism in Latin America? Globalisation, Agribusiness and Land Re-concentration in Chile", *Journal of Peasant Studies* 33(4): 646-77. Routledge.

Patnaik Utsa (1987). *Peasant Class Differentiation: A Study in Method with Reference to Haryana*. Delhi: Oxford University Press.

Saha, Anamitra and Swaminathan, Madhura, (1994). "Agricultural Growth in West Bengal in the 1980s: A Disaggregation by Districts and Groups", *Economic and Political Weekly* 29(13).

Sen, Abhijit, and Ranja Sengupta (1995). "The Recent Growth in Agricultural Output in Eastern India, with Special Reference to the Case of West Bengal", Paper Presented at a Workshop on *Agricultural Growth and Agrarian Structure in Contemporary West Bengal and Bangladesh*. Calcutta. January.

Sen, Amartya (1966). "Peasants and Dualism with and without Surplus Labour", Journal of Political Economy 74.

This study has been conducted out of the financial support from UGC in the form of Major Resarch Project entitled 'Market Economic Reform and Agrarian West Bengal [ref no. F-5-99/2003]. There are many data used in this study has not been shown in the tables because of space constraint.

APPENDIX A-18.1

Patnaik's E-Criterion

The labour-exploitation index seeks to give an empirical approximation to the analytical concept of the class status of the household. The latter is determined by the extent of the use of outside labour, or the degree to which the peasant family works for others, relative to self-employment on its own holdings. Under certain simplifying assumptions, the labour-exploitation index identifies the extent to which surplus labour is appropriated from the workers concerned.

$E = X/Y = \{(Hi\text{-}Ho) + (Lo\text{-}Li)\}/F$

where Hi = Labour-days hired on the operational holding of the household.

Ho = Family labour days hired out to others.

Li = Labour days worked on leased in land (whether by family or hired labour).

Lo = Labour days similarly worked on land leased out by the household.

F = Labour days worked by household workers on the operational holding.

The following limits are specified to the value of E in order to classify households into a set of mutually exclusive and all-exhaustive categories (sub-categories not specified here are not ruled out).

Appendix Table A-18.1

Patnaik's E-Criterion

Class	*Defining Characteristics*	*Value of E=X/F & Reason*	*Remarks*
1. Landless Labourers	No Self-employment; working entirely for others	$(E \rightarrow -\infty)$ $F=0$, $X<0$ & large	Exploited
2. Poor Peasants	Working for others exceeds self-employment	$(E \leq -1)$, $F>0$, $X<0$, $\lvert X \rvert \geq F$	
3. Small Peasants	Self-employment exceeds working for others	$(0 \geq E > -1)$, $F>0$ $X \leq 0$, $\lvert X \rvert < F$	Self-employed
4. Middle Peasants	Self-employment dominates combined with the small amount of others' lab	$(1 > E > 0)$, $F>0$ $X>0$, $X<F$	
5. Rich Peasants	At least as large an employment of others' lab as selfemployment	$(E \geq 1)$, $F>0$ $X>0$, $X \geq F$	Exploiter
6. Landlords	No manual lab expanded in the form of self-employment	$(E \rightarrow \infty)$, $F=0$ $X>0$ & Large	

Source: Patnaik (1987).

19 How Remunerative is Indian Agriculture?

Focus on Maharashtra

Sangeeta Shroff

Backdrop

India, in the 21st century has emerged as a major economic power marked by rising per capita incomes and impressive performance in growth rate of gross domestic product (GDP). However, this remarkable achievement does not seem to be confined to all sectors of the economy. During the Tenth Five Year Plan period (2002-2007) while the industry and service sector grew at a little more than 9 per cent per annum, that of the agricultural sector was a miniscule 2.3 per cent per annum. In the Eleventh Five Year Plan period (2007-2012) also, agriculture could not achieve its target growth rate of 4 per cent and while the overall growth rate of the economy was 7.9 per cent, that of agriculture was 3.3 per cent. Deceleration in growth rate in the agricultural sector is coupled by dominance of workforce in this sector with no major reduction over the decades. The causes for the underperformance and stagnation of Indian agriculture are many, such as lack of irrigation facilities, technology fatigue in areas which contributed to the granary of the country, declining public investment and size of holding in agriculture, etc. These factors have culminated to agriculture being an unremunerative economic activity and in the ultimate analysis resulted into agrarian distress.

Although the agricultural sector has shown dismal performance, it may be noted that the country has made remarkable progress in this sector since independence. Prior to the green revolution, India was "Ship to Mouth" in response to meeting the food shortages being faced by the country. However, the new seed-fertiliser technology signaled the beginning of a more dynamic agriculture which brought about not only self sufficiency in food but the country was in a position to maintain a buffer stock and also export. The green revolution was followed by the white revolution which brought about a revolution in the dairy sector and further the country also went through a yellow revolution with the adoption of new varieties of oilseeds catalysed by Technology Mission of Oilseeds. The blue revolution aimed at enhancing fish production and of late the National Horticulture Mission aims at ushering a golden revolution by increasing production of fruits and vegetables. Progress in agriculture was made possible because of technology as well as the institution of planning, price policy measures, land reforms, etc. However, this sector is again showing signs of deceleration with failure to achieve targeted growth rate.

In view of the above, an attempt is made in this paper to highlight critical factors responsible for dismal performance of agricultural sector, observe profitability of Indian agriculture, by taking the case of Maharashtra, explain the causes of agrarian distress in Maharashtra and finally suggest policy issues which need to be addressed.

Growth Rate of Area, Production and Yield of Major Crops

In Table 19.1, the growth rates of area, production and yield of major crops in the country have been indicated. It is observed that the growth rate of area under food grains during the period 2001-2010 has declined as compared to the 1971-1980 period. However, that under pulses, oilseeds and cotton, though fluctuating, has registered an increase in the post-2000 period. Area under sugarcane has been declining continuously. This pattern is also observed in case of production growth rates. It can be clearly observed that the growth rate in yield of food grains has shown a marked decline in the period 2001-2010. The growth rate in yield of food grains which was 2.97 per cent per annum in 1981-1990 declined to 1.60 per cent per annum in 2001-2010. Rice and wheat which are the major food grains presented a worse picture. The growth rate of yield of rice which was 3.19 per cent per annum in 1981-1990 declined to 1.62 per cent per annum in 2001-2010 while in case of wheat the growth rate declined from 3.10 per cent per annum to 0.68 per cent per annum, during the corresponding period.

Table 19.1

Growth Rate of Area (A), Production (P) and Yield (Y) of Major Crops

Crop		*1971-1980*	*1981-1990*	*1991-2000*	*2001-2010*
Food grains	A	0.46	-0.26	-0.08	0.29
	P	2.07	2.73	2.09	1.9
	Y	1.6	2.97	2.17	1.6
Rice	A	0.88	0.38	0.67	-0.03
	P	1.9	3.62	2.02	1.59
	Y	1.01	3.19	1.34	1.62
Wheat	A	2.39	0.46	1.72	1.24
	P	4.31	3.58	3.57	1.89
	Y	1.87	3.1	1.82	0.68
Coarse Cereals	A	-0.87	-1.34	-2.11	-0.76
	P	1.11	0.35	-0.01	2.43
	Y	2	1.71	2.14	3.22
Pulses	A	0.59	-0.1	-0.6	1.17
	P	-0.39	1.49	0.67	2.68
	Y	-0.98	1.59	1.27	1.5
Oilseeds	A	0.4	2.44	0.17	2.48
	P	0.74	5.45	2.25	5.15
	Y	0.33	2.95	2.07	2.6
Cotton	A	0.42	-1.26	2.71	2.13
	P	3.68	2.8	2.3	13.59
	Y	3.3	4.1	-0.41	11.22
Sugarcane	A	1.9	1.46	1.67	0.78
	P	2.56	2.71	2.73	0.92
	Y	0.64	1.23	1.05	0.15

Source: Calculated from data obtained from GoI (2010). *Agricultural Statistics at a Glance.*

Four states in India, namely, West Bengal, Andhra Pradesh, Uttar Pradesh and Punjab account for more than half, i.e., 53.4 per cent of area under rice. However, while yield of rice in Punjab is 4022 kg per hectare, the yield in Uttar Pradesh is nearly half that of Punjab. West Bengal which has highest area under rice has a yield which is only 0.63 times that of Punjab. Orissa and Bihar which account for 12 per cent area under rice have yields as low as 1500 kg per hectare which is 0.37 times of that of Punjab and 0.70 times that of All India (Government of India, 2010).

In case of wheat, about 68 per cent of the output is produced by Uttar Pradesh, Punjab and Haryana though these states contribute to 56 per cent of area. Punjab again has highest yield of about 4462 kg per hectare in 2008-09. In contrast, the All India average is about 2907 kg per hectare which indicates that there is huge scope to increase yield levels. For example, in Madhya Pradesh, which contributes 13.64 per cent of area to the country, there is potential to increase yield by 54.25 per cent (Report of the Commission for Agricultural Costs and Prices, GoI, 2009-10). However, the yield levels in Punjab have more or less reached a plateau and other states have to increase productivity so as to maintain the level of food production that is required to feed the population.

There are a number of region specific factors causing low productivity of food grains. In Punjab, Haryana and Western Uttar Pradesh, ground water depletion, soil salinity, micro nutrient deficiency and decreasing total factor productivity (TFP) are major constraints for low growth rate in productivity. The possible reason for degradation in soil is due to imbalance in fertiliser use. For example, in Punjab the N:P:K ratio is 18.4:5.9:1 which clearly indicates excessive use of nitrogen in proportion to phosphorous and potash. In the North East region where a number of schemes are being introduced to extend the green revolution, severe floods/water logging, improper drainage and soil acidity inhibit increase in yield levels. Further, low seed replacement rate, poor input delivery, non-availability of electricity, and improper infrastructure, are also factors leading to low productivity.

Cost of Cultivation and Net Returns to Farmers: Case of Maharashtra

As mentioned earlier, in the recent past there has been deceleration in growth rate in Indian agriculture, declining input efficiency, degradation of natural resources. Further, evidence of the worsening situation of farming households is observed from the 59th round of NSSO (2005; 2005a) on the "Situation Assessment of Farmers", which indicates that 48.6 per cent of the farmers' households in India are in debt and about 41 per cent farmers' households in the country did not like farming because it is not profitable, is risky and it lacks social status.

In view of the above, an attempt is made in this section to observe net returns to farmers in Maharashtra with respect to major crops cultivated in the *kharif* and *rabi* seasons. The *kharif* crops included are cotton, soyabean, *kharif jowar* and *tur*, while the *rabi* crops selected are *rabi jowar*, wheat and gram. In addition, sugarcane which is an important crop in Maharashtra has also been included. The crops selected are presented in Table 19.2 and Table 19.3 and together constitute 72.84 per cent of the gross cropped area (GCA) in the state

Table 19.2

Cost of Production and Net Returns of Major Kharif Crops in Maharashtra (₹ per Quintal) 2010-11

Item	*Cotton*	*Soyabean*	*Jowar*	*Tur*
Paid out cost	2144	1274	867	1628
Interest on fixed capital	62	37	22	36
Rental value of land	497	417	189	664
Transport cost	45	45	45	45
Marketing cost	134	111	48	169
Total paid out cost	2323	1430	960	1842
Total cost	2882	1879	1171	2542
Yield in quintals per hectare	12.51	13.04	18.39	12.36
Gross return per quintal	3000	2435	1143	3998
Net return over paid out cost	856	1005	183	2156
Net return over total cost	118	556	-28	1456
Net income (₹ per hectare) considering only paid out cost	10709	13105	3365	26648
Net income (₹ per hectare) considering total cost)	1476	7250	-515	17996

Source: Same as Table 19.3.

It can be observed from Table 19.2 that for crops cultivated in the *kharif* season, the net income per hectare considering only paid out cost varied from ₹ 3365 for *kharif jowar* to ₹ 26648 for *tur*. However, if total cost is considered, the net returns per hectare are substantially reduced and are infact negative for *kharif jowar*. In case of cotton, the net returns after considering total cost are only ₹ 1476/hectare. Maharashtra has highest area under cotton in the country which constitutes 18.69 per cent of GCA.

Table 19.3

Cost of Production and Net Returns of Major Rabi Crops in Maharashtra (₹ per Quintal) 2010-11

Item	*Rabi Jowar*	*Wheat*	*Gram*	*Sugarcane*
Paid out cost	1427	958	1468	68
Interest on fixed capital	79	88	96	3
Rental value of	209	198	353	13
Transport cost	74	67	61	28
Marketing cost	49	60	97	-
Total paid out cost	1550	1085	1626	96
Total cost	1838	1371	2075	112
Yield (in quintals) per hectare	8.84	20.52	7.5	863.2
Gross return	1271	1206	2138	145
Net return over paid out cost,	-279	121	512	49
Net return over total cost	-567	-165	63	33
Net income (₹ per hectare considering only paid out cost)	-2466	2483	3840	42296
Net income (₹ per hectare considering total cost)	-5012	-3386	472.5	28486

Source: Agricultural Price Committee Cell, Government of Maharashtra, (Unpublished data).

The situation with respect to *rabi* crops is even worse. *Rabi jowar* constitutes 12.55 per cent of GCA and net income per hectare is negative considering both paid out cost as well as total cost. In case of wheat, also when total cost is considered, the net returns are negative. In case of gram, the net returns per hectare considering paid out cost is ₹ 3840 per hectare while it is only ₹ 472.5 per hectare when total cost is considered.

Other important issues related to the state of Maharashtra are that the cropping intensity in the state is only 128.6 per cent due to lack of irrigation facilities. Presently, only 18 per cent of GCA is irrigated. Hence, farmers mostly depend upon rainfed farming and about 71 per cent area is under *kharif* crops.

Besides the high cost of cultivation, it can also be observed that not only is the average size of holding in Maharashtra declining over the years, but there is also marginalisation of farmers. This is indicated in Table 19.4.

Table 19.4

Average Size of Holding in Maharashtra (Hectare)

Size Group	*1970-71*	*1980-81*	*1990-91*	*2005-06*
Marginal	0.46 (25.05)	0.48 (28.07)	0.49 (34.58)	0.46 (44.6)
Small	1.47 (17.60)	1.49 (22.44)	1.47 (28.80)	1.26 (30.25)
Semi-medium	2.87 (21.82)	2.85 (24.57)	2.8 (22.45)	2.50 (17.87)
Medium	6.3 (24.65)	6.04 (20.29)	5.85 (12.37)	5.28 (6.74)
Large	16.6 (10.38)	14.9 (4.63)	15.3 (1.80)	13.39 (0.51)
All Sizes	4.28	3.11	2.2	1.46

Note: Figures in brackets are percentages to total farmers in the respective category.

Source: Calculated from *Economic Survey of Maharashtra* (various issues), Directorate of Economics and Statistics, Government of Maharashtra.

It can be observed from Table 19.4 that the average size of holding which was 4.28 hectares in 1970-71 has shown decline over the years and is presently 1.46 hectares. A more disturbing feature is that the number of marginal and small holdings which constituted 42.65 per cent of total holdings has increased over the years and reached 74.85 per cent in 2005-06. Thus, three-fourth of the holdings fall in the marginal and small category while the share of marginal holdings is 44.6 per cent. This could therefore lead us to say that the returns of marginal farmers who own an average holding size of 0.46 hectare is a little less than half of the net returns per hectare indicated in Table 19.2 and Table 19.3. The problem is further aggravated because this net returns from cultivation is spread over a period of 12 months. Since cropping intensity is only 128.6, it clearly indicates that most farmers take only one crop in a year and barely 22 per cent of the area is cropped more than once.

Agrarian Crisis in Maharashtra

A large number of factors have been responsible for agrarian distress in Maharashtra, which has ultimately culminated into suicides of farmers. The situation reached alarming proportions in 2006 in Maharashtra, when the suicide rate was 1427 per lakh. Vidarbha region of Maharashtra which is often termed as the suicide belt reported maximum suicides. The Vidarbha Jan Andolan Samiti, a farmer's movement, reported that between June 2005 and December 2006, 1158 farmers had committed suicide in

Vidarbha. A major reason for agrarian distress in Maharashtra is poor irrigation facilities and only 18 per cent of GCA is irrigated, leading to low productivity. By and large, one-third of the state falls under rain-shadow region, with scanty rainfall. The state has the distinction of having more than 24 per cent of drought prone area of the country and faces major problems associated with rainfed agriculture, such as low yields, crop uncertainties and failures due to erratic rainfall.

Further, an important observation regarding farmers' distress was that they were concentrated mainly in regions where cotton is an important crop. A brief backdrop of Maharashtra's agricultural economy shows that it has highest area under cotton in the country, which is about one-third of the country's area and this area is mostly concentrated in Vidarbha. Nearly 97 per cent of the cotton cultivated in Maharashtra is unirrigated.

The state-wise cotton yields during the period 2003-04 to 2005-06 indicate that Maharashtra had the lowest yield in 2004-05 and 2005-06 among all major cotton producing states. Its average yield of 122 kg per hectare for the period 2003-04 to 2005-06 was only 18 per cent of India's leader Punjab. Gujarat experienced rapidly rising yields. An international comparison revealed that Maharashtra's yield was 10 per cent that of China. The cotton farmer in Maharashtra was thus not efficient enough to compete in the international markets which became accessible after globalisation. Opening up of the economy also encouraged cotton cultivation by increasing area under the crop. In 1993-94 while area under cotton was 7.4 million hectares, it gradually increased to 12.12 million hectares in 2011-12 with fluctuations depending upon not only domestic prices but also international prices. In Maharashtra also where cotton is a major crop, area increased from 2.48 million hectares to 3.9 million hectares in the corresponding period. But low yields in the state have been accompanied by a fall in price of cotton. The wholesale price index of cotton declined from 181 in 2003-04 to 144 in 2005-06. The international prices also declined and Cotlook Index A declined from 69.1 US cents per lb to 57.04 US cents per lb during this corresponding period.

In 2004-05, the cost of production for cotton in Maharashtra according to the Commission on Agricultural Costs and Prices (CACP) was ₹ 2216 per quintal but the minimum support price (MSP) was ₹ 1960 per quintal. Further, a study (Mitra and Shroff, 2007) showed that the real cost of cultivation of cotton in Maharashtra from the period 2000-01 to 2005-06 increased by 124 per cent leading to a decline in profits per hectare. It was only Maharashtra that had a negative profit per hectare in 2006. The size of this negative surplus was substantial and over ₹ 1000 per hectare at 2000-01 prices.

Thus the cotton economy in Maharashtra continued to be characterised by low yield, low prices and rising cost of cultivation. The suicide epidemic among farmers in Maharashtra has also been attributed to widespread adoption of area under BT cotton especially in Vidarbha. The notable feature of BT cotton is that it contains a gene which gives protection to the crop against pest attack (American Bollworm). These seeds are however very costly and require irrigation. In Vidarbha however, farmers lack irrigation facilities and depend upon rainfall. Across the country while average cost of cultivation is little more than ₹ 16,000 per hectare with yield of 460 kg per hectare, in Vidarbha the cost of cultivation could go well beyond ₹ 20,000 per hectare with average productivity of 146 kg per hectare (*www.dnaindia.com*).

Thus, in light of the above factors the yield of cotton declined making it unprofitable. Other states such as Gujarat benefited from use of BT cotton seeds because of which they experienced enhanced yields. The farmers in Vidarbha are thus pushed to debt with more than 90 per cent of them borrowing from private traders and money lenders. This explains the agrarian crisis in Vidarbha leading to the extreme step of suicides, especially in 2006.

The above mentioned situations have reoccurred. The year 2010-11 was an exceptionally good year for cotton farmers and the Cotlook Index A increased from 77.99 in 2009-10 to 165.12 in 2010-11 while the wholesale price index (WPI) increased from 218 to 312. Production of cotton in the country was also high at 29.5 million bales and farmers benefited from high output as well as high price due to increase in international demand. In response to high profits made in 2010-11, area under cotton increased from 11.1 million hectares to 12.19 million hectares in 2011-12. Area under cotton in Maharashtra was around 4.1 million hectares in 2010-11 and 2011-12. However, the high prices in 2010-11 could not be sustained and the prices began to fall in 2011-12 due to less international demand following the euro-zone crisis, modest imports from China, comfortable world cotton situation, etc. International prices declined and Cotlook Index A sharply declined from 165.12 in 2010-11 to 105.4 in 2011-12. Accordingly prices in domestic market also began declining. Further, in March 2012, the government issued a notification imposing a ban on cotton exports which further depressed prices. Due to opposition from farmers, the government partially relaxed the restriction on export ban by allowing shipments for consignments which had already been registered. Such decisions of the government make the country lose its credibility as a regular supplier in the international market and international buyers turn to other markets for their requirements. Export policy of the government is therefore against the interests of the farmers as it is bringing about a downward pressure on prices. The important point to note is that in a closed economy, farmers face only yield risk. If yield is low then farmers are compensated by higher prices. However, in an open economy, farmers face both yield risk and price risk. A low output may not be compensated by higher price increase if prices in the international market are ruling low. Thus in India when trade barriers were dismantled, farmers had to face global competition and were therefore susceptible to price shocks. As mentioned earlier, since cotton is an important crop in Maharashtra, especially in Vidarbha, factors such as poor irrigation facilities, technology, price shocks after liberalisation and government policy, all played a role in making the cotton crop unremunerative in several years.

Public Expenditure in Agriculture in Maharashtra

Investment in agriculture is essential for increasing as well as sustaining the overall growth of agriculture. Development planning in Maharashtra began with the Third Five Year Plan in 1961. The sectoral distribution of plan expenditure in Maharashtra is given in the Table 19.5. It can be observed that from Third Plan to Tenth Plan period, the share of agriculture and allied services has sharply declined from 30.87 per cent in Third Plan to 3.47 per cent in Tenth Plan in Maharashtra. Unlike the agriculture and allied services, the plan expenditure in irrigation and flood control has sharply increased from 14.87 per cent in Third Plan to nearly 37 per cent in Ninth Plan and then decreased to about 28 per cent in Tenth

Table 19.5

Sectoral Distribution of Plan Expenditure in the Maharashtra

(₹ in Crore)

Sr. No	Five Year Plan Schemes	Agriculture and Allied Services	Rural Development	Irrigation and Flood Control	Power Development	Industry and Minerals	Transport and Communication	Others	Total (3 to 9)
1	3rd Plan (1961-66)	134.20 (30.87)	@	64.64 (14.87)	94.71 (21.79)	14.82 (3.41)	37.45 (8.61)	88.91 (20.68)	434.73 (100.00)
2	Three Annual Plans (1966-69)	111.20 (28.77)	@	57.69 (14.92)	115.46 (29.87)	10.64 (2.75)	35.19 (9.10)	56.40 (14.59)	386.58 (100.00)
3	4th Plan (1969-74)	221.71 (22.07)	@	165.96 (16.52)	283.64 (28.24)	39.62 (3.94)	81.08 (8.07)	212.50 (21.15)	1004.51 (100.00)
4	5th Plan (1974-79)	343.32 (12.91)	190.70 (7.17)	491.74 (18.49)	910.47 (34.22)	158.69 (3.97)	220.08 (8.27)	345.13 (12.97)	2660.13 (100.00)
5	Annual Plan (1979-80)	64.38 (7.93)	89.45 (11.01)	150.00 (18.47)	299.90 (36.92)	30.03 (3.70)	64.04 (7.88)	114.43 (14.09)	812.23 (100.0)
6	6th Plan (1980-85)	393.38 (6.01)	894.94 (13.66)	1437.44 (21.95)	1734.76 (26.49)	225.31 (3.44)	422.81 (6.46)	1440.75 (21.99)	6549.39 (100.00)
7	7th Plan (1985-90)	614.39 (5.56)	1575.86 (14.27)	2239.53 (20.28)	2835.39 (25.67)	420.25 (3.81)	585.77 (5.30)	2773.02 (25.11)	11044.21 (100.00)
8	8th Plan (1992-97)	1470.60 (5.71)	2836.44 (11.01)	5668.34 (22.01)	5864.09 (22.77)	941.89 (3.66)	2015.36 (7.83)	6954.94 (27.01)	25751.66 (100.00)
9	9th Plan (1997-2002)	1486.38 (3.33)	4190.88 (9.38)	16383.32 (36.69)	6977.79 (15.63)	474.63 (1.06)	4331.74 (9.70)	10811.43 (24.21)	44656.17 (100.0)
10	10th Plan (2002-2007)	1918.86 (3.47)	7148.09 (12.92)	15238.14 (27.54)	5300.58 (9.58)	886.24 (1.60)	6442.63 (11.65)	18389.19 (33.24)	55323.72 (100.0)
	Total (3rd to 10th Plan)	6758.42 (4.55)	16926.36 (11.39)	41896.8 (28.19)	24416.79 (16.43)	3202.12 (2.15)	14236.15 (9.58)	41186.7 (27.71)	148623.33 (100.0)

Note: @ included under agriculture and allied services; figures in parenthesis percentage to total.

Source: GoM (2009). *Annual Plan 2008-09*, Planning Department, Government of Maharashtra.

Plan. However, the extent to which investment in irrigation has translated into irrigated area in the state is still a gray area in the state.

The Way Forward

From the above discussions it is clear that India made considerable progress in moving out of an era of severe food shortage and dependence on imports to self sufficiency in food by late 1980s when yield rates in food grains showed marked increase. The transformation of the food grain economy was largely due to success of the green revolution in the late 1960s which was concentrated mainly in regions having resources such as irrigation. However, as noted earlier, agriculture has begun to show deceleration in growth rates especially in the food grain sector. The yield rates of food grains during the period 2001-2010 declined to 1.6 per cent per annum from 2.17 per cent per annum in the previous decade. In case of wheat, the yield rate declined to 0.68 per cent per annum during 2001-2010 from 1.82 per cent per annum in the previous decade. Cultivation of commercial crops such as cotton is also proving to be unremunerative and being an important cause of agrarian distress especially in states like Maharashtra. In the light of this, certain policy issues must be addressed.

In view of the increase in population, it is necessary to further increase food production. The per capita availability of food grains which was 186.2 kgs in 1991 has declined to160 kgs in 2010. It is therefore necessary to spread the green revolution to the eastern region of the country. Tremendous opportunities to utilise untapped groundwater for enhancing productivity exist in these regions as they receive 2-3 times more rain than North-west regions. The Eastern and North Eastern regions have the combined potential to meet the increasing demand of food grains and enable the country to ensure food security. The North West region presently serving as the backbone of the country's food granary is at a critical juncture due to fast depleting groundwater resources. If the green revolution is extended to Eastern India by increasing area and yield, India need not resort to short term measures such as ban on exports of rice to contain inflation and food security.

In order to salvage the agrarian distress in Maharashtra, it is necessary to increase area under irrigation which has remained at about 18 per cent since long. Maharashtra is a pioneering state in drip irrigation and this must be popularised alongwith watershed strategies so as to achieve yield potential.

Besides increasing productivity in agricultural sector, rural non-farm employment and other skill development programmes must be initiated, so as to combat structural employment in other sectors and create job opportunities for the workforce concentrated in the agricultural sector.

References

Government of India. (2010). *Agricultural Statistics at a Glance*. New Delhi: Directorate of Economics and Statistics, Department of Agriculture and Co-operation, Ministry of Agriculture.

———. (2011-12). *Economic Survey*. New Delhi: Ministry of Finance.

———. (2010). *Report of the Commission for Agricultural Costs and Prices-2009-10*. New Delhi: Commission for Agricultural Costs and Prices, Department of Agriculture and Co-operation, Ministry of Agriculture.

NSSO (2005). *Situation Assessment Survey of Farmers, Some Aspects of Farming*, Report No. 496, 59th Round (January-December 2003). New Delhi: National Sample Survey Organization, Ministry of Statistics and Programme Implementation, Government of India, December.

———. (2005a). *Situation Assessment Survey of Farmers, Indebtedness of Farmers Households*, Report No. 498, 59th Round (January-December 2003). New Delhi: National Sample Survey Organization, Ministry of Statistics and Programme Implementation, Government of India, May.

Mitra and Shroff (2007). " Farmers' Suicides in Maharashtra", *Economic & Political Weekly:* 73-77. December 8.

Websites:

www.cotcorp.gov.in

www.cotlook.com

http:// indiabudget.nic.in

(www.angelcommodities.com)

www.dnaindia.com